HEARTLAND & HINTERLAND

Third Edition

HEARTLAND & HINTERLAND

A Regional Geography of Canada

EDITED BY

Larry McCann
University of Victoria

Angus Gunn
University of British Columbia

Prentice Hall Canada Inc.
Scarborough, Ontario

Canadian Cataloguing in Publication Data

Main entry under title:

Heartland and hinterland : a regional geography of Canada

3rd ed.
Includes index.
ISBN 0-13-839671-X

1. Regionalism - Canada.   2. Canada - Economic conditions.
I. McCann, L. D. (Lawrence Douglas), 1945-   .   II. Gunn, Angus M., 1920-   .

HC113.H42   1998   330.971   C97-931511-5

ISBN 0-13-839671-X

Publisher: Pat Ferrier
Acquisitions Editor: Sarah Kimball
Developmental Editor: Maurice Esses
Marketing Manager: Tracey Hawken
Copy Editor: Jenifer Ludbrook
Production Editor: Marjan Farahbaksh
Production Coordinator: Deborah Starks
Photo Research: Karen Taylor/Alene McNeill
Cover Design: Monica Kompter
Cover Images: Barrett & MacKay Photography Inc.
Page Layout: Leanne Knox/ArtPlus Limited
Map Art: Donna Guilfoyle, Adele Webster/ArtPlus Limited

Printed in Canada
28 29 30 CP 15 14 13
All maps are based on information taken from the National Atlas Digital Base Maps, and more particularly described as Canada at the scale of 1:7.5 Million. ©1997 Her Majesty the Queen in Right
of Canada with permission of Natural Resources Canada.

To our Native peoples—
Canada's first "geographers"

Brief Contents

Table of Contents

Maps, Figures, and Tables

Maps

Figures

Tables

Preface

Regional geography has come a long way from its former practice of recording unique quantities of encyclopedic data for any and every place. In the modern phase, which originates from French and German traditions, regions are studied selectively. In particular, the French tradition, with its descriptions of *pays* — small French rural regions — as blends of human and physical elements forming regional personalities, set the stage for the more sophisticated conceptual frameworks of today. That method of defining places as personalities is still found in Canada in regional texts for schools and universities.

Early work in Canada focussed on the natural environment to divide the country into a number of regions, each based on the uniform distribution of one or more physical elements. In recent times these gave way to a more balanced choice of human and physical factors. Then, with the rapid growth of urban areas in the second half of the twentieth century, a new type of region emerged, one that dealt with the interactions between urban cores and their peripheral areas. Heartland and hinterland is an excellent example of this conceptual approach. It examines the symbiotic links between large, influential urban centres and their lower-density outlying areas. It is a valuable and flexible perspective: all the regions of the country can be studied as a single, interconnected, functioning unit, for example. Alternatively, patterns of interaction may be investigated at regional and interregional scales.

In Chapter 1, Larry McCann points out that heartlands and hinterlands are found in all countries. Heartlands are centres of power and status, places where people, wealth, and political influence are concentrated; hinterlands are scenes of marginality in all three of these conditions. This is not to say that all peripheral areas are poor, or that poverty and other forms of marginality are restricted to peripheral areas. On the contrary, they are found in all large cities. Rather, it is the preponderance of power in its many forms that identifies a heartland.

While the heartland–hinterland paradigm is as powerful as ever in interpreting Canada's regional structure, there are three new developments to be examined under this rubric, changes that have profoundly affected Canadian life since the previous edition of this book was published. The new relationship with Native peoples is one of these changes. After centuries of being shunted aside as of no consequence, then treated as nuisances or subjected to various attempts at assimilation, aboriginal peoples are finally being accorded equality coupled with a new respect for their values.

The globalization of business, industry, and investment is the second new development, and the growing tension over Québec's desire for independence is the third. Both are little short of revolutionary, without precedent in the scale and intensity of their influences on Canadian economic life. The implications of globalization are examined at various points throughout the book. The tension about Québec happens to reflect a rare form of the kind of regionalism that is found in many parts of the world, rare because it involves the aspirations of both an aboriginal people and a modern society. Chapter 5 examines the problem in some depth and Cole Harris in Chapter 11 analyses some of its roots and current forms.

As in the two previous editions, contributors have been guided by the heartland–hinterland framework in their descriptions and analyses. For each region, development is assessed in terms of economic, social, political, and cultural factors, as each one of these is influential within the given historical and physical milieu. While the contemporary scene is the principal focus, it is important to remember that there were older heartlands and hinterlands shaping Canadian life and leaving a legacy for today. One of these was the cultural hearth of Québec in the eighteenth century, and its scattered hinterland of French settlers across the continent. A more significant and longer-lasting one was the settlement pattern of the Native peoples, where heartlands and hinterlands were well defined by the presence or absence of resources for food and shelter. The physical environment that determined the lifestyles of these first settlers is still a vital element in the delineation of present-day heartlands and hinterlands, and John Stager spells this out in Chapter 2.

Two chapters on the Industrial Heartland highlight the concentration of economic power in what Maurice Yeates calls the "Windsor–Québec City Axis." Donald Kerr describes and documents the rise and development of this region from mid-eighteenth century to the post World War II period. Maurice Yeates follows with a study of the contemporary scene and future trends. While, in one sense, it is acceptable to identify a single economic heartland extending from Windsor to the city of Québec, the reality, as Yeates points out, is a steady drift of the principal elements of the heartland westward toward southwest Ontario.

Atlantic Canada is today very much a hinterland, although it was once an economic centre of power. Graeme Wynn explains these developments in his chapter on the four provinces, New Brunswick, Newfoundland, Nova Scotia, and Prince Edward Island. Major industries of the past are in decline and population growth rates are among the lowest in the country. At the other end of the country, British Columbia's strong economy stands in sharp contrast to Atlantic Canada. Lewis Robinson traces some of the factors that gave rise to present conditions. Asia-Pacific trade is one. The youthfulness of settlement on the West Coast is another; industry does not have to contend with the inertia of infrastructures and plants that were designed for a very different age.

All three provinces of the Western Interior share many characteristics: prairie, parkland, and boreal forest landscapes; a share of the Canadian Shield's terrain and climate; historical reliance on agriculture as the main source of economic strength; and a rich mix of population, the result of large-scale immigration from several European countries. John Lehr assesses the present centres of power and the areas of poverty while looking ahead to the growing influence of individual provinces. Across the broad expanse of what is often called the Near North lies the Canadian Shield. From its earliest times it has been a region of exploitation, harvesting furs, timber, and minerals, to meet demands for these products from Europe and the United States The result today, as Iain Wallace explains, is a place defined by its physical environment and characterized by a resource-based economy, frequently associated with single-enterprise settlements that have difficulty coping with a highly competitive global marketplace.

The North, a place described by Peter Usher as "one land, two ways of life," has a particular fascination for Canadians because of the heroic ventures in search of a northwest passage and the drama of gold rushes in the Yukon at the end of the nineteenth century. For most of the twentieth century, little thought was given to the impacts of imported enterprises on the livelihoods and cultures of the Native peoples. Now, with new political and territorial structures in place, old tensions between development and the survival of traditional ways of life among Inuit and Dene have largely faded.

Beyond the confines of the Canadian map, on the world scale, ideas of core and periphery provide a useful framework for the study of Canada's place among the trading nations of the world. On the one hand, as an affluent country with several multinational corporations and significant political power, Canada could be regarded as a heartland. On the other hand, Canada could just as readily be deemed a hinterland if consideration is given to its dependence on foreign capital, the substantial volume of primary products in its exports, and the overshadowing influence of a neighbourhood world metropolis, New York.

In this edition, I have changed the boundaries of the major regions in order to align them with traditional perceptions of national life — the Atlantic region, St. Lawrence Lowlands, Québec, Canadian Shield, Western Interior, British Columbia, and the North. These seven entities are well-known to individual Canadians and are frequently accorded local and national recognition. Recently, federal recommendations proposed constitutional vetoes for the five most populous regions of the country: Atlantic provinces, Québec, Ontario, three western interior provinces, and British Columbia.

Chapters have been revised to incorporate some of the enormous changes already indicated. The maps in this edition have also been revised extensively. They are now larger and in two colours for better readability.

SUPPLEMENTS

The following supplements have been carefully prepared for instructors to accompany this new edition:

- An *Instructor's Manual*, which includes a bank of test questions and transparency masters of all the maps, figures, and tables in the textbook.
- *Electronic Transparencies (for Windows)*, consisting of all the maps, figures, and tables in the textbook. This item is available upon request.

ACKNOWLEDGMENTS

We are grateful to the many people who offered such helpful suggestions and recommendations for the new edition. We would particularly like to thank the following instructors for providing formal reviews of the manuscript:

Suzanne Greaves (The University of Western Ontario)
William Norton (University of Manitoba)
Harun Rasid (Lakehead University)
Michael Ripmeester (Brock University)
Tod Rutherford (University of Waterloo)
Robert Summerby-Murray (Mount Allison University)

I have greatly enjoyed working with such a fine group of scholars, some of whom were colleagues or mentors in earlier years. It is a particular pleasure to see Lewis Robinson, dean of Canadian regional geographers, involved. My thanks go to the staff of Prentice-Hall for their help, particularly to Maurice Esses.

Angus M. Gunn
West Vancouver,
British Columbia

Contributors

Angus Gunn, formerly a Professor of Social and Educational Studies at the University of British Columbia, is the author or editor of a large number of geographical books and articles including *Patterns in World Geography, High School Geography Project: Legacy for the Seventies, Man's Physical Environment*, and *Human Settlements in an Urban Age*.

Cole Harris, Professor of Geography at the University of British Columbia and Fellow of the Royal Society of Canada, is the editor of Volume I of the *Historical Atlas of Canada*, and author of numerous books and articles on the historical geography of Canada including *The Seigneurial System in Early Canada, Canada Before Confederation* (with John Warkentin), and *The Resettlement of British Columbia: Essays on Colonialism and Geographical Change*.

Donald Kerr, formerly Professor and Chair of the Department of Geography at the University of Toronto and past-president of the Canadian Association of Geographers, remains actively involved in research activities relating to Canada's geography. He is co-editor of Volume III of the *Historical Atlas of Canada*.

John Lehr, Professor of Geography at the University of Winnipeg, is a graduate of the University of Alberta where he wrote his dissertation on Ukrainian settlement in the Western Interior. His present research focusses on Ukrainian settlement in Canada and Latin America, and on heritage tourism in Western Canada.

Larry McCann is a Professor of Geography at the University of Victoria. From 1987 to 1992 he was Director of the Centre for Canadian Studies and the Edgar and Dorothy Davidson Professor of Canadian Studies at Mount Allison University. He has written widely on Canadian urban and regional development and is at work on a book about the urbanization of Canada and its regions for Cambridge University Press.

Lewis Robinson is a former Professor and Head of the Department of Geography at the University of British Columbia. He has conducted field studies in every part of Canada, and his lengthy publishing record on Canadian geography includes *Concepts and Themes in the Regional Geography of Canada*.

John Stager is a former Professor of Geography at the University of British Columbia where he specialized in physical geography. His publications deal with physical, historical, social, and environmental issues affecting the Arctic. He is co-author, with Harry Swain, of *Canada North: A Journey to the High Arctic*. He served for two terms as a member of the Canadian Polar Commission.

Peter Usher has a Ph.D from the University of British Columbia and has conducted field research for many years in the Canadian North. Since 1977 he has operated an independent consultancy based in Ottawa, providing research on northern land issues and aboriginal claims. At present he is Chair of the Wildlife Management Advisory Committee in the Western Arctic.

Iain Wallace is a Professor of Geography at Carleton University and a member of the Board of Governors of the Royal Canadian Geographical Society. He earned his doctorate in geography at the University of Oxford. He is the author of *The Global Economic System* and has written widely on aspects of the economic geography of Canada and on its relation to environmentalism.

Graeme Wynn is Professor and Head of the Department of Geography at the University of British Columbia where, from 1990 to 1996, he was Associate Dean of Arts. In 1988–90 he was recipient of the I. W. Killam Research Fellowship, and in 1998 was appointed Senior Visiting Research Fellow at St. John's College, Oxford. He is the author of many books and articles on Canadian geography and in recent years has contributed to *The Illustrated History of Canada* and *A Scholar's Guide to Geographical Writing on the American and Canadian Past.*

Maurice Yeates is the author of many articles and books on the urban geography of Canada and the United States including *Main Street: Windsor to Québec City* and *The North American City*. A former Dean of Graduate Studies and Research and Chairman of the Department of Geography at Queen's University, he is now a Senior Research Fellow at the Centre for the Study of Commercial Activity at Ryerson Polytechnic University, Toronto. He is a Fellow of the Royal Society of Canada.

chapter 1

Interpreting Canada's Heartland and Hinterland

Larry McCann

In a recent and provocative book, Richard Gwyn, a journalist for *The Toronto Star*, labelled Canada as the first "postmodern Dominion." For Gwyn, Canadian society at the close of the twentieth century was very different in "texture and temper" from even a decade or so ago, when *Heartland and Hinterland* was first published. Canada was no longer the "old anglophone and francophone *deux nations* plus aboriginals." A recent burst of immigrants from old and new source areas, particularly Asia and the Caribbean, but also from the Middle East and South America, was making the country a "microcosm of all the world's peoples." The old economy of natural resource and automobile production was still evident — even flourishing — but the country's external business connections, sparked by a communications revolution, were more global and diverse than ever before, as the post industrial wave of producer service industries became critical for creating wealth and new jobs (see Table 1.1). Gwyn also argued that these social, economic, and technological changes were having a profound spatial impact. Canada was becoming more "decentred...no longer just into our historic regions and provinces but more and more now into all our new 'identity' communities." The "city states" of Toronto, Montréal, and Vancouver were becoming increasingly important in the lives of Canadians.[1]

Whether or not Canada is indeed the world's first postmodern country — with its society, economy, and spatial divisions more fragmented — is a matter for debate, but Gwyn is certainly correct in pointing to the substantial forces that are currently remaking the regional geography of Canada. Like all countries in the

TABLE 1.1   Canada's Share of Global Exports: Telecommunications and Aerospace Products, 1980 and 1990

Telecommunications (%)				Aerospace (%)			
1980		1990		1980		1990	
Germany	16.7	Japan	28.4	USA	47.6	USA	50.3
Sweden	15.3	USA	15.9	Britain	19.7	France	17.5
USA	10.9	Sweden	6.9	Germany	9.1	Britain	7.7
Japan	10.3	Germany	6.5	France	6.0	Germany	4.1
Netherlands	9.3	Canada	4.7	Canada	4.4	Canada	4.0
Belgium	7.4						
France	6.5						
Canada	5.1						

Source: Handbook of International Economic Statistics, 1993

[handwritten margin note: main forces remaking the regional geography of Canada]

modern world-system, Canada is currently experiencing processes of change that have become global in scope and impact. In earlier periods, mercantile trade, territorial expansion, industrialization, and urbanization successively and gradually shaped and reshaped the Canadian landscape.[2] More recently, ever-expanding flows of foreign capital and immigrant labour, increasing political realignment and technological advances, and widespread industrial restructuring and urban transformation are being implemented with incredible speed, forcing seemingly continuous change. Throughout this metamorphosis, however, many features of Canada have proven remarkably stable, not least the unmistakable — and enduring — spatial pattern of heartland and hinterland regions. This pairing of terms, also expressed as the metropolis–hinterland, core–periphery, centre–periphery, or centre–margin dichotomy,[3] is the fundamental geographic framework — a product of capitalism itself — through which the ever-changing character of Canada's regions has been shaped and reshaped.[4]

By blending theory and empiricism, this introductory chapter outlines the heartland–hinterland approach as a meaningful framework or paradigm for regional analysis, and introduces some important and integrating themes that are addressed throughout the book. The chapter also emphasizes that to understand Canada's regional geography, an approach stressing change and development over time is essential. A region's contemporary economic and cultural character, like that of the country as a whole, is an expression of both past and present geographies. By focussing on geographical change since the late nineteenth century, we are also able to identify the full and accumulated impact of the heartland–hinterland process in shaping Canada's regional geography. As Richard Gwyn suggests, Canada might very well be entering a new phase where our traditional national identity is being challenged, even redefined. But as we consider just what this identity is, it is essential to remember that past landscapes are still very much a part of the country's contemporary geography.

HEARTLAND AND HINTERLAND:
Landscapes of Contrast

The heartland–hinterland model has been defined by John Friedmann as a "general theory of polarized growth" that is applicable at all geographical scales, including international, national, and (in Canada's case) provincial.[5] Just as the trading areas of central places overlap in hierarchical fashion, so also do the spheres of influence of heartlands and hinterlands, as well as their cities within. The world-system now comprises several metropolises that are the control points for the world economy — New York, London, and Tokyo.[6] Toronto and Montréal, while important in Canada and displaying a global outreach in some economic sectors, fall very much under the financial shadow of these global cities, especially New York.[7] In turn, provincial cities like Halifax and Moncton in the Maritimes, or Winnipeg and Regina in the Western Interior, feel the decision-making strength of Toronto. But these provincial places nevertheless exert a force over their local regions.

Regardless of scale, the meaning of heartland and hinterland regions can be clearly defined by reference to certain economic, social, settlement, and political processes, as well as geographic factors. Heartlands usually develop in areas possessing favourable physical qualities and facilitating good access to markets. They

New York City – skyline

display a diversified profile of secondary, tertiary, and quaternary industries, and they are characterized by a highly urbanized and concentrated population which participates in a well-integrated system of cities. They are well advanced along the development path, and possess the capacity for "innovative change;"[8] and they are able to influence and usually control — primarily through metropolitan actions — economic, social, and often political decisions of national importance.

heartlands

Hinterlands are characterized by the obverse; an emphasis on primary resource production; scattered population and weakly integrated urban systems; limited innovative capacity; and restricted political prowess. Hinterlands, therefore, are all the regions lying beyond the heartland whose growth and change is largely determined by their weaker status and dependency relationships with the heartland.

hinterlands

Population patterns across the country support the heartland–hinterland scheme. A little more than 30 000 000 people inhabit Canada's 9 970 000 square kilometres. About one-third of them live in metropolitan Toronto, Montréal, and Vancouver. Only one-eighth of the country is effectively occupied. There are still large areas where no human has yet set foot (Map 1.1). Agricultural societies emerged early in areas fringing the lower Great Lakes and parallelling the St. Lawrence River, and later appeared across the prairie and parkland zones of the Great Plains. Still, only about four per cent of Canada is cultivated, and it is likely that little more will ever be useful agricultural land. In the Cordilleran region and across the massive Canadian Shield, population is scattered, hugging valley bottoms, or staying close to transportation routes. Settlements appear in isolation — logging camps, lumbering towns, mining communities — fixed in space by the availability of forest and mineral resources. These settlement patterns are characteristic of the great resource-producing hinterlands of the country.

Daniel Johnson Dam,
Manicouagan River,
Québec

Map 1.1 The Canadian Ecumene and the Distribution of Urban Population, 1996

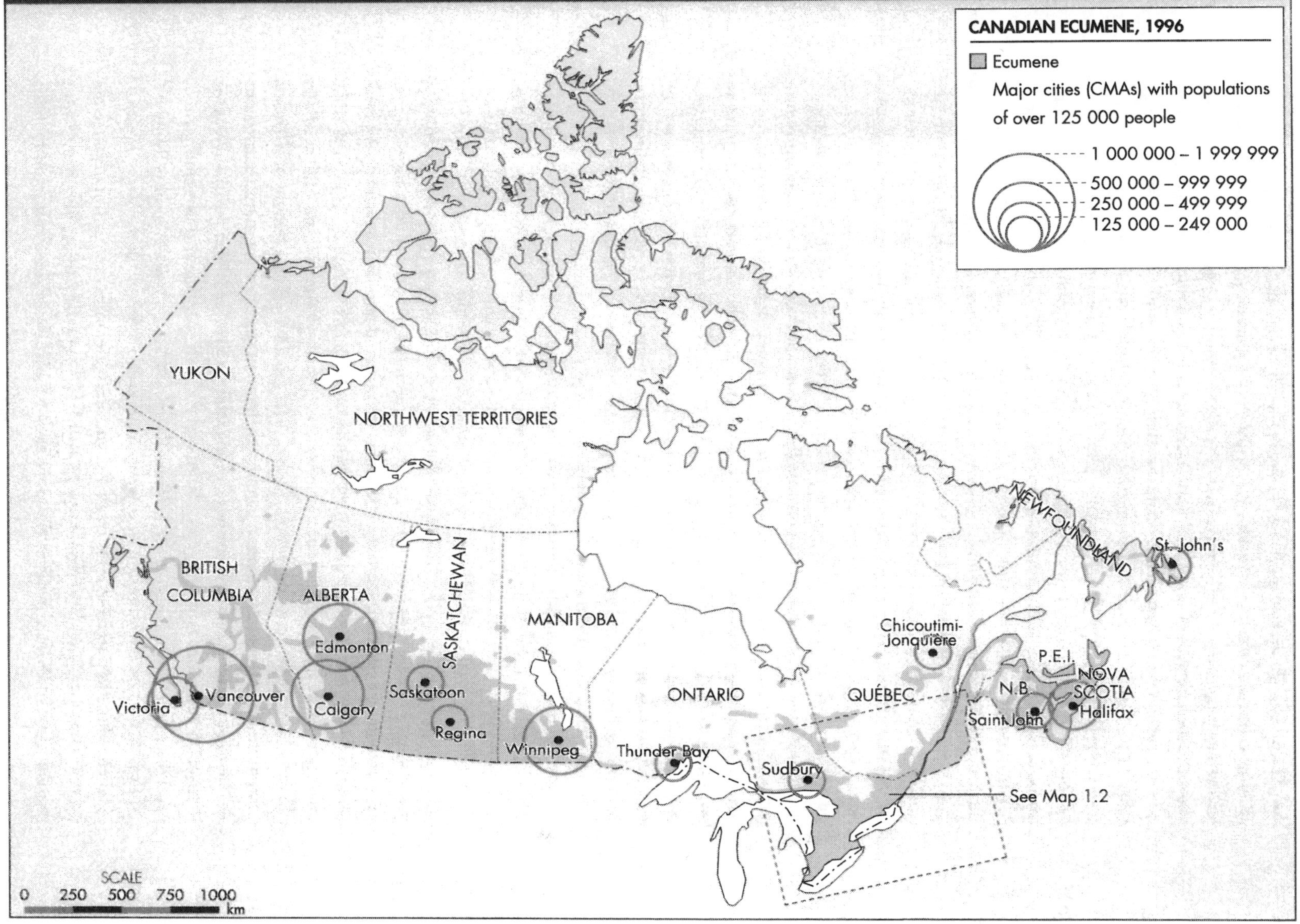

See Map 1.2

CANADIAN ECUMENE, 1996

Ecumene

Major cities (CMAs) with populations
of over 125 000 people

1 000 000 – 1 999 999
500 000 – 999 999
250 000 – 499 999
125 000 – 249 000

St. John's

NEWFOUNDLAND

NOVA SCOTIA

P.E.I.

N.B.

Halifax

Saint John

Chicoutimi-Jonquière

QUÉBEC

Sudbury

ONTARIO

Thunder Bay

MANITOBA

Winnipeg

SASKATCHEWAN

Saskatoon

Regina

ALBERTA

Edmonton

Calgary

BRITISH COLUMBIA

Vancouver

Victoria

YUKON

NORTHWEST TERRITORIES

SCALE

0   250   500   750   1000
km

physical environment influences heartland – hinterland structure

The physical environments of Canada — the immense physiographic divisions, the broad sweeps of climatic variations, the location of soils, the diversity of vegetation types — are all firmly etched in the Canadian psyche, influencing our perception of heartland and hinterland structure. It is obvious that Canada's economic core, lying in southern Ontario and Québec and called the Industrial Heartland, coincides with favourable physical characteristics (Map 1.2). It occupies the Great Lakes–St. Lawrence Lowlands physiographic unit, which is endowed with rich grey-brown podzolic soils and the growth-inducing temperatures and precipitation of a humid continental climate. The hinterland regions generally experience harsher and often rather limiting physical traits. The Cordillera, Interior Plains, Canadian Shield, and the Appalachian physiographic regions yield tremendous resource wealth, but their soils, vegetation types, and climatic patterns do not favour wide distributions of population and concentrated development. Across the Shield, the severe climate and rugged terrain — interspersed with lakes, muskeg, and forest — have made settlement difficult. In southern districts of the vast Western Interior region, the natural landscape is, for the most part, gently undulating. Together with a continental climate of cold winters and dry summers, this environment favours an agricultural economy based mainly on grain and cattle production. Variability of precipitation in Palliser's Triangle, however, creates seasons of drought. Elsewhere, early fall frosts or summer hail can spell disaster. In the Maritimes, by contrast, resource production is affected by the marine environment. For example, offshore storms and icebergs will be a threat to the burgeoning oil and natural gas industry; and unless ocean water temperatures are near 5°C, cod do not spawn in inshore waters.

In the final analysis the natural environment is no mere given, no neutral constant, no passively endured condition. Rather, the natural environment is an integral part of our life-world, as deeply shaped by social and economic conditions as social and economic conditions are mediated by it. To this end, and standing in sharp contrast to the bulk of the Canadian landscape, several densely populated regions have emerged as core areas where people have concentrated in large cities to manufacture products, provide tertiary services, and offer producer service goods.[9] These are the control or pivot points of Canada's regional geography that reach across the country, integrating the space economy (Map 1.1 and Map 1.2). In the west, Victoria and Vancouver define the Georgia Strait region of British Columbia, Calgary and Edmonton anchor an intensifying development corridor in the heart of Alberta, and Winnipeg, which once controlled entry to the Western Interior, is Manitoba's primate city. On the east coast, a string of towns and cities, stretching from Saint John through Moncton to Halifax, supports the development aspirations of the Maritimes, and in Newfoundland, St. John's alone plays the pivotal role. But standing above all others in national importance is the highly urbanized Windsor to Québec corridor, the core of the Canadian ecumene, where the economy is diversified, population concentrated, and settlements interconnected.

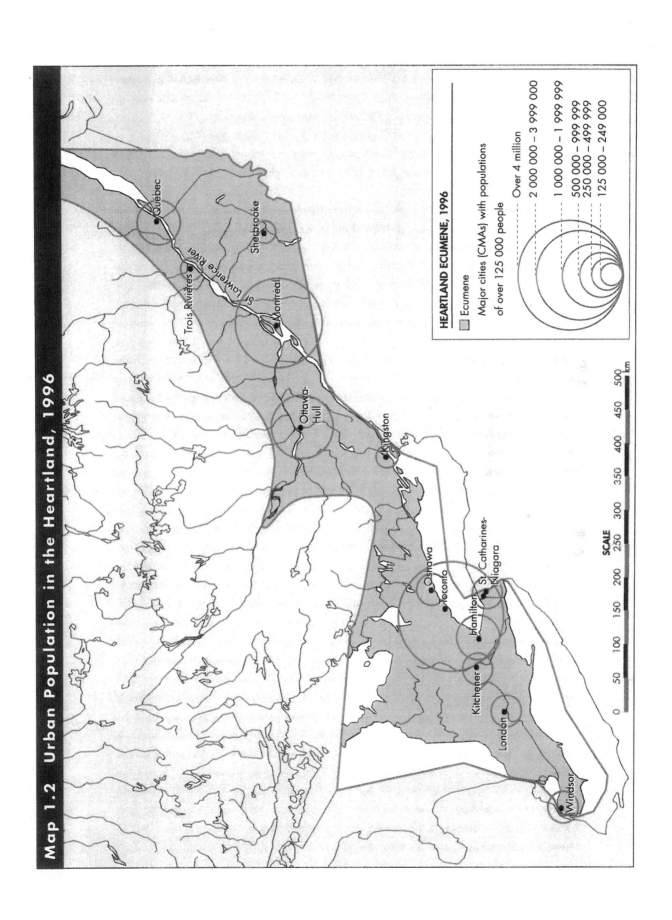

Map 1.2 Urban Population in the Heartland, 1996

HEARTLAND ECUMENE, 1996

Ecumene

Major cities (CMAs) with populations of over 125 000 people

- Over 4 million
- 2 000 000 – 3 999 000
- 1 000 000 – 1 999 999
- 500 000 – 999 999
- 250 000 – 499 999
- 125 000 – 249 000

Québec

Sherbrooke

Trois Rivières

St Lawrence River

Montréal

Ottawa-Hull

Kingston

Oshawa

Toronto

Hamilton

St Catharines-Niagara

Kitchener

London

Windsor

SCALE

0   50   100   150   200   250   300   350   400   450   500
km

The spatial structure of heartland and hinterland regions — economically rooted, socially constructed, environmentally conditioned, and expressive of cultural values — is a most notable feature of the Canadian landscape.

THE HEARTLAND–HINTERLAND PROCESS

In the heartland–hinterland system, regions interact with, or relate to, each other in a variety of important ways to shape their geographic character. The most important is usually economic, exemplified by exchanges from core to periphery of financial capital, good and services, and government transfer payments, or by the shipment of staple commodities from a resource periphery to an industrial core. The cultural transfer of architectural forms, foods, ethnic groups, languages, religious ways, and laws — to cite only a few cultural traits that might move from cultural hearth to a newly settled area — is also important for shaping regional character. The same can be said for state policies, corporate initiatives, technological innovations, and social trends. Economically, the manner of and degree to which these interactions take place measure the quality and strength of regional integration and character in a core–periphery system, resulting in such differentiating traits as regional equality or inequality, self-reliance or dependency. Culturally, the transfer of values, customs, or traits can lead to national similarities or more locally or regionally expressed traditions. Politically, interaction between centres of power and marginal areas can induce intense feelings of alienation and regionalism (or regional consciousness) that further define regional character. In these ways, the interaction between regions, particularly as expressed through the heartland–hinterland process, is a critical force in shaping a region's geographical character.

The basic forms of economic interaction in a heartland–hinterland system distinguish a staple-producing periphery from an industrial core.[10] It is the industrial core (or heartland) that creates the demand for staple commodities, supplying the resource periphery (or hinterland), in turn, with capital, labour, technology, and entrepreneurship. These factors of production are essential for the initial growth and sustained development of a hinterland. The heartland–hinterland model thus offers a framework for examining, at various geographic scales, the movement of people, goods and services, investment capital, and technology from one region to another, and of internal and external change over time. It also suggests that, because the source of demand rests at the centre, the periphery is dependent on the centre not only for supplies of capital or technical expertise, for example, but also for patterns of living, social organization, and well-being. The theoretical implication is clear: that power — the ability to innovate and control — usually belongs to the heartland. To make this theoretical claim, however, is not to deny the empirical observations discussed throughout this book that hinterlands can — and do — successfully challenge core power.

Interaction between heartland and hinterland regions, whether of an economic, social, cultural, or political nature, therefore has considerable force in shaping regional and national character. The following examples illustrate this point. On a world scale, the seventeenth- and eighteenth-century migrations of French and British peoples from European hearth areas established settlement patters across eastern Canada that have persisted to the present.[11] Later, the nineteenth-century emigration of Maritimers to the "Boston States" recorded not only the powerful attraction of the American heartland, but also a declining regional economy. Even today, Canada is characterized as a hinterland source of industrial materials, exporting billions of dollars worth of staple commodities (raw materials or resource-intensive goods such as grains or wood pulp) to world markets — wheat to China and Russia, copper, coal, and pulp to Japan, hogs, lumber, and paper to the United States, potash and zinc to India — but (excepting automobiles and auto parts) comparatively few manufactured products or specialized services.

This imbalance is slowly changing, in part because of recent free-trade agreements such as NAFTA (the North American Free Trade Agreement), in part because of the need to adapt new technologies to traditional industries, and in part because of the growth of producer service industries spurred on by global economic restructuring. Free trade and the necessity of technological change, for example, have both encouraged and forced high-tech manufacturers such as Northern Telecom and firms that re-manufacture specialty wood products in British Columbia to re-tool to win over global markets. Reflecting post-industrial trends, Canadian architectural, engineering, and construction firms have won major contracts in Asia where the need for an industrial infrastructure is expanding greatly. Canada's position in the world-economy is clearly paradoxical, for it does send abroad a limited range of manufactured goods and producer services; and it does invest capital directly in many foreign economies.[12]

Heartland–hinterland interaction can be a complex matter. Canada's hinterland areas, for example, ship many of their staple commodities directly to foreign customers, so that the goods neither pass through the national core nor use its business organizations. It can thus be argued that, to a degree, the Canadian heartland depends on the development of the periphery, not the other way around. Without the hinterland's demand for the industrial core's goods and services, or the political core's right to tax the international activities of hinterland firms, would central Canada be as prosperous or as important as it is today?[13] On the other hand, when the shipment of manufactured goods is considered, industrially important central Canada stands in marked contrast to the resource-oriented hinterlands. The Industrial Heartland is still, after all, the country's chief source of manufactured goods. In addition, the internal movement of manufactured products within the Heartland (including semi-finished materials such as steel, which are destined for further upgrading) is an indication of regional integration, giving added importance to the Windsor-Québec corridor as a highly specialized, functional region.[14]

Regional structure, then, is the product of the interaction between heartland and hinterland areas at various geographic scales, and of the processes giving definition to each type of area. Locational forces and even policy decisions of a political nature draw secondary manufacturing and service activities, as well as a skilled labour force, to core areas. The further concentration of corporate headquarters and financial institutions there, encouraged by government regulations and supported by substantial agglomerative economies, also causes a flow of profits from the hinterland to the heartland, making it difficult to raise development capital within the periphery. Even in the current period of greater capital mobility at a global scale, such occurrences still lie at the root of hinterland dependency and the underdevelopment problem, and are even more difficult to overcome without state intervention. Under the terms of NAFTA, of course, the ability of the federal government to provide regional development incentives is greatly restricted. Even government assistance through transfer payments (fast reducing as all governments pursue programs of fiscal restraint) is no guarantee that disparities between core and peripheral regions will narrow, let alone disappear.

If disparities are to be diminished, it seems more likely — and more essential than ever before — that hinterland regions must develop generally according to the ways in which heartland areas have always developed, although the specific growth factors need not, nor would they likely, be the same. For a region to advance toward heartland status and therefore be capable of innovating change and wielding power, it must progress beyond the phase of merely supplying staples to an industrial core, generating, instead, through the multiplier process, a diversified industrial profile in most economic sectors (Figure 1.1). By this process, linked industries emerge as spinoffs from the principal staple exports. The region must also become highly urbanized and develop an integrated city system. These advances will only come about when capital, derived from the export sector, is reinvested directly into a program of regional economic diversification, providing, of course, that local or external markets warrant such a program. At this point, the emerging core area will begin to attract additional supplies of labour, investment capital, and technical expertise, not just to the export sector, but also to residential industries serving an internal market. As the regional market expands and region-serving activities proliferate, conditions may develop for self-reinforcing growth, and new internal factors — a decisive entrepreneurial group, for example — become all-important in determining regional growth. Internal growth and spatial concentration will be reinforced as more manufacturing or producer service industries are attracted to the core by the external economies associated with overhead capital and the agglomeration or clustering of industries.[15]

The effect of such economic and population growth is to create a settlement system that increasingly displays visible signs of regional integration. Highway construction, rail and air systems, and communication networks will connect urban places and show features of the rank-size rule of urban population distribution. An internal core–periphery design is the end product. This process of settlement development is most advanced in the Windsor–Québec corridor, Canada's Industrial

FIGURE 1.1 The Process of Urban Growth in a Heartland and Hinterland Economic System

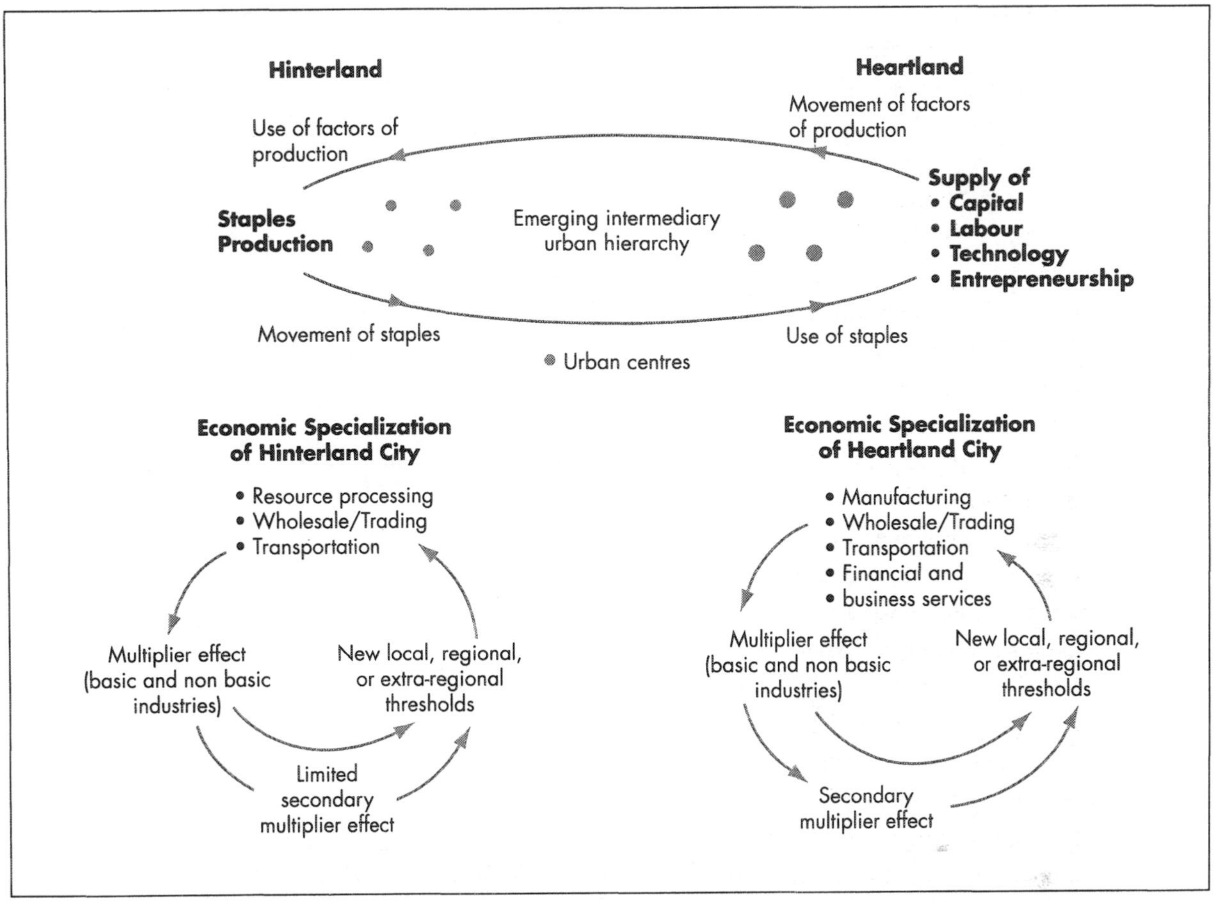

Heartland. Over the past few decades, it has emerged in dramatic fashion in British Columbia and Alberta, but is much less featured in the Maritime provinces and in Newfoundland, Manitoba, or Saskatchewan.

Table 1.2 records a century and more of progress in the development of Canada's provinces, as well as the changing structural relationship between heartland and hinterland. Employment statistics for key economic sectors and urbanization levels record succinctly the process of regional transformation accompanying Canada's transition from industrial to post-industrial nation. The course of economic and population development in most Canadian regions depended initially on the exploitation of a number of staple products and their export to world markets. Secondary manufacturing in time played a significant role. Now, producer services and more general service activities are of particular importance to some regions. Increasingly, these are gravitating toward a few large metropolitan centres like Toronto, Montréal, and Vancouver. The provincial values of Table 1.2 mask this growing urban concentration. Thus, we need to remember the importance of cities in shaping the space economy.

TABLE 1.2 Economic Development and Urbanization in Canada, By Regions 1891-1995

| Region | Gainfully Occupied by Sectors, 1891 | | | | | | | | | | Population, 1891 | | | |
| | Percentage of Regional Total | | | Total Regional Labour Force (000s) | Percentage of Canadian Total | | | Total Regional Population | Percentage of Canadian Total | Percentage of Regional Population Urban |
	Primary	Secondary	Tertiary		Primary	Secondary	Tertiary			
HEARTLAND										
Ontario	48.0	26.9	25.1	724.1	43.7	48.4	46.2	2 114 321	43.8	38.7
Québec	48.6	26.8	24.6	449.6	27.5	29.9	27.9	1 488 535	30.9	33.6
HINTERLAND										
Nova Scotia	54.2	20.3	25.5	156.5	10.6	10.1	10.1	450 396	9.3	17.1
New Brunswick	52.2	24.3	23.5	107.2	7.0	6.5	6.4	321 263	6.6	15.2
Prince Edward Island	65.1	17.7	17.2	35.0	2.8	1.6	1.5	109 078	2.3	12.8
Maritimes	*54.8*	*21.4*	*23.8*	*298.7*	*20.4*	*15.6*	*18.0*	*880 737*	*18.2*	*16.5*
Manitoba	64.6	13.4	22.0	53.7	4.3	1.9	3.0	152 506	3.1	26.8
Saskatchewan	64.0	10.9	25.1	21.1	1.7	0.7	1.3	–	–	–
Alberta	–	–	–	–	–	–	–	–	–	–
Western Interior	*64.4*	*12.7*	*22.9*	*74.8*	*6.0*	*2.6*	*4.3*	–	*2.0*	–
British Columbia	40.7	29.6	29.7	47.2	2.4	3.5	3.6	98 173	2.0	39.9
Canada	50.0	25.3	24.7	1 595.5	797.7	403.2	394.6	4 833 239	–	31.8

TABLE 1.2 (continued)

| Region | Gainfully Occupied by Sectors, 1929 | | | | | | | Population, 1931 | | |
| | Percentage of Regional Total | | | Total Regional Labour Force (000s) | Percentage of Canadian Total | | | Total Regional Population | Percentage of Canadian Total | Percentage of Regional Population Urban |
	Primary	Secondary	Tertiary		Primary	Secondary	Tertiary			
HEARTLAND										
Ontario	25.2	32.2	42.6	1 423.8	25.6	44.2	36.3	3 431 683	33.1	58.7
Québec	26.5	31.4	42.1	1 082.0	20.5	32.6	27.2	2 874 662	27.7	58.6
HINTERLAND										
Nova Scotia	47.4	17.4	35.2	197.9	6.7	3.3	4.2	512 846	4.9	43.5
New Brunswick	44.9	18.3	36.8	142.3	4.6	2.5	3.1	408 219	3.9	31.1
Prince Edward Island	63.9	9.0	27.1	33.2	1.5	0.3	0.5	88 038	0.9	19.5
Maritimes	47.9	17.0	35.1	373.4	9.4	6.1	7.8	1 009 103	9.7	36.3
Manitoba	37.7	17.4	44.9	276.8	7.5	4.6	7.4	700 139	6.8	42.0
Saskatchewan	61.0	6.9	32.1	335.5	14.6	2.2	6.4	921 785	8.9	20.3
Alberta	55.0	9.6	35.4	286.6	11.2	2.7	6.0	731 605	7.1	31.1
Western Interior	53.3	9.5	37.1	898.9	33.3	9.5	19.8	2 353 529	22.8	30.1
British Columbia	51.8	10.9	44.3	337.9	7.8	7.6	8.9	694 263	6.7	55.4
Canada	34.0	25.3	40.7	4 116.0	1 398.7	1 039.7	1 677.6	10 376 786	—	49.7

TABLE 1.2 (continued)

Gainfully Occupied by Sectors, 1956

Region	Percentage of Regional Total			Total Regional Labour Force (000s)	Percentage of Canadian Total			Population, 1956		
	Primary	Secondary	Tertiary		Primary	Secondary	Tertiary	Total Regional Population	Percentage of Canadian Total	Percentage of Regional Population Urban
HEARTLAND										
Ontario	11.2	38.8	49.0	2 148.2	24.6	44.8	37.6	5 404 933	33.6	75.9
Québec	16.1	37.9	46.0	1 598.2	24.3	32.5	26.3	4 628 378	28.8	70.0
HINTERLAND										
Nova Scotia	22.6	24.3	53.1	242.8	5.1	3.1	4.6	694 717	4.3	57.4
New Brunswick	25.3	25.1	49.6	180.4	4.3	2.4	3.2	554 616	3.5	45.8
Prince Edward Island	40.4	16.4	43.2	36.6	1.4	0.3	0.6	99 285	0.6	30.3
Maritimes	*25.1*	*24.0*	*50.9*	*459.8*	*10.8*	*5.8*	*8.4*	*1 348 618*	*8.4*	*50.6*
Manitoba	24.5	22.1	53.4	328.4	7.5	3.9	6.3	850 040	5.3	60.1
Saskatchewan	46.6	10.4	43.0	323.7	14.1	1.8	4.9	880 665	5.5	36.6
Alberta	34.6	17.7	47.7	387.4	12.6	3.7	6.6	1 123 116	7.0	56.6
Western Interior	*35.2*	*16.8*	*48.0*	*1 039.5*	*34.2*	*9.4*	*17.8*	*2 853 821*	*17.8*	*51.5*
British Columbia	13.6	29.0	57.4	483.0	6.1	7.5	9.9	1 398 464	8.7	73.3
Canada	18.6	32.6	48.8	5 728.7	1 066.9	864.7	2 797.1	16 080 791	—	66.6

TABLE 1.2 (continued)

Region	Percentage of Regional Total[a]			Total Regional Labour Force (000s)	Percentage of Canadian Total			Total Regional Population	Percentage of Canadian Total	Percentage of Regional Population Urban
	Primary	Secondary	Tertiary		Primary	Secondary	Tertiary			
HEARTLAND										
Ontario	4.4	28.1	62.8	4 548	24.2	43.0	36.0	8 625 110	35.4	81.7
Québec	4.3	25.5	62.8	3 100	16.1	26.6	25.0	6 438 400	26.4	77.6
HINTERLAND										
Newfoundland	9.4	22.6	60.1	232	2.6	1.8	1.8	567 680	2.3	58.6
Nova Scotia	7.0	20.3	67.0	388	3.3	2.7	3.3	847 445	3.5	55.1
New Brunswick	7.7	22.5	63.1	308	2.9	2.3	2.5	696 405	2.9	50.7
Prince Edward Island	15.6	17.7	61.3	57	1.1	0.3	0.5	122 510	0.5	36.3
Maritimes	*12.6*	*20.9*	*64.9*	*753*	*7.3*	*5.3*	*6.3*	*1 666 360*	*6.9*	*52.9*
Manitoba	10.0	18.6	67.2	510	6.2	3.2	4.4	1 026 245	4.2	71.2
Saskatchewan	21.7	12.6	61.3	461	12.1	2.0	3.6	968 310	4.0	58.2
Alberta	13.2	19.1	64.0	1 213	19.3	7.8	10.0	2 237 725	9.2	77.2
Western Interior	*14.2*	*17.6*	*64.1*	*2 184*	*37.6*	*13.0*	*18.0*	*4 232 280*	*17.4*	*71.4*
British Columbia	6.9	21.5	66.5	1 412	11.8	10.2	12.0	2 744 470	11.3	78.0
Yukon	11.5	9.1	71.1	13	0.2	0.04	0.12	23 150	0.1	64.0
Northwest Territories	12.6	7.1	74.1	19	0.3	0.05	0.19	45 740	0.2	48.1
Canada	6.8	24.2	63.6	12 267	—	—	—	24 343 190	—	75.7

Gainfully Occupied by Sectors, 1981 | Population, 1981

[a] These data do not total 100 per cent because they exclude the share of the unspecified labour force.

TABLE 1.2 (continued)

| | Gainfully Occupied by Sectors, 1994 | | | | | | | Population, 1996 | | |
| | Percentage of Regional Total | | | Total Regional Labour Force (000s) | Percentage of Canadian Total | | | Total Regional Population | Percentage of Canadian Total | Percentage of Regional Population Urban |
Region	Primary	Secondary	Tertiary		Primary	Secondary	Tertiary			
HEARTLAND										
Ontario	3.0	23.0	74.0	5 160	22.1	43.9	38.6	10 753 573	37.3	83.3
Québec	3.6	22.5	73.9	3 156	16.4	26.3	22.6	7 138 795	24.7	78.4
HINTERLAND										
Newfoundland	9.2	12.3	78.5	195	2.6	0.9	1.5	551 792	1.9	56.9
Nova Scotia	7.1	16.0	77.9	380	3.3	2.3	3.0	909 282	3.1	54.8
New Brunswick	6.2	16.6	77.2	307	2.7	1.9	2.4	738 133	2.5	48.8
Prince Edward Island	7.1	16.1	76.8	56	0.6	0.3	0.4	134 557	0.5	44.2
Maritimes	7.9	15.4	77.7	938	6.6	4.5	5.8	2 333 764	—	52.2
Manitoba	8.6	16.6	74.8	511	6.3	3.1	3.9	1 113 898	3.9	71.8
Saskatchewan	17.7	10.3	72.0	457	11.5	1.7	3.3	990 237	3.4	63.3
Alberta	11.7	14.5	73.8	1 337	22.2	7.2	10.0	2 696 826	9.3	79.5
Western Interior	12.2	14.1	73.7	2 305	40.0	12.0	17.2	4 800 961	—	75.0
British Columbia	4.7	19.4	75.9	1 733	11.5	12.5	13.3	3 724 500	12.9	82.1
Yukon	—	—	—	—	—	—	—	30 766	0.1	60.0
Northwest Territories	—	—	—	—	—	—	—	64 402	0.2	42.5
Canada	5.2	21.0	73.8	13 292	—	—	—	28 846 761	—	77.9

Sources: Labour force data have been derived from information published in Alan G. Green, *Regional Aspects of Canada's Economic Growth* (Toronto: University of Toronto Press, 1971), Appendix C. pp. 102-7; Statistics Canada, *Labour Force Annual Averages 1975-1994* (Ottawa: 1979-1995); and Statistics Canada, *Census of Canada*, 1981 (Ottawa: 1983). Green's analysis for 1891, 1929, and 1956 excludes Newfoundland. Population data have been compiled from information in *Census of Canada, 1891-1991*; and Statistics Canada, CANSIM (Ottawa: 1996), Matrix 6231.

More than a century ago, when rural settlements sheltered most of Canada's population, over half the country's labour force was still engaged in agriculture, lumbering, fishing, and mining activities, mostly in eastern Canada (Table 1.2). These resource-based industries supplied a growing demand for lumber, fish, coal, and foodstuffs in Europe, the United States, the West Indies, and elsewhere. Since the late nineteenth century, there has been a steady westward progression of resource production because foreign demand continued to increase, a national transportation system was put in place, and new staples replaced exhausted supplies. This spatial progression across Canada — part of the country's so-called Great Transformation[16] — reversed the rural-urban balance and expanded the core–periphery pattern. It did so by attracting European and American investment and immigration, by stimulating population shifts and manufacturing expansion, and by fostering the build-up of cities and social well-being — all themes developed in subsequent chapters.

In the context of our theoretical discussion of regional development, staple exports promoted the initial development of the Canadian economy through spread effects. Income from the staple-based economy was used to manufacture consumer and producer goods and to foster specialized tertiary services, creating the basis for the emergence of a national industrial and financial core embracing Montréal and Toronto. There, in the Industrial Heartland, merchants invested heavily in a variety of manufacturing activities and founded banking and insurance companies throughout Ontario and Québec. Over the course of the past one hundred or so years, various factors — including transport and route developments, the accumulation of agglomeration economies, entrepreneurial behaviour (particularly combination and merger and oligopolistic competition), factor immobility, and initial advantages in the guise of situation, relative accessibility, labour and capital availability, and tariff protection — all have placed central Canada at a distinct advantage over peripheral regions.

As the twentieth century closes, labour force and population data confirm that Canada's traditional pattern of heartland and hinterland remains firmly in place, despite weakening incursions (Table 1.2). The Industrial Heartland, especially the group of cities forming the so-called Golden Horseshoe around Toronto, continues to attract the majority of foreign investment funds in secondary manufacturing and the business operations of multinational corporations. Where the United States once clearly dominated investment in this area, Japanese and other Asian firms are now major participants. Despite this Asian presence, the core's relative share of population and economic production has been slipping over the last few decades. In Canada's goods-producing industries, Ontario and Québec's share of census value-added products declined by about 10 percentage points between 1961 and 1995, or from about 70 to 60 per cent. The slack has been taken up especially by Alberta, but also (and increasingly) by British Columbia (19 per cent to 27 per cent). Within Canada's Industrial Heartland region, Ontario has maintained its almost 50 per cent share of the country's manufacturing. The province is attempting, somewhat successfully, to restructure its manufacturing base by promoting new and technologically advanced industries. Certainly, the province's manufacturing sector has not declined at the hands of NAFTA as drastically as some forecasters initially predicted. By contrast, manufacturing in Québec, which has relied traditionally on labour-intensive industries, continues to lose ground proportionately, from 33 per cent in 1961 to about one-quarter in 1995.[17]

This loss has been captured by Alberta's petroleum products and petrochemical industries, and by British Columbia's share of lumber, plywood, and pulp and paper production, now totally over half the Canadian total. Neither western province is immune from the world-economy's negative shocks. In British Columbia, as forest reserves dwindle, as new technologies emerge, as the United States market fluctuates, and as Japan wins concessions to import raw logs, the forest products industry has been forced to restructure and downsize, engaging increasingly, for example, in the re-manufacturing of specialized wood products (e.g. building materials) or in closing down outdated mills. Despite broadening its economy somewhat through an expanding pulp industry, Alberta, too, has suffered intermittent setbacks. Fluctuating provincial revenues in the oil and gas sectors have been profound, prompting substantial government cutbacks after a dramatic expansionary era and infrastructural build-up during the late 1970s and early 1980s. The cities of both provinces — but especially Vancouver and Calgary — nevertheless remain magnets for migrants from across Canada who seek better, if not steady, employment. Foreign immigrants, especially from Asian countries also target these cities. Fuelled in part by the entrepreneurial efforts of new immigrants, Alberta and British Columbia have diversified beyond staple production. Their service industries now comprise between 60 and 70 per cent of gross provincial domestic product. With their new-found wealth and global connections, especially in the western United States and throughout the Pacific Rim, western corporations are increasingly investing abroad, just as a core area would.[18]

Regions vary widely in their capacity to achieve full development, and certainly in their ability to attain heartland status. Few regions ever achieve a position of dominance in a core–periphery system. Not only is it difficult to overcome the cumulative advantages of an existing heartland, but a diversified profile of economic functions is unlikely to develop in a region which does not have, for example, strong access to large external markets. Few regions are so blessed. All of those functions that are influenced by internal and external scale economies, such as manufacturing and producer services, would be denied to such a region. The region would have to continue its growth and expansion through more limited but specialized activities and remain a hinterland. Its people might enjoy a high standard of living, as in Alberta and British Columbia, but fully diversified development is unlikely. With out concern in this book for regional development, this is an important point to keep in mind, particularly when interpreting the evolving geography of Canada's peripheral regions.

HEARTLAND AND HINTERLAND CITIES

While the industrial revolution in Canada, as elsewhere, benefitted urban places both large and small, the shift to a post-industrial economy is becoming more and more concentrated in the largest metropolitan centres. Toronto, Montréal, and Vancouver, and to a lesser extent Calgary, Edmonton, Winnipeg, the city of Québec, and Ottawa, continue to absorb a disproportionate share of the new investment and economic activity associated with post-industrial activity. Their skyscraper skylines host not only the multinational corporations participating in the changing global economy, but also symbolize status as leading metropolitan centres within Canada's national and regional space economies.

More than ever before, these cities are playing a fundamental role in shaping the heartland and hinterland character of the country. While this role is chiefly economic, metropolitan centres also disperse — amongst other things — technological, cultural, and social innovations to lesser urban places.[19] From a heartland-hinterland perspective, therefore, it is useful to think of cities as intermediaries, whose economic bases and functional roles stem from their handling of the factors of production as these move between core and periphery. Such interaction creates cities whose livelihoods depend on trade, transportation, manufacturing, or financial and specialized business activities (Map 1.3 and Map 1.4).[20] Although both heartland and hinterland cities will perform these functions, the degree of specialization and the composition of economic sectors in each type of city will vary. Consider these differences. It is quite unlikely that a hinterland city will have a fully diversified economic base. Depending upon the type and distribution of resources found within the periphery, cities there will function primarily as resource towns (forest products, mining, fishing, hydroelectricity), as central places (agriculture), as break-in-bulk or transshipment points, or, in those exceptional situations where location or circumstance favours diversification, as multifunctional settlements.

In most heartland cities, by contrast, manufacturing is a favoured activity because the core's accessibility to national markets creates an initial advantage, making possible the manufacture of a wide range of primary and consumer and producer goods. Such production in turn stimulates considerable regional employment and population growth. Because cities of the industrial core can usually supply national markets more efficiently and economically, manufacturers of consumer and producer goods are minimal in most hinterland cities, just as they are in peripheral regions. In a similar way, the economic base of a metropolitan centre is marked increasingly by highly specialized financial, management, and research activities — the very quaternary activities that are stimulating the post-industrial revolution.

Clearly, the attributes of cities — size, function, and regional settlement patterns — tell us a great deal about the geographic character of heartland and hinterland regions.[21] Without doubt, Canada's largest and most economically diverse cities are located in Ontario and Québec. Toronto and Montréal lead the way, but there are other important centres, including Ottawa, Québec, Hamilton,

Cities as intermediaries

Ottawa –
Rideau Canal
and Parliament
Buildings

St. Catharines, Kitchener, Waterloo, London, and Windsor (Map 1.4). Many of the core's other urban places are specialized manufacturing centres now in a state of transition. Cambridge, Brantford, Drummondville, Granby, Magog, Oshawa, Stratford, and Woodstock are examples of places that have experienced a sequence of industrialization, de-industrialization, and restructuring since World War II. While their past growth and prosperity were supported partially by tariff protection and branch plants from the Untied States, they are now being forced to adapt to new economic conditions imposed by NAFTA's regulations. The Québec textile towns are most vulnerable to decline. Ontario's Auto Pact participants are most likely to make the transition.

In the peripheral regions, by contrast, urban places are often dependent upon the external demand for one or two staple products, as in the fishing and pulp and paper towns of British Columbia, the agricultural service centres of the Western Interior, the mining communities of the Canadian Shield, or the fishing settlements of the Maritimes and Newfoundland. As fish stocks diminish on both coasts, fishing communities have suffered greatly in the 1990s.[22] The situation is quite different in several of the larger hinterland centres — Vancouver, Calgary, Edmonton, Saskatoon, Winnipeg, and Halifax, for instance. Here, a combination of resource,

Map 1.3 Economic Specialization of Urban-centred Regions, 1980s*

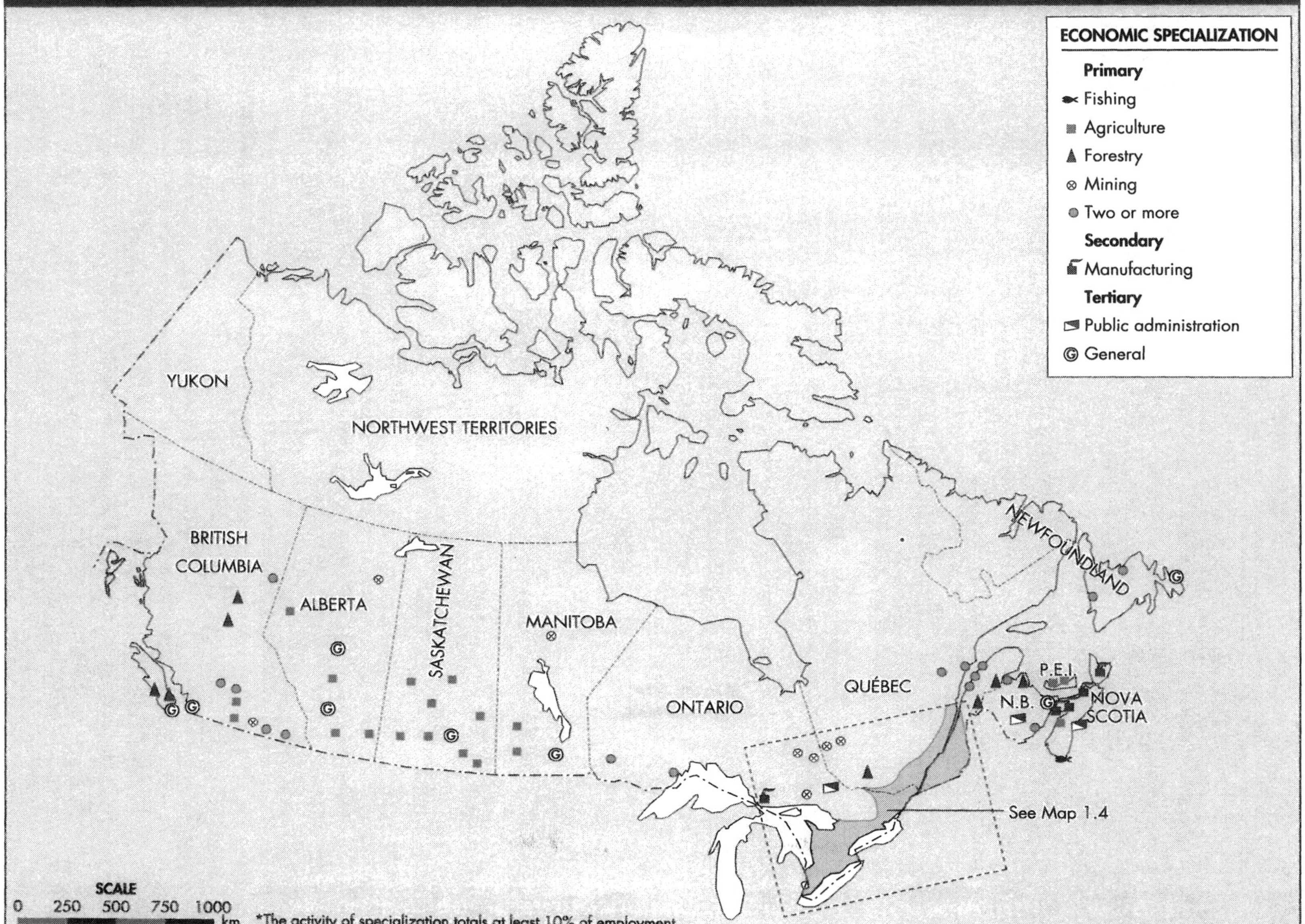

ECONOMIC SPECIALIZATION

Primary
- 🐟 Fishing
- ▦ Agriculture
- ▲ Forestry
- ⊗ Mining
- ● Two or more

Secondary
- ◣ Manufacturing

Tertiary
- ▱ Public administration
- Ⓖ General

YUKON

NORTHWEST TERRITORIES

BRITISH COLUMBIA

ALBERTA

SASKATCHEWAN

MANITOBA

ONTARIO

QUÉBEC

NEWFOUNDLAND

P.E.I.

N.B.

NOVA SCOTIA

See Map 1.4

SCALE

0   250   500   750   1000
km

*The activity of specialization totals at least 10% of employment.

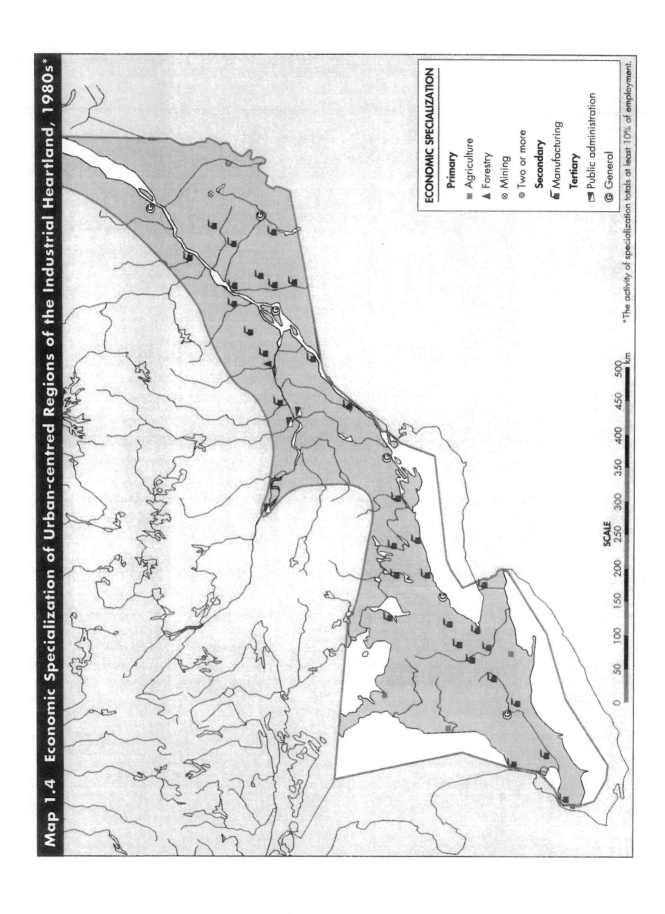

Map 1.4 Economic Specialization of Urban-centred Regions of the Industrial Heartland, 1980s*

ECONOMIC SPECIALIZATION

Primary
- Agriculture
- ▲ Forestry
- ⊗ Mining
- Two or more

Secondary
- Manufacturing

Tertiary
- Public administration
- Ⓖ General

*The activity of specialization totals at least 10% of employment.

SCALE

0   50   100   150   200   250   300   350   400   450   500
km

public sector, and manufacturing activities, as well as large population size, support the designation of regional metropolitan centre. Despite the presence of these centres there is still little clustering of cities in the periphery, for the dispersed pattern of resources there usually encourages the wide separation of urban places.

THE ROLE OF THE METROPOLIS

By the sheer weight of their importance, there are several Canadian cities that validate using the metropolitanism concept. The historian Maurice Careless has outlined the relevance of the term for understanding Canada's urban development in particular and the course of the country's geographical history in general:

> *Metropolitanism is at root a socio-economic concept... [implying] the emergence of a city of outstanding size to dominate not only its surrounding countryside but [also] other cities and their countrysides, the whole area being organized by the metropolis, through control of communication, trade, finance, into one economic and social unit that is focussed on the metropolitan 'centre of dominance' and through it trades with the world. Political activity, too, many often become centred on the metropolis.*[23]

In Canada, this definition of metropolitanism certainly applies to both Toronto and Montréal, and possibly describes the emerging power of Vancouver and Calgary, and less so Edmonton, Winnipeg, and other urban places. Metropolitan centres have considerable bearing on the functioning of a heartland–hinterland system. They organize and give definition to a region. Because of its concentration of people, wealth, and economic activity, a metropolis takes a leading role in shaping the economy, society, culture, and sometimes political life of a nation. A metropolis operates from a position of strength, achieved gradually through the acquisition of trading, manufacturing, transportation, and financial and specialized business roles. The order in which these functions is acquired is not important for determining leadership; rather, a controlling role measured by economic power is what determines dominance and binds a heartland–hinterland system together. Collectively, the more cities within a region that display this type of power, the greater the strength of the region as a whole for influencing change.

If this power is all-pervasive, it should be possible to measure the social and economic features of metropolitan strength, demonstrating that in such corporate sectors as finance, manufacturing, service, and resource concentration, the metropolis reigns supreme (Tables 1.3 and 1.4). In Canada, this remains as true today as ever before. Toronto has consolidated its function as the main control point of the heartland–hinterland system, a process that actually began before World War I, became quite obvious a few decades ago, and increased tremendously in the 1990s. Since the 1970s, Toronto has consistently retained control of nearly 50 per cent of the non-financial assets of Canadian corporations and, following the deregulation of the financial sector in the 1980s, it has far surpassed Montréal in this all-important sector.[24] For international banks and other foreign-controlled corporations in search of a Canadian headquarters site Toronto simply offers more locational advantages than Montréal or any other Canadian city (Table 1.4). As post-industrial growth in financial and producer services continues apace, Toronto has benefitted in a cumulative way from agglomeration economies associated with access to financial intermediaries, business contacts, and other economic benefits of urbanization.

Besides succumbing to Toronto's financial challenge, the fading Québec metropolis, Montréal has lost some important corporate battles. Recently, it did manage to retain headquarters for the world's largest newsprint conglomerate when Abitibi Price and Stone Consolidated joined together in 1997. The reasons were reportedly economic: Montréal has long been prominent in the pulp and paper sector, offering strong research facilities and business ties. But it did lose the head offices of Canadian Pacific and key officers of the Bank of Montréal to Calgary, where various amenities and access to skilled technical personnel and information were decisive relocation factors. In fact, for a small but growing number of corporations on the move, hinterland cities like Calgary possess locational advantages that far outweigh the higher costs of Montréal and Toronto. Moreover, while Toronto is still the overall financial leader and ranks first in revenues earned by manufacturing, service, and resource industries, metropolitan control in these non-financial sectors is more dispersed. For example, Vancouver is Canada's leading corporate centre for the forest products industry, Calgary heads over 80 per cent of the Canadian-based exploration and production companies in the oil and gas

Toronto – skyline at waterfront

TABLE 1.3   Canadian Corporations and Metropolitan Dominance in Canada's Economy, 1989

	Financial Sectors				Non-Financial Sectors[a]		
Rank	Location of Headquarters	Total Assets ($ million)	Percentage of Total	Rank	Location of Headquarters	Total Revenues ($ million)	Percentage of Total
1	Toronto	620 258	47.98	1	Toronto	306 303	44.14
2	Montréal	359 282	27.79	2	Montréal	154 120	22.21
3	Québec	50 000	3.87	3	Calgary	46 592	6.72
4	Ottawa	35 426	2.74	4	Vancouver	43 359	6.25
5	Calgary	35 222	2.72	5	Winnipeg	22 862	3.30
6	London	34 112	2.64	6	Hamilton	12 177	1.76
7	Vancouver	32 335	2.50	7	Windsor	11 753	1.68
8	Winnipeg	28 406	2.20	8	London	10 975	1.58
9	Halifax	19 502	1.51	9	Ottawa	10 332	1.49
10	Edmonton	19 444	1.50	10	Edmonton	8 677	1.25
11	Kitchener	17 534	1.36				
12	Stratford	14 051	1.09				
	TOTAL	1 292 688	97.9		TOTAL	693 820	90.38

[a] Manufacturing, service, and resource

Source: R. Keith Semple, "Quaternary Places in Canada," in John N. H. Britton ed. Canada and the Global Economy (Montréal: McGill-Queen's University Press, 1996), pp. 356 and 357.

TABLE 1.4 Foreign-Controlled Corporations and Metropolitan Dominance in Canada's Economy, 1989

Rank	Location of Headquarters	Financial Sectors Total Assets ($ million)	Percentage of Total	Rank	Location of Canadian Headquarters	Non-Financial Sectors[a] Total Revenues ($ million)	Percentage of Total
1	Toronto	91 190	59.11	1	Toronto	155 157	58.73
2	London	33 693	21.84	2	Montréal	35 278	13.36
3	Montréal	13 383	8.68	3	Calgary	20 688	7.83
4	Vancouver	6 138	3.98	4	Vancouver	13 843	5.24
5	Ottawa	4 749	3.08	5	Windsor	11 251	4.26
6	Kitchener	3 321	2.15	6	Hamilton	4 414	1.67
				7	Winnipeg	3 588	1.36
				8	London	2 996	1.13
	TOTAL	154 267	98.84		TOTAL	264 154	93.58

[a] Manufacturing, service, and resource.

Source: R. Keith Semple, "Quaternary Places in Canada," in John N. H. Britton, ed., *Canada and the Global Economy* (Montréal: McGill-Queen's University Press, 1996), pp. 261 and 262.

sector, and tiny Florenceville, New Brunswick, is home to McCain's, one of the world's largest food processing multinationals.[25]

This evidence raises two questions that are important for understanding the changing roles of heartland and hinterland regions. First, how do cities grow in size and economic status to become metropolitan centres? Second, are we justified in asserting that Vancouver, Calgary, and possibly other places warrant metropolitan status? Both questions are significant because the emergence of a metropolis can lead to the formation of a regional core–periphery system capable of innovating change on a broad geographic scale, and to the restructuring of long-standing dependency relationships between heartland and hinterland regions.

Cities grow in size and prominence through natural population growth and immigration or by attracting new investment and employment opportunities. Over time, growth can become self-perpetuating and expansionary, particularly when income earned by local entrepreneurs from export activities is reinvested in the urban economy, spawning a more diverse profile of specialized manufacturing, tertiary, and quaternary services that supply and control external markets. But it is only when the emerging metropolitan centre becomes innovative, transmitting social values, shaping cultural patterns, displaying political prowess, and wielding economic control well beyond its traditional hinterland into a larger arena — in effect, challenging and recasting traditional dependency relationships in its favour — that metropolitan status is achieved. Such a proposition surely applies to Vancouver and Calgary, for each is increasingly drawing some financial and corporate control away from central Canada, and each maintains dominance over key resource sectors. While important in the Canadian context, these two places stand well behind Toronto and Montréal, and remain as rather minor players in the global chess game of urban economic growth and development (Tables 1.3 and 1.4).[26]

becoming a metropolis

The metropolis is a powerful agent of change, and an examination of its deleterious actions adds to our understanding of Canada's regional geography, particularly the notion of regionalism. Regionalism is the expression, by the people or institutions of an area, of the values, interests, and concerns shared by the region's population. From an urban perspective, regionalism can be, on the one hand, an internal behavioural response to forces emanating from an external metropolis and, on the other, the use of internal metropolitan power to express regional character outside of the region. At one level, then, corporations based in a metropolis, or more broadly speaking, in a core region, can cause underdevelopment or at least de-industrialization in the hinterland. For example, plant closures can be viewed as either rational economic processes or seemingly unjust policy decisions. Unfortunately, while the same metropolitan corporations were once capable of acting in partnership to help adjust levels of inequality by assisting governments in the implementation of regional development programs, now, under the terms of NAFTA, this once powerful equalization tool is no longer available.

Several examples illustrate the nature of these metropolitan-based actions. Some Maritimers would argue that one policy decision in particular — the withdrawal of freight rate concessions immediately after World War I — was critical in creating the subsequent and long-standing depressed condition of the economy in the Maritimes. Explaining the woeful state of well-being in the Maritimes is much more complex, of course. But the continued regional anger and distrust voiced against metropolitan interests in central Canada as a consequence of changing freight rates illustrates that regionalism is a powerful and enduring element in shaping regional consciousness — in short, of forging a collective identity or creating a unified regional character.[27] Similar expressions by Québec separatists and Newfoundland fishers, or by Alberta's stand on energy issues and British Columbia's distrust sometimes of federal resource management policies, have also played a role in forging strong regional identities. On an international level, the critics of American penetration of the Canadian economy by giant multinationals — of the actions of an external industrial core or metropolis — are responding from the same emotional base. The only difference is that the term "nationalism" rather than "regionalism" is used when questioning issues of an international dimension. Often these responses are bounded by political space, and although the use of provincial boundaries for regionalization purposes can be questioned, it has been argued that "for all practical purposes 'province' and 'region' [in Canada] are now synonymous."[28] If this is so, one reason surely relates to the fact that individual provinces have become increasingly unified over time in their expression of regionalism.

Beyond economics, the metropolis diffuses social and cultural processes, but with conflicting effects on regional character and regionalism. With today's focus on the mass media and use of computer technologies, metropolitan-based values and norms spread rapidly. The corporate operations of Hollis Street in Halifax, Portage and Main in Winnipeg, or Howe Street in Vancouver differ in scale but not in purpose from those of Bay Street in Toronto or Rue St. Jacques in Montréal. Even in suburbia, there is little to differentiate the recent development of cities across Canada. The same house type plans, conforming to national housing standards and often imitating American models, are available through nationally distributed magazines and are found in suburbs everywhere. Notwithstanding the ethnicity of Canada's largest metropolitan centres, nor the long-standing tensions between English and French, Canadian cities have, in a sense, coalesced in social character. Putting pure political issues to one side, there is little substantive evidence of regional protest against the social and cultural values of the metropolis, at least not at the urban level. Regionalism based on social and cultural issues tends to be more strongly rural in origin, and in these circumstances the protest can be directed as much against a regional city as it can against a national or international metropolis.[29]

Regionalism of this type suggests two possible consequences for interpreting the geography of Canada: the rural and urban components of a region are drawn together; or they are pushed apart. Throughout its development, Cape Breton

Kidd Creek mine,
Timmins, Ontario

Island has at times displayed an uneasy tension with mainland Nova Scotia, but shared problems of economic disparity on a national scale now override past issues and pull the region together. Mainland Chinese economic interests, courted by Ottawa and Halifax politicians, currently have a direct say in running Cape Breton's Sydney steel plant. Their funds are essential for the plant's continued operation. Ontario illustrates the other consequence of regions diverging, or pulling apart. The gulf dividing northern Ontario, a resource hinterland for both national and international markets, from southern Ontario, a mature, post-industrial society, runs much deeper than the physical barrier of the Shield–Lowlands interface that separates them. Differences in social make-up, economic well-being, and political orientation are profound.[30] At this level, the force of the national metropolis is supplemented by regional divergence on a provincial scale.

THE EVOLVING PATTERN OF HEARTLAND AND HINTERLAND

One of the striking features of Canada's regional geography is that the roles of major regions have, at times, been significantly altered. The heartland–hinterland approach provides a meaningful framework for analysing regions in evolution. For this analysis, we will incorporate some of the themes discussed to this point in an overview of the changing pattern of heartland and hinterland in Canada, and thereby establish a context for the regional development issues analysed in the chapters that follow.

This first sign of an industrial heartland in Canada, with a national metropolis, integrated industrial complex, concentrated population, and political strength, appeared following the federation of some of the British North American colonies in 1867. The pre-Confederation colonies, characterized by staple economies and a settlement system of rural farmsteads and scattered and loosely integrated staple entrepôts, garrison outposts, and commercial towns, were essentially hinterlands of a European world-economy controlled from London. Urban settlements in the colonies were chiefly intermediaries in an Atlantic trading system. Through them, and on to the frontier, flowed the immigrants, investment capital, and goods and services sent in exchange for commodities of the sea, forest, soil, and mine. There were some important urban places, to be sure — Halifax, Saint John, Québec, Montréal, and Toronto — but only Montréal's influence was felt widely beyond its immediate hinterland. Moreover, because most people lived on farms or in small villages, rural localism and ties to the Atlantic economy were the most characteristic patterns. Although there were significant regional differences in the pre-Confederation period, the basic distinction was between the land-based economies of the Canadas and the maritime orientation of Newfoundland, New Brunswick, Nova Scotia, and Prince Edward Island. No region had advanced to the stage in its development where it could stand alone as a heartland.[31]

With Confederation came increased social and economic interaction between regions, forming the basis for regional integration — railroad construction, inter-regional trade, a uniform legal structure, centralized banking — and with integration, in turn, came the centring force of a metropolitan presence (Map 1.5). The localism of communities tied to a trans-Atlantic hearth was partly undone and replaced by linkages to central Canadian leadership and innovation. New political and resource peripheries were developed across Canada. Provincial status was achieved successively by Manitoba (1870), British Columbia (1871), Prince Edward Island (1873), and Saskatchewan and Alberta (1905). Mining and lumbering thrust into both British Columbia and the Canadian Shield, and agriculture expanded across the Western Interior. Following these developments, the hinterland regions became attached primarily to Montréal and Toronto. Throughout the Great Transformation of urban-industrial expansion, integration was spurred by various federal government plans (for example, colonization schemes, central banking regulations, manufacturing incentives or port development), by the diverse activities of railroad companies, and by the branch business operations of the major banks, insurance and trust companies, resource companies, manufacturers, and wholesale and retail corporations.[32]

Within the world-economy at large, however, even central Canada remained peripheral. For example, its regional economy was still strongly dependent on British investment capital, European and United States manufacturing technology, and skilled immigrant labour from abroad. Nevertheless, the Canadian pattern of heartland and hinterland continued to intensify. Building on income derived from

Map 1.5 Heartland and Hinterland, 1871

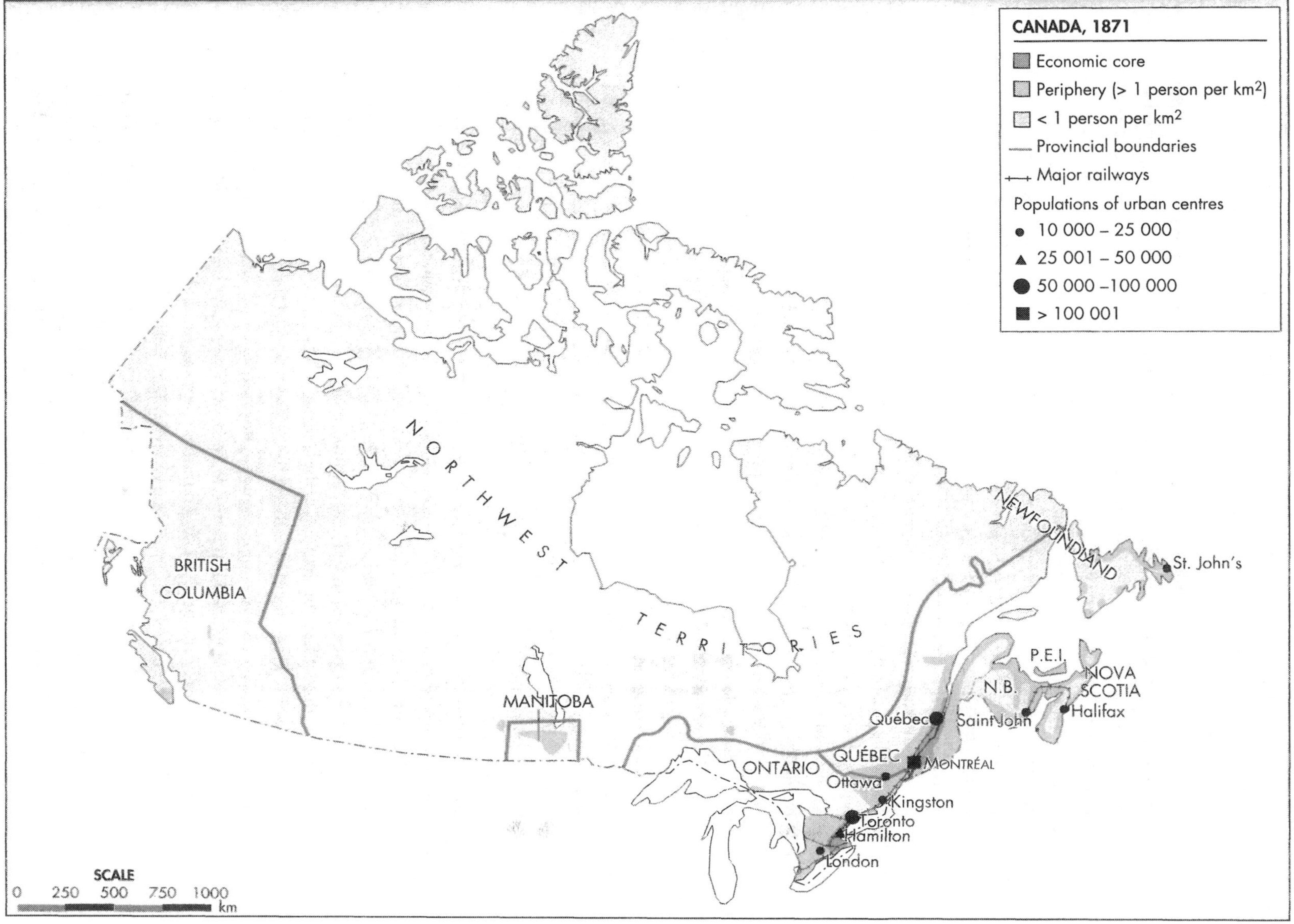

CANADA, 1871

Economic core
Periphery (> 1 person per km²)
< 1 person per km²
Provincial boundaries
Major railways

Populations of urban centres
- • 10 000 – 25 000
- ▲ 25 001 – 50 000
- ● 50 000 –100 000
- ■ >100 001

NEWFOUNDLAND

St. John's

NOVA SCOTIA

Halifax

P.E.I.

N.B.

Saint John

MONTRÉAL

Québec

QUÉBEC

Ottawa

Kingston

ONTARIO

Toronto

Hamilton

London

NORTHWEST   TERRITORIES

MANITOBA

BRITISH COLUMBIA

SCALE

0   250   500   750   1000

km

a relatively rich agricultural base and the staple trades, Ontario and Québec entrepreneurs used the National Policy of industrial incentives, first introduced in the late 1870s, to establish further strength in their manufacturing, transportation, and financial sectors. As Canada entered the twentieth century, factors of cumulative advantage — geographic situation, large local markets, the nexus of transportation routes, and political power — continued to reinforce the heartland status of central Canada. An intensive period of industrial-led urbanization shaped an expanding urban system in southern Ontario and southern Québec. Acting in unison as a core region to wield substantial control over the developing resource hinterlands of the Shield and the west, the power of central Canadian cities also brought about, in part, the demise of a short-lived industrial revolution in the Maritimes. Although a national core region was therefore firmly in place by the eve of World War I, the lines of ultimate control still extended beyond Canada to reach an international metropolis, including London and New York, a pattern which has remained remarkably stable over time (Map 1.6 and Map 1.7).[33]

As subsequent chapters will show, this pattern of core and periphery remained little changed during years of war and depression and even during the expansionary era of the 1950s and 1960s (Map 1.8). Central Canada consolidated its heartland leadership through the expanding participation of American branch plants in its manufacturing sector, by controlling wartime industrial production, by gaining strength in distribution and retailing activities, and through the financial and corporate management of the rush for a new generation of staples. The Maritimes remained entrenched in difficult times, its population earning incomes well below the national average, and none of the western provinces showed signs of significant diversification around traditional export sectors. On the other hand, the external and internal features of the pattern had become more complex. Canada's links to the United States were more diverse and stronger than ever before, and new ties with Japan and other nations had been forged. Internally, as resource exploration and the limits of settlement expanded, cities of heartland and hinterland grew in size and intensified their characteristic functions.

There were, nonetheless, portents of change in the evolving pattern of heartland and hinterland. The precursors of change were chiefly political and economic in nature. The expansion of federal administrative responsibilities into the Yukon and the Northwest Territories, particularly after 1912 and during the 1920s, and the joining of Newfoundland to Canada in 1949, bolstered the political make-up of the hinterland, giving Ottawa heightened political status. More recently, the thrust of the Parti Québécois onto the political stage in Québec in the 1970s was a warning of the possible fracture of Canadian unity, a fracture that still seems possible.

On an economic level, new resource potentials were being realized that would alter traditional metropolis–hinterland relationships. Increased world demand for lumber and pulp and paper and new calls for metallic ores, coal, and hydroelectricity have enriched British Columbia's economy. The province now sends over 90 per cent of its extra-regional exports to foreign countries, making it increasingly inde-

Map 1.6 Heartland and Hinterland, 1901

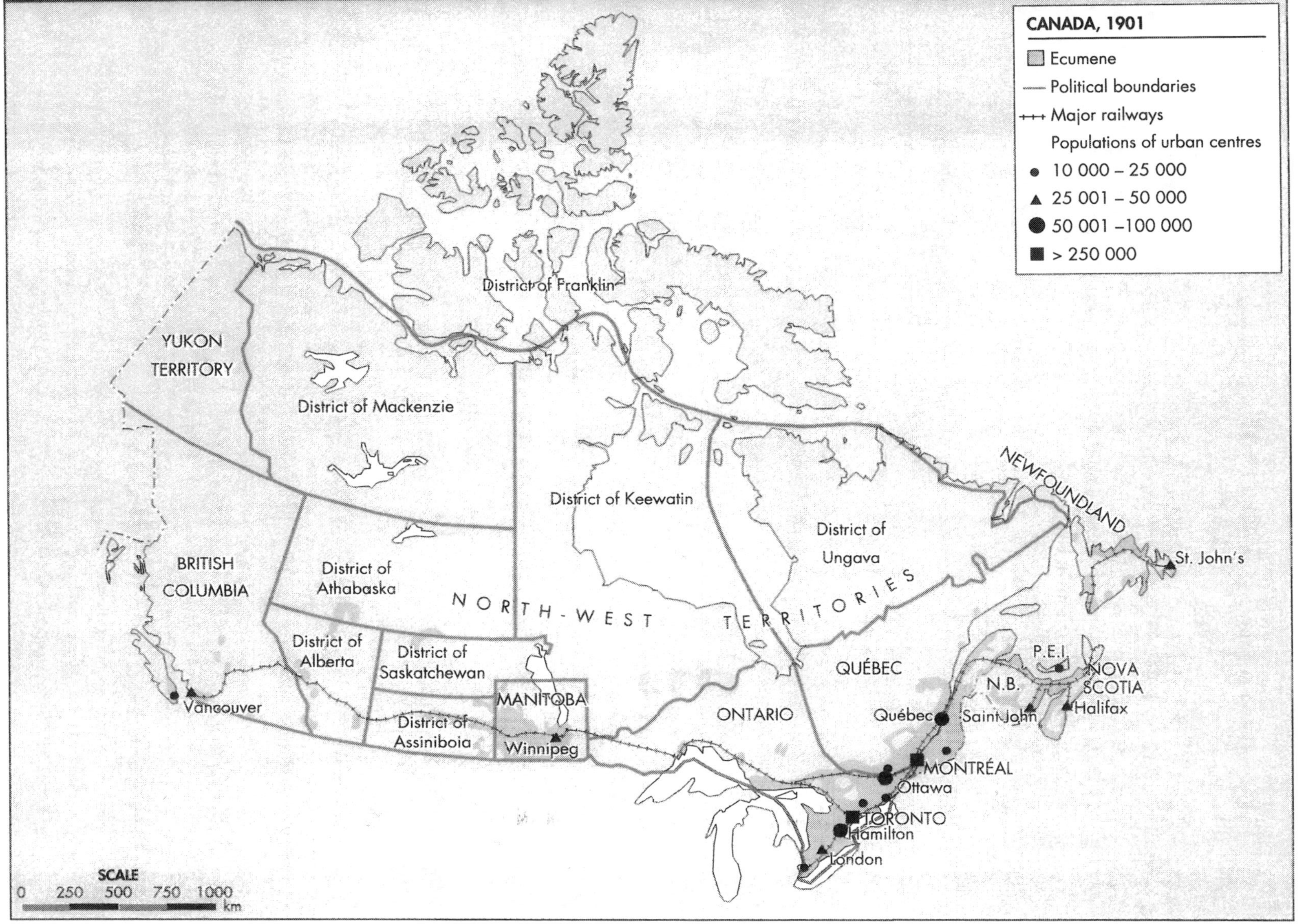

CANADA, 1901

- Ecumene
- —— Political boundaries
- +++ Major railways

Populations of urban centres
- • 10 000 – 25 000
- ▲ 25 001 – 50 000
- ● 50 001 –100 000
- ■ > 250 000

St. John's

NEWFOUNDLAND

P.E.I.

NOVA SCOTIA

Halifax

N.B.

Saint John

QUÉBEC

District of Ungava

NORTH-WEST TERRITORIES

District of Keewatin

District of Franklin

District of Mackenzie

MONTRÉAL

Ottawa

Québec

TORONTO

Hamilton

London

ONTARIO

District of Athabaska

District of Saskatchewan

District of Alberta

District of Assiniboia

MANITOBA

Winnipeg

YUKON TERRITORY

BRITISH COLUMBIA

Vancouver

SCALE

0   250   500   750   1000

km

Map 1.7 Heartland and Hinterland, 1921

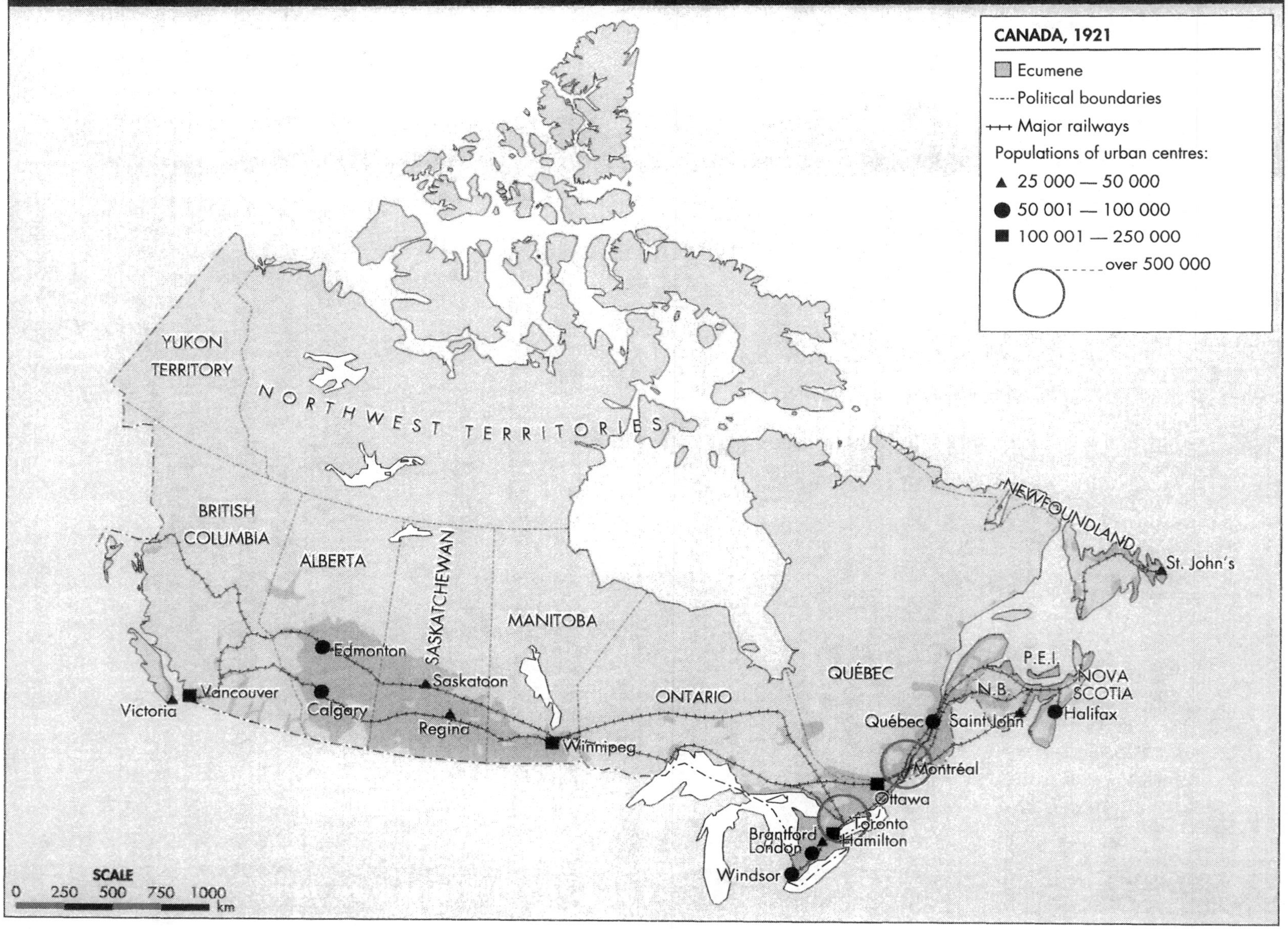

CANADA, 1921

- Ecumene
- ----- Political boundaries
- +++ Major railways

Populations of urban centres:

- ▲ 25 000 — 50 000
- ● 50 001 — 100 000
- ■ 100 001 — 250 000
- ⊙ over 500 000

YUKON TERRITORY

NORTHWEST TERRITORIES

BRITISH COLUMBIA

ALBERTA

SASKATCHEWAN

MANITOBA

ONTARIO

QUÉBEC

NEWFOUNDLAND

P.E.I.

N.B.

NOVA SCOTIA

St. John's

Halifax

Saint John

Montréal

Québec

Ottawa

Toronto

Hamilton

Brantford

London

Windsor

Winnipeg

Regina

Saskatoon

Calgary

Edmonton

Vancouver

Victoria

SCALE

0   250   500   750   1000

km

Map 1.8 Heartland and Hinterland, 1951

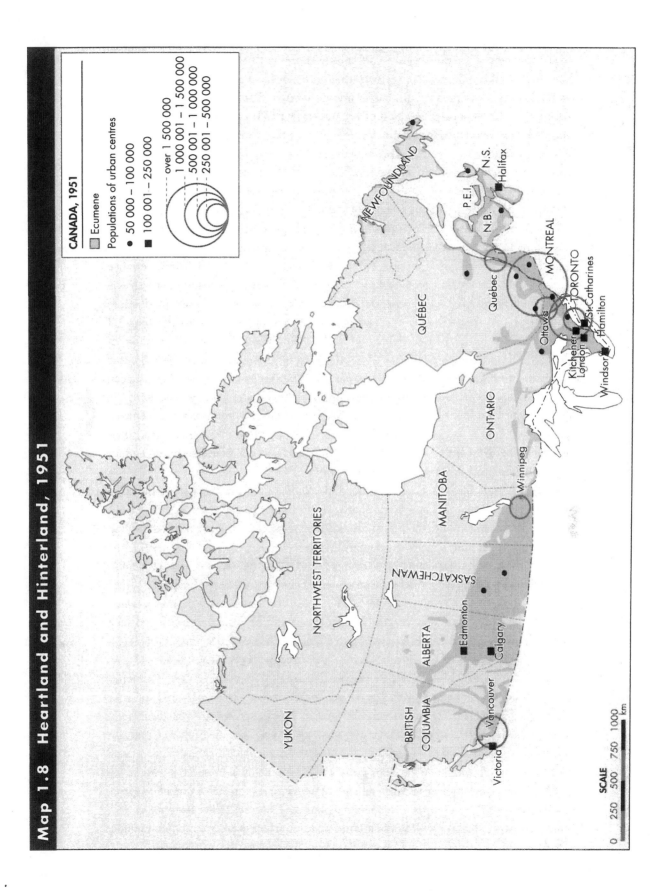

CANADA, 1951

Ecumene
Populations of urban centres

● 50 000 – 100 000
■ 100 001 – 250 000

over 1 500 000
1 000 001 – 1 500 000
500 001 – 1 000 000
250 001 – 500 000

YUKON

NORTHWEST TERRITORIES

BRITISH COLUMBIA

ALBERTA

SASKATCHEWAN

MANITOBA

ONTARIO

QUÉBEC

NEWFOUNDLAND

Victoria

Vancouver

Edmonton

Calgary

Winnipeg

Québec

MONTREAL

Ottawa

TORONTO

Kitchener
London
St. Catharines
Hamilton
Windsor

P.E.I.
N.B.
N.S.
Halifax

SCALE

0   250   500   750   1000
km

pendent of other Canadian regions and thus reinforcing its own pattern of metropolis and hinterland in microcosm. This external outreach has taken on new connections with Vancouver's increased financial involvement in Pacific Rim countries. For Alberta, the international energy crisis of the early 1970s generated windfall profits from oil and natural gas development, creating for that province an unprecedented position of economic strength. This boom was short-lived, but some of the slack has been taken up by petroleum and natural gas exports, especially to the United States. More recently in the late 1980s and 1990s, economic development was expanded somewhat by the massive investment of Japanese pulp and paper multinationals in Alberta's boreal forest. These trends have consolidated a highly integrated settlement system between Edmonton and Calgary, making Alberta all the more distinguishable from Saskatchewan and Manitoba within the Western Interior.[34] Saskatchewan also claims a more diversified economy and improved well-being as a result of potash and uranium development and rising profits from grain, livestock, and pork production. Increasingly, specialized financial services, corporate offices, and a wide range of lesser businesses are being established in the "New West," including some former Industrial Heartland corporations. More alarming for the core, at least initially, was the withdrawal of American branch plants from Ontario as the United States began to consolidate its industrial empire early in the 1990s under the aegis of, first, the Free Trade Agreement (FTA) and later, NAFTA. While the final outcome of these changes on Ontario's and Québec's space economy is still unclear, it is certain that restructuring will continue to influence the role and survival of American manufacturing subsidiaries in Canada's industrial core.[35]

Elsewhere in the hinterland, traditional dependency relationships are being challenged. The potential for massive resource development in the territorial North has drawn that region closer to the industrial core and in the process has nurtured a strong sense of political unity among the region's Native peoples. For example, in 1977, the Berger Commission, respecting local demands as well as the pleas of environmentalists influenced a ten-year moratorium on pipeline construction in the Mackenzie Valley. Because the cost of developing Arctic petroleum has escalated, calls on this resource have not been renewed. More recently, the eastern Arctic has won the right to a degree of self-government, a process that will take effect in 1999 with the official proclamation of Nunavut. This type of development is occurring in part because the federal government is increasingly devolving its responsibilities to meet provincial and territorial demands for more autonomy. Not all provinces appreciate the extent of federal cutbacks in transfer payments. This is particularly true in Newfoundland and the Maritimes, where income levels have been converging only slowly on the national average. Prospects in Newfoundland for further and quickened advance have now been placed on oil revenues flowing from the development of Hibernia's offshore reserves, which are almost ready to be shipped to North American markets. As the 1990s come to a close, and even with the country's provincial regions and largest cities developing new roles, Canada's spatial pattern of heartland and hinterland is still firmly entrenched.

Map 1.9 Traditional Pattern of Canadian Regions

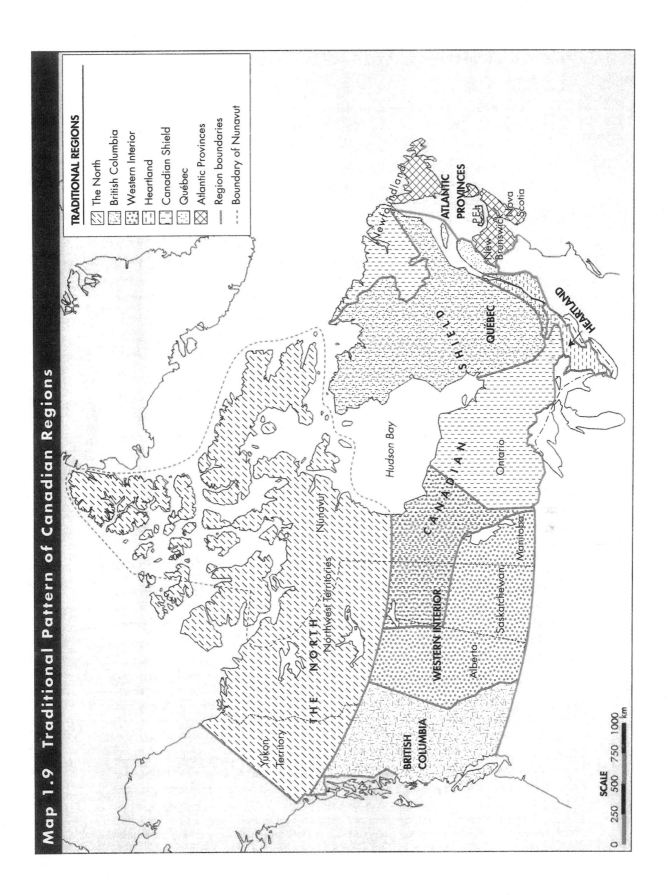

TRADITIONAL REGIONS

- The North
- British Columbia
- Western Interior
- Heartland
- Canadian Shield
- Québec
- Atlantic Provinces
- Region boundaries
- Boundary of Nunavut

THE NORTH

Yukon Territory

Northwest Territories

Nunavut

BRITISH COLUMBIA

WESTERN INTERIOR

Alberta

Saskatchewan

Manitoba

CANADIAN SHIELD

Ontario

Hudson Bay

QUÉBEC

Newfoundland

ATLANTIC PROVINCES

New Brunswick

PEI

Nova Scotia

HEARTLAND

SCALE

0   250   500   750   1000   km

ORGANIZATION OF THIS BOOK

The chapters of this book have been written to capture the essence of the evolving pattern of heartland and hinterland in Canada and the factors which have shaped regional character (Map 1.9). In this edition, physical or environmental factors are discussed in much more detail, returning this important topic to its traditional position of strength when considering the changing regional geography of Canada. In the regional schema followed in the text, all regions fall either into heartland or hinterland status. There is the national core — the Industrial Heartland — centred in southern Ontario and southern Québec. Then there are the provincial hinterland regions, some joined to create a larger whole but each possessing its own heartland–hinterland system of varying importance: Atlantic Canada, the Western Interior, and British Columbia. As well, there are the resource and political hinterlands — the Canadian Shield and the territorial North — which lack core areas, but stand alone because physical differentiation, economic specialization, or political dependency mark them as distinctive regions. Finally, because Québec is a classic illustration of a worldwide trend — so much so, that its people are considering political separation from the rest of Canada — a separate chapter on this province has again been included to promote debate on this important question. As before, the primary objective of this book remains intact: to stimulate discussion and interpretation of Canada's evolving regional geography.

NOTES

[1] Richard Gwyn, *Nationalism Without Walls: The Unbearable Lightness of Being Canadian* (Toronto: McClelland and Stewart, 1994), p. 244.

[2] Since the last edition of *Heartland and Hinterland*, students of Canadian geography now have access to an invaluable source of information on the geographical development of the country — the three volumes of the *Historical Atlas of Canada*. The hundreds of maps produced in these volumes record the transformation of Canada in comprehensive and intricate detail. The volume which coincides most closely with this book is Donald Kerr et al., eds. *Addressing the Twentieth Century*, Vol. III of *Historical Atlas of Canada* (Toronto: University of Toronto Press, 1990).

[3] The literature on the heartland–hinterland paradigm is huge, but several of the more important theorists who have written on the subject are Fernand Braudel, *The Perspective of the World*, Vol. III of *Civilization and Capitalism, 15th–18th Century*, trans. S. Reynolds (New York: Harper and Row, 1984); John Friedmann, *Urbanization, Planning, and National Development* (Beverly Hills: Sage, 1973); and Immanuel Wallerstein, *The Modern World-System, I, II, and III* (New York: Academic Press, 1974, 1978, and 1989).

[4] The notion of a metropolis or industrial heartland holding sway over Canada's regions is exemplified well in the writings of Harold Innis, *The Fur Trade of Canada: An Introduction to Canadian Economic History* (New Haven: Yale University Press, 1930); J. M. S. Careless, *Frontier and Metropolis: Regions, Cities, and Identities in Canada Before 1914* (Toronto: University of Toronto Press, 1989); D. Michael Ray, *Dimensions of Canadian Regionalism*, Geographical Paper No. 49 (Ottawa: Canada, Department of Energy, Mines and Resources, 1971); and *idem*, "Canadian

Regions: A Hierarchy of Heartlands and Hinterlands," in David Knight, ed. *Our Geographic Mosaic*, (Ottawa: Carleton University Press, 1985), pp. 49–58.

[5] John Friedmann, *Urbanization, Planning, and National Development*, pp. 45 and 51.

[6] Saskia Sassen, *The Global City: New York, London, Tokyo* (Princeton: Princeton University Press, 1992).

[7] Stephen Meyer and Milford Green, "Foreign Direct Investment from Canada," *The Canadian Geographer*, 40 (1996), 266-31.

[8] John Friedmann, who has done more than anyone else to conceptualize core and periphery regions in a spatial context, defines innovation as the introduction of ideas or artifacts perceived as new into a social system, and further argues that it is the ability to innovate, above all else, that distinguishes core from peripheral areas: "Major centres of innovative change will be called core regions." Friedmann, *Urbanization, Planning, and National Development*, p.45.

[9] On the importance of cities to Canada's regional geography, see the chapters by James W. Simmons, "The Urban System," and Larry McCann and Peter J. Smith, "Canada Becomes Urban: Cities and Urbanization in Historical Perspective," in Trudie Bunting and Pierre Filion, eds. *Canadian Cities in Transition* (Toronto: Oxford University Press, 1991), pp. 69–99 and 100–25.

[10] For the initial theoretical discussion of the core–periphery interaction and regional economic development that are still relevant to the Canadian context, see R. E. Caves, "Vent for Surplus Models of Trade and Growth," in R. E. Baldwin et al., eds. *Trade, Growth and the Balance of Payments* (Chicago: Rand McNally, 1965), pp. 95–115; and M. H. Watkins, "A Staple Theory of Economic Growth," *The Canadian Journal of Economics and Political Science*, 28 (1963), 141–58. These and other models are summarized in O. F. G. Sitwell and N. R. M. Seifried, *The Regional Structure of the Canadian Economy* (Toronto: Methuen, 1984).

[11] R. Cole Harris, ed. *From the Beginning to 1800*, Vol. I of *Historical Atlas of Canada* (Toronto: University of Toronto Press, 1987).

[12] Canada's newly emerging position in the changing world-economy is well documented in the various chapters that comprise John N. H. Britton, ed. *Canada and the Global Economy: The Geography of Structural and Technological Change* (Montréal: McGill-Queen's University Press, 1996).

[13] This point is argued in J. Tait David, "Some Implications of Recent Trends in the Provincial Distribution of Income and Industrial Product in Canada," *The Canadian Geographer*, 24 (1980), 221–36. See also Trevor Barnes, "External Shocks: Regional Implications of an Open Staple Economy," In John N. H. Britton, *Canada and the Global Economy*, pp. 48–68.

[14] Maurice Yeates, *Main Street: Windsor to Québec City* (Toronto: Macmillan, 1975).

[15] Harvey S. Perloff, *How a Region Grows* (New York: Committee for Economic Development, 1963).

[16] This theme is central to *Addressing the Twentieth Century*, Vol. III of *Historical Atlas of Canada*, especially Plates 5–39.

[17] Data in this and the following paragraph have been derived from various annual estimates published by Statistics Canada, *Provincial Gross Domestic Product by Industry* (Ottawa: Queen's Printer, 1962–1996).

[18] Trevor Barnes et al., "Vancouver, the Province, and the Pacific Rim," in Graeme Wynn and Timothy Oke, eds. *Vancouver and Its Region* (Vancouver: University of British Columbia Press, 1992), pp. 171-99; Edward Chambers and Michael Percy, *Western Canada In the International Economy* (Edmonton: University of Alberta Press, 1992); Larry Pratt and John Richards, *Prairie Capitalism: Power and Influence in the New West* (Toronto: McClelland and Stewart, 1979); and

idem and Ian Urquhart, *The Last Great Forest: Japanese Multinationals and Alberta's Northern Forests* (Edmonton: NeWest Press, 1994).

[19] William J. Coffey, *The Evolution of Canada's Metropolitan Economies* (Montréal: Institute for Research on Public Policy, 1994).

[20] Theoretical perspectives on the heartland–hinterland character of Canadian cities can be found in L. D. McCann, "Urban Growth in a Staple Economy: The Emergence of Vancouver as a Regional Metropolis, 1886-1914," in L. J. Evenden, ed. *Vancouver: Western Metropolis*, Western Geographical Series, Vol. 16 (Victoria: Department of Geography, University of Victoria, 1978), pp. 17–41; idem, "Staples and the New Industrialism in the Growth of Post-Confederation Halifax," *Acadiensis*, 8 (1979), 47–79; and idem and Smith, "Canada Becomes Urban."

[21] See, for example, L. S. Bourne and Mark Flowers, *The Canadian Urban System Revisited: A Statistical Analysis*, Research Paper 192 (Toronto: Centre for Urban and Community Studies, University of Toronto, 1996); Coffey, *The Evolution of Canada's Metropolitan Economies*; W. K. D. Davies and D. P. Donoghue, "Economic Diversification and Group Stability in the Urban System: The Case of Canada, 1951–1986," *Urban Studies*, 30 (1993), 1165–86; and M. Yeates, "The Core-Periphery Model and Urban Development in Central Canada," *Urban Geography*, 6 (1985), 101–21.

[22] J. Randall and G. Ironside, "Communities on the Edge: An Economic Geography of Resource-Dependent Communities in Canada," *The Canadian Geographer*, 40 (1996), 17–35.

[23] Maurice Careless, "Frontierism, Metropolitanism, and Canadian History," *Canadian Historical Review*, 35 (1954), 17. See also Don Davis, "The Metropolitan Thesis' and the Writing of Canadian Urban History," *The Urban History Review*, 14 (1985), 95–114; and Donald P. Kerr, "Metropolitan Dominance in Canada," in John Warkentin, ed. *Canada: A Geographical Interpretation* (Toronto: Methuen, 1967), pp. 531–55.

[24] For a comparison between 1977 and 1989, see the tables in R. Keith Semple and W. Randy Smith, "Metropolitan Dominance and Foreign Ownership in the Canadian Urban System," *The Canadian Geographer*, 25 (1981), 9, 22; and R. Keith Semple, "Quaternary Places in Canada," in Britton, ed. *Canada and the Global Economy*, pp. 261 and 262.

[25] Data in this paragraph are derived from R. Keith Semple, "Quaternary Places in Canada," in Britton, ed. *Canada and the Global Economy*, p. 273.

[26] The development of these metropolitan centres is discussed in Wynn and Oke, eds. *Vancouver and Its Region*; L. J. Evenden, ed. *Vancouver: Western Metropolis*; and B. M. Barr, ed. *Calgary: Metropolitan Structure and Influence*, Western Geographical Series, vol. 12 (Victoria: Department of Geography, University of Victoria, 1975). For their current metropolitan outreach overseas, see Meyer and Green, "Foreign Direct Investment from Canada."

[27] For the development of Maritimes regionalism in the 1920s, see Ernest R. Forbes, *The Maritimes Rights Movement, 1919–1927: A Study in Canadian Regionalism* (Montréal: McGill-Queen's University Press, 1979). For a fuller historical explanation, see the various chapters in E. R. Forbes and D. A. Muise, eds. *The Atlantic Provinces in Confederation* (Toronto: University of Toronto Press, 1993).

[28] Garth Stevenson, "Canadian Regionalism in Continental Perspective," *Journal of Canadian Studies*, 15 (1980), 18.

[29] I demur slightly on this issue, for there are times when metropolitan culture takes on a rather surreal quality. While travelling recently in Newfoundland, I was in need of money. Stopping in Lewisport, a service and fish processing centre on the northeastern coast, I entered a branch of the Canadian Imperial Bank of Commerce which, of course, is headquartered in Toronto. By using my Mastercard, I was subject to Toronto-based banking policies. Moreover, playing clearly in the back-

ground was the music of a Toronto FM-radio station, beamed by satellite to this distant and remote community. Apart from the accents of tellers, the local vernacular was nowhere in evidence.

[30] A very persuasive argument for distinguishing these two Ontario regions on both economic and political grounds is authored by G. R. Wellar, "Hinterland Politics: The Case of Northwestern Ontario," *The Canadian Journal of Political Sciences*, 10 (1977), 727–54.

[31] R. C. Harris and J. Warkentin, *Canada Before Confederation* (Toronto: Oxford University Press, 1974). See also Harris, *From the Beginning to 1800*, Vol. I of *Historical Atlas of Canada* and R. Louis Gentilcore, ed. *The Nineteenth Century*, Vol. II of *Historical Atlas of Canada* (Toronto: University of Toronto Press, 1993).

[32] McCann and Smith, "Canada Becomes Urban," pp. 86–9.

[33] The development of a core–periphery pattern in the United States is best described by David Ward, *Cites and Immigrants: A Geography of Change in Nineteenth Century America* (New York: Oxford University Press, 1971), pp. 11–49.

[34] P. J. Smith, "Alberta Since 1945: The Maturing Settlement System," in L. D. McCann, ed. *Heartland and Hinterland: A Geography of Canada*, 2d ed. (Scarborough: Prentice-Hall, 1987), pp. 378–82.

[35] Ian MacLachlan, "Organizational Restructuring of U.S.-based Manufacturing Subsidiaries and Plant Closure," in Britton, ed. *Canada and the Global Economy*, p. 213.

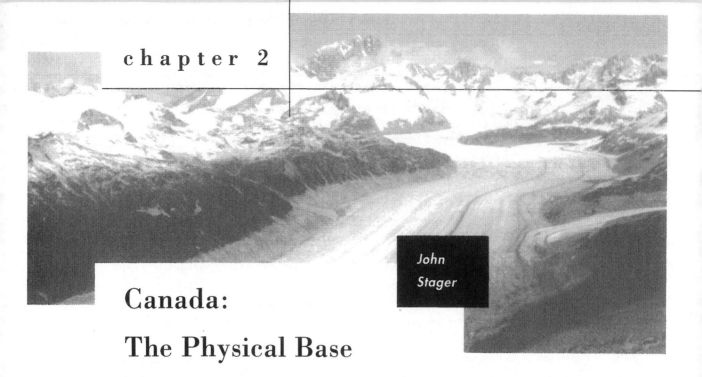

chapter 2

John Stager

Canada:
The Physical Base

In a book dealing with Canada's heartland and hinterland regional structure, a concern for the country's rocks, air masses, and tundra may, at first glance, seem remote in a world of globalization, economic competitiveness and growth, or from the historical and cultural roots of Canadian society. Yet all this activity takes place on a stage — the land of Canada. This chapter provides a physical geographical background, which describes and explains the nature of the land surface, the prevailing climatic conditions and the patterns of natural vegetation. The Native peoples of this part of the world had high respect for the land, for they were part of it. The opportunities and limitations of this same natural environment are reflected in the ways in which subsequent settlers reordered it.

Topography, temperature, forest, and plain affect everyday life. Mountain roads are expensive to build and maintain, the great distances from one end of the country to the other challenge communications and transportation, and climate makes Canada a cold country, demanding heating, adequate clothing, and snow removal. Global atmospheric circulation links us with other countries, including their industrial wastes. These are carried to us as pollutants which, because of our climate, tend to accumulate in the Arctic regions, posing threats to plants, animals, and the Native peoples who live there. Our forests have great commercial value and underpin large parts of the Canadian economy. But they also call up concerns for protection and preservation that are important to environmental sustainability — an issue that is commanding more and more attention from people worldwide.

This chapter, then, is both a reminder of our dependence on the land, water, and air that sustain life, and a means of equipping us to better understand Canada's natural stage, and the drama that is being played out on it.

GEOLOGICAL PROVINCES

Canada was built, through geologic time, in a series of blocks presently exposed and described as geological provinces (Map 2.1). At the centre of Canada and the rest of the continent is the Precambrian Canadian Shield, ancient rocks, some over 3500 million years old, that underwent long periods of building, deformation, and erosion, and now rest as the stable backbone for all of North America. This Shield extends at depth all the way to the western mountains, and south almost to Mexico.

Overlying the buried margins are the following more recent platforms of sedimentary rocks, layers of limestone and dolomite from ancient seas, and sandstones and shales and evaporites accumulated during periods of submergence and emergence: the Interior Platform of the Canadian plains, the Arctic Platform of the southern Arctic islands, the Hudson Platform below Hudson Bay and exposed in the lowlands at the south end of the bay, and the St. Lawrence Platform along the southeast margin of the Shield. The last-mentioned is divided by the Frontenac Axis into two sections, the Southern Ontario platform and the Ottawa–St. Lawrence Valley part. The axis crosses the St. Lawrence River at the east end of Lake Ontario.

Butted up against the platforms and the buried Shield are mountain belts or orogens. Once regions of massive accumulation of eroded sediments and other deposits, these orogen belts were uplifted during episodes of mountain building that lasted for millions of years and still continue today. The oldest is the Appalachian orogen of the eastern provinces, formed between 300 million and 500 million years ago during the Palaeozoic era. The Cordilleran orogen is a wide band of mountains and plateaus all along the western edge of Canada, formed and reformed throughout the Mesozoic and Cenozoic eras 50 million to 150 million years ago. The Innuitian orogen in the Far North is the youngest. It dates from the Tertiary period of the Cenozoic era, 20 million to 60 million years ago, and shows a simpler pattern of orogenic folding with low mountains and ridges (Table 2.1).

The only other geological province of major extent is the Arctic Coastal Plain stretching from Yukon's northern slopes across the Mackenzie Delta to the outer western edge of Banks Island and other high-Arctic islands. This plain is an area of active accumulation and deposition of eroded sediments from the uplands behind. Pleistocene glaciation is the last chapter in our survey of Canada's geologic history. Within the last million years Canada was covered by ice caps 1.5 km or more in depth. In four major periods of spreading growth and melting retreat, they have left a footprint on the land that largely determines its present character.

Map 2.1 Major Geological and Coastal Shelf Regions

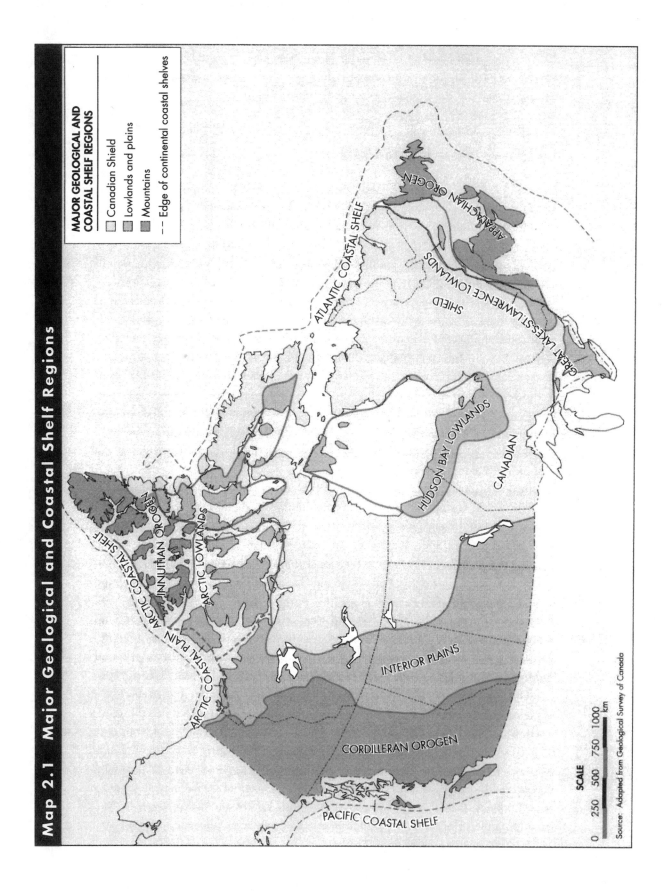

MAJOR GEOLOGICAL AND COASTAL SHELF REGIONS

- ☐ Canadian Shield
- ▨ Lowlands and plains
- ▨ Mountains
- -- Edge of continental coastal shelves

APPALACHIAN OROGEN

ATLANTIC COASTAL SHELF

GREAT LAKES-ST. LAWRENCE LOWLANDS

SHIELD

HUDSON BAY LOWLANDS

CANADIAN

INNUITIAN OROGEN

ARCTIC LOWLANDS

ARCTIC COASTAL SHELF

ARCTIC COASTAL PLAIN

INTERIOR PLAINS

CORDILLERAN OROGEN

PACIFIC COASTAL SHELF

SCALE

0   250   500   750   1000
km

Source: Adapted from Geological Survey of Canada

TABLE 2-1   Geological Time Table

Eon	Eras	Periods and Systems	Epochs	Millions of Years to Beginning of Geological Stage
		Quaternary	Holocene	
			Pleistocene	1.6
	Cenozoic		Pliocene	
			Miocene	
		Tertiary	Oligocene	
			Eocene	
			Paleocene	65
Phanerozoic	Mesozoic	Cretaceous		144
		Jurassic		208
		Triassic		245
		Permian		286
		Carboniferous		360
	Palaeozoic	Devonian		408
		Silurian		438
		Ordovician		505
		Cambrian		570
Proterozoic	Precambrian			4000
Archean				

Canadian Shield

Rocks with the longest and most tumultuous geologic history are found in the Canadian Shield, and considering its past, the present surface is rather subdued. Composed of granites, gneisses, ancient sediments, and volcanics, the Shield in its worn-down state is shaped like a saucer, highest around the outer rim and submerged in the middle at the Hudson Bay. The western margin along the Interior Platform is a broad edge extending from a high of 600 m in the north at Great Bear Lake and dropping to 300 m in southern Manitoba. The high rim is readily visible along parts of Lake Superior.

In the Laurentians, summits are higher than 600 m and north of Québec City they rise to more than 900 m. In the Torngat Mountains of Labrador they are as high as 1200 m. In the north the Shield is exposed at the surface on the southern and northeastern sections of Baffin Island. It underlies the eastern end of Devon Island and extends north along Ellesmere Island halfway to its northern tip. The

Shield in this location is the high eastern rim of the Arctic. It forms mountainous terrain with elevations increasing toward the north from 1000 m to over 1800 m between south Baffin Island and Ellesmere Island. The surface bears all the traces of Pleistocene glaciation, particularly its well-developed alpine forms, and it includes one of the most beautiful series of fiords anywhere in the world.

Much of the ground is bedrock except where remnant ice caps cover the higher ground. An undulating section of the Shield is found west of Hudson Bay, its boundary intersecting with the Arctic coast at Coppermine. There are two finger-like northward extensions, one of which is the backbone of Boothia Peninsula and west Somerset Island. The other includes Melville Peninsula and crosses to Baffin Island. The surface elevations increase from Hudson Bay outward to the outer margins and, except for a couple of broad domelike rises over 300 m high northwest of Hudson Bay, the surfaces are generally 100 and 200 m above sea level.

The Shield was the platform for the great continental ice sheets of the Pleistocene era. Labrador–Ungava and the Keewatin area west of Hudson Bay were centres of accumulation, so ice moved radially outward from these location. This ice severely eroded the Shield, leaving smooth bare rock along the edges when the ice melted. A thin mantle of ground moraine and other melting-edge landforms are visible signs of the past. Stretching back from the shores of Hudson Bay, like a flight of steps, are old beach lines that mark past shores of meltwater bodies and seas (Map 2.2). They became dry ridges when the land gradually rose as it was released from the weight of glacial ice. Patterns of lineation on the ground from ice movement, fanning outward from the centres, are characteristic of Shield terrain. Another feature is the thousands of lakes, poorly integrated into an immature drainage pattern, filling the depressions caused either by structures and joints of ancient formations, or as a result of recent erosion and deposition. On the western half of the Shield, almost one-fifth of the area is covered with water. The eastern half is not much different.

Interior Platform

The level layers of bedrock that underlie the Interior Platform are increasingly younger as they approach the west, and are rarely exposed at the surface because of the liberal cover of glacial drift. The plateau consists of three levels: the first, over 300 m above sea level, forms a belt, 200 km wide, known as the Manitoba Lowland, stretching north along the edge of the Shield; farther west, the Saskatchewan Level at 600 m above sea level reaches northwest as far as the Peace River area of British Columbia; the Alberta Level, at 900 m elevation, includes all of southern Alberta from the Swan Hills to the southwestern corner of Saskatchewan. The Saskatchewan and Assiniboine rivers drain the southern half of this level, while the Peace, Athabasca, Slave, and Mackenzie rivers carry surface waters from the northern half to the Arctic Ocean.

Map 2.2 Glacial Lacustrine and Marine Deposits

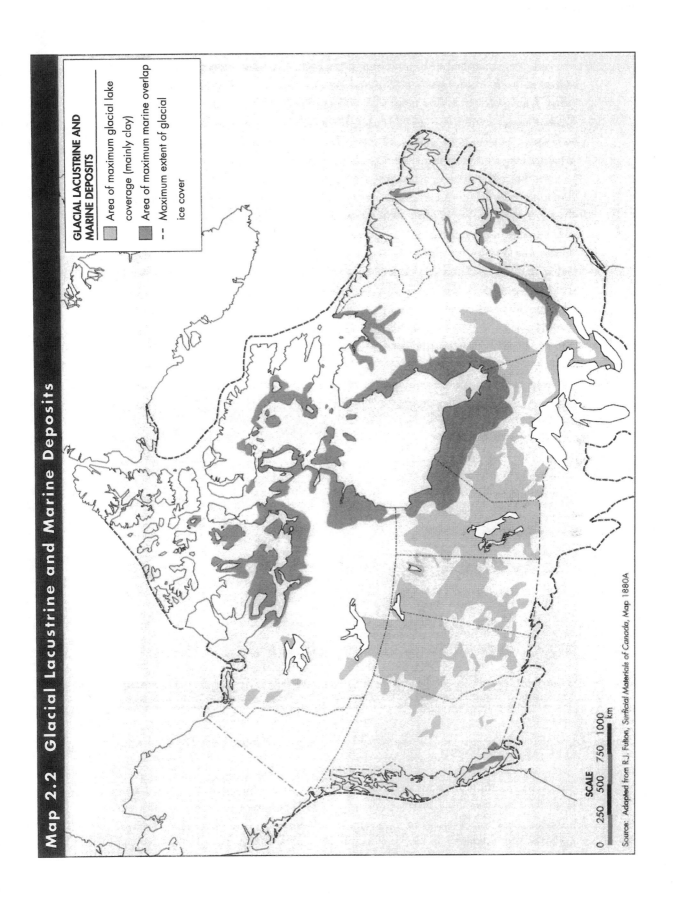

GLACIAL LACUSTRINE AND MARINE DEPOSITS

Area of maximum glacial lake coverage (mainly clay)

Area of maximum marine overlap

-- Maximum extent of glacial ice cover

SCALE

0 250 500 750 1000 km

Source: Adapted from R.J. Fulton, *Surficial Materials of Canada*, Map 1880A

In a few places bedrock surfaces to produce the Manitoba Escarpment, a series of 150 m block "mountains" in the Pembina Hills and the Riding and Duck mountains. One peak reaches a height of 400 m. The Cypress Hills on the Alberta-Saskatchewan border rise as high as 1400 m above sea level. During the glacial meltback, thousands of square kilometres were inundated by ponded meltwater, creating major proglacial lakes (Map 2.2).

Southern Manitoba is noticeably flat, and much of southern Saskatchewan is a flat, silt or clay lakebed surface almost entirely free of stones. The mantle of glacial debris is everywhere, in places hundreds of metres thick. Surface till forms irregular mounds, depressions, and ponds, leaving the impression that glacially carried debris was dumped in place as the glaciers died. Drainage meltwater sorted and redistributed gravel and sand into flats, some of which were subsequently blown by wind into a dune landscape. Traversing the platform are great channels that once carried meltwaters, creating broad and deep valleys which today are either abandoned or occupied by beaded lakes. To the north the Platform narrows along the Mackenzie River, where it is covered by till.

Sand dunes, Kettle Lake, Manitoba

Arctic Platform

The Arctic Platform overlaps and in places abuts on the Shield. It is a broad geologically stable zone of layered sedimentary rock. Where rock is exposed, cliffs and hillsides show successions of limestones, dolomites, and shales. The Platform underlies the islands between the mainland coast and the Parry Channel, and crosses north of the channel in Cornwallis and Devon islands. To the east, it is found surrounding Foxe Basin and under the Brodeur Peninsula at the north end of Baffin Island. The land surface has some rock outcrops but it is mainly weathered bedrock and glacial deposits. Elevations are low, generally less than 100 m, but with a few high points in the west and northeast. Cliffed coastlines can be seen along Lancaster Sound.

Hudson Bay Platform

The Hudson Bay Platform is visible as the lowlands of Hudson Bay, extending from Churchill to the south of James Bay. On the north side it is exposed on Southampton and a couple of other islands. The whole of Hudson Bay is underlain by Palaeozoic limestone, with its present surface low and gently sloping. In the late stages of glacial retreat, the rock surface was covered with till and lake muds from a large glacial lake, then later with marine clays that were deposited during a marine advance. The rivers that cross the lowland pass across an otherwise poorly drained organic terrain that covers much of this landscape. In the north, the limestone rock is exposed as fractured or crushed rock and elsewhere it is covered with mosses and lichens.

Great Lakes — St. Lawrence Platform

There are two separate segments of the Great Lakes—St. Lawrence Platform: the Ontario peninsula from Windsor to Kingston, and the St. Lawrence River valley along with part of the lower Ottawa River Valley. In the St. Lawrence estuary, Anticosti Island is part of this geological province, as is part of the Long Peninsula of Newfoundland. All this area has early Palaeozoic limestones, dolomites, and shales as basement rock, but the topographic expression is unique in each part of the platform.

The Shield boundary lies along a fault line with the Ottawa–St. Lawrence Valley as a downthrow block, which left an abrupt northern scarp edge 50 to 100 m high. The original sediments that were confined by the Shield and the Appalachian orogen were generally level, built up and re-excavated by ancient rivers, yet remaining quite stable.

The last episode of landscape shaping came at the end of glaciation when the retreat of the ice front caused a lake to form on the plain. Melting of the ice opened the valley to the sea, and the Champlain Sea flooded the region. That was about 11 800 years ago (Map 2.2) and it created beaches, which can still be seen. It also lined the valley floor with marine clays and silts.

The release from the weight of ice brought rebound of the land, so in later stages before the sea withdrew entirely, rivers from the Shield flushed great quantities of gravel straths and sandy deltas into the marine body. For example, the St. François and St. Maurice rivers built large deltas. Much clastic material was

Landslide at Saint-Jean Vianney, Québec, 1971

dumped on the clays so that sand frequently overlies the clay. Consequently the marine clays have a history of instability, causing landslides with much damage and occasional loss of life. There are several hills that contrast with the more level ground, and there are resistant Shield rocks like Oka Mountain near Montréal. A series of intrusive plugs, the Monteregian Hills, much eroded but with their own Champlain Sea beaches ringing them, stretches from Mount Royal in Montréal in a line across the plain to Mount Brome in the Eastern Townships.

The early Palaeozoic limestones, dolomites, sandstones, and shales play a more prominent role in the shape of the Ontario peninsula. The sediments lie in overlapping layers gently dipping from the Shield edge around Muskoka southwest to Windsor. There are three main layers forming escarpments toward the east and ridged elsewhere. The most prominent is the Niagara Escarpment, which appears in the Niagara Peninsula in the clifflike form into which the Niagara River eroded a canyon upstream to the present location of Niagara Falls. The escarpment curves around the end of Lake Ontario forming Hamilton's "Mountain," and then cuts diagonally across the peninsula to become the backbone of Bruce Peninsula and Manitoulin Island. The escarpment is highest in the Blue Ridge area to the south of Georgian Bay. Other escarpments are only visible locally, all pretty much masked by the glacial drift that shaped most of the topography of southern Ontario.

During ice advance, the Great Lakes were deepened, and their shorelines stood higher as proglacial lakes formed at the retreating edge of continental ice. Meltwaters spilled south, first to the Mississippi and Hudson rivers, and then across Lake Nipissing to the Ottawa River before the St. Lawrence re-entrant was uncovered and the present-day pattern established. As a result, a layer of shallow water was laid down in the Windsor-Leamington region. There are old beach lines like Dundas Street in Toronto, which was the line of the Governor's Road east and west of the provincial capital. In the late stages of ice retreat, lobes of ice stood in Lake Huron, Georgian Bay, and Lake Ontario and large moraines were deposited around the margins. They formed a horseshoe shape, open to the southwest and ringing the core of the Ontario peninsula. These moraines, along with some spill-ways north of Toronto, produce a very irregular pattern of hills and depressions forming a coarse rolling landscape. Despite the roughness, moraines have proved fertile and a large amount of their surfaces is under cultivation. Toward the centre of the peninsula, the till cover forms a smooth undulating surface, and here and there around Peterborough there are some unusual glacial deposits like the drumlin fields and whaleback hills.

Appalachian Orogen

In the Appalachian Orogen, mountain-building activities in the Pleistocene era pressed succeeding mountain ranges against one another in a southwest–northeast trend. At some point the sediments were thrust forward along a fault called Logan's Line that is now the boundary of the St. Lawrence Lowlands. In the Eastern Townships of Québec, the upland piedmont is the erosional remnant of past mountains. Fluvial erosion exploited the softer sediments to produce valleys as deep as 100 m, parallel to the main mountain orientation. Glaciation overtopped all, and widened some valleys to fill them with fluvioglacial drift. The rivers that originally drained off the piedmont toward the St. Lawrence River have been consolidated into a few major streams as a result of river capture.

To the northeast, from the city of Québec to the Gaspé Peninsula, there is a tightly folded ridge rising eastward to more than 1200 m. In New Brunswick there is an upland rising to 600 m in places and forming a broad hump from south of the Bay of Chaleur to the western side of the province. There are parallel hills like the Caledonia Highlands along the north side of the Bay of Fundy. Eastern New Brunswick, the flats at the head of the Bay of Fundy, the Chignecto isthmus in Nova Scotia, and Prince Edward Island are all part of a general lowland, glacially smoothed and mantled with drift. Prince Edward Island is noticeably different because the underlying red sandstones and shales present a distinctive red soil.

The backbone of Nova Scotia is a till-covered rocky upland running the length of the province. Behind a volcanic ridge along the Bay of Fundy runs the narrow Annapolis Valley, widening toward Kentville where soil and climate favour fruit growing. The Cobequid Hills on the north side of Minas Basin are aligned with the Cape Breton Highlands on the northern half of that island. This block of high land stands 300 m above sea level in places, affording dramatic scenery. On Newfoundland, the same general alignment of all the orogen is present in the Long Range of 600 m-high mountains, running northward from the southwest corner of the island onto the Long Peninsula. In contrast, the northeast coast, the French Shore, is a low country, till covered and sometimes rocky, but poorly drained. To the southeast, peninsulas like the rocky Burin Peninsula and the rough ground of the Avalon Peninsula have the same basic trends of the Appalachian folding.

Cordilleran Orogen

The mountain-building forces of the Cordilleran Orogen have built a longitudinal series of mountains and plateaus from the early Palaeozoic era that are now part of the whole of North America. There are three systems that best describe the complexity of the region. First, there is an eastern system that includes the Rocky Mountains, the Foothills, and the Rocky Mountain Trench in the south, and the Mackenzie, Franklin, and Richardson mountains in Yukon and Northwest Territories. These mountains and foothills were formerly level strata of sedimentary rocks that were thrust-faulted. Later, with shearing, older rocks were placed over younger ones. The end result is a splendid array of flat, inclined, folded, and in places almost vertical layers of sediments.

What is visible today, however, is the result of a Cordilleran ice sheet that over-topped all but the highest peaks, and, in the process of alpine glaciation, straight-ened river valleys, created cirques and sharp ridges near peaks, and formed hanging valleys in U-shaped troughs. The alpine landscape exhibits many peaks over 3000 m with Mount Robson rising still higher. In the valleys the local relief varies from

Volcanic cone near Telegraph Creek, BC

1500 to 1800 m. The Rocky Mountain Trench is an unusual feature that continues with some interruptions into the Tintina Trench in Yukon. It is several kilometres wide with steep sides as high as 900 m. Glaciation has buried the basement rock and flattened the floor with fluvioglacial products. Both the Columbia and Kootenay rivers rise in the Trench near Canal Flats. In Yukon, the folded Mackenzie Mountains lie against the Selwyn Mountains, another fold belt occasionally intruded with igneous materials. These mountains are much lower, rarely over 200 m, but they still exhibit alpine glacial landforms. Farther north, glaciation was less intrusive because there was not enough moisture to nourish ice caps.

Second, the western system marks the coastal edge of the Cordilleran Geological Province. The core of the system is the Coast Mountains, which begin as the Cascades south of the Fraser River and continue northward through the Alaska Panhandle to Yukon. Much of this core is underlain by a plutonic granitic batholith, the roots of an ancient mountain system that was finally exposed after long periods of erosion. It too was glaciated and alpine landforms with jagged peaks and ridges create the current profile. The coast has many fiords where glaciers descended and deepened the valleys, which are now flooded by the sea. West of the Coast Mountains across a coastal trough are the Insular Mountains forming the spine of Vancouver Island, part of the Queen Charlotte Islands, and the St. Elias Mountains in Yukon.

Between the two outer systems is the third of the Cordilleran geological provinces, the Interior Province. It is mainly plateau country, interrupted by mountain masses. In the southeast are the Columbia Mountains with peaks of 2000 to 2700 m, a group of four separate ranges separated by glacially deepened lakes. The Okanagan lakes are along the west side, Upper and Lower Arrow lakes fill the valley between the Monashee and Selkirk ranges, and Kootenay Lake is in the southern part of the main valley between the Selkirks and Purcell ranges. Volcanic activity is found in their rocks but their surfaces were shaped by glaciers, creating many cirques in the mountains, and ridges and moraines in the valleys. The plateau region is broken about latitude 56°N by the Skeena–Omineca mountains, which stretch north to Yukon west of the Trench. These mountains, geologically complex with sediments and intrusives, reach over 2000 m. The present surface is a glaciated one.

The plateau itself is in two units, the Interior Plateau of British Columbia, and a series of plateaus and plains north from the Stikine River into Yukon. This northern part is about 1800 m high except for the Liard Plain, and the land surface is rolling but frequently broken by isolated hills and groups of low mountains. Glaciation here was less intense, and low precipitation resulted in many rivers cutting deeply into the plateau, causing wide gravel straths and braided streams. The Interior Plateau is a gently rolling upland that belies the complex geologic history, which includes erosion, uplift, and sufficient volcanic activity to produce layers of basalt over an area of 35 000 square kilometres between Kamloops and Prince George. The surface is generally above 1000 m and is covered by glacial deposits, ground moraine, and water-worked materials.

During deglaciation, meltwater was ponded in an extensive series of lakes that linked the lowlands of Okanagan Valley to the basin of the Shushwap lakes, the Thompson River valleys and the Nicola basin. The silt terraces of the Okanagan, so favoured for fruit and vine, were built by sediments washed from the slopes against a block of wasting ice that was left in the centre of the valley. The glacial lakes retreated in a complex series of stages that left flights of terraces and old beach lines plainly visible around places like Kamloops. The valley of the Fraser River was filled with glacial products only to be re-excavated by modern streams, thus creating more terraces. The Fraser River cuts across the Interior Plateau and Coast Mountains to the sea where it has built a large delta. This delta, the Lower Mainland, and the Fraser Valley from Vancouver to Hope form a fluvial plain that is densely settled and has intensive agriculture.

The Cordilleran Orogen, as a mainly mountainous province, has a high percentage of bare rock surfaces. It also has, more than anywhere else in Canada, most of its land in sloping ground, frequently steeply sloping. Thus there is a lack of stability for both earth and rock surfaces, and landslides or avalanches are common experiences. They create risks for valley dwellers and expensive maintenance for valley communications.

Innuitian Orogen

The Innuitian Orogen covers most of the Queen Elizabeth Islands. It is a mountain-building zone where accumulated sediments were folded, faulted, and elevated at the edge of the Shield. From the northern half of Ellesmere Island in a broad belt that lies south and west across the islands, the grain of the country is oriented in successive ridges parallel to the axis of the belt. The western two-thirds of the area has a surface covered with unconsolidated materials, partly glacial, and elevations that approach 200 m. There are some high hills that exceed 300 m on Victoria and Banks islands. The eastern one-third has more bedrock exposed, and because there is little or no vegetation, the folded structures form colourful, textbook examples of synclines and anticlines. This is the mountainous part where, beyond elevations of 1100 m, peaks are submerged in ice caps. Alpine glacial landforms with long straight trenches, hanging valleys, and active glaciers reach into coastal fiords, offering spectacular scenery.

Arctic Coastal Plain

The Arctic Coastal Plain is the narrow seaward edge of the low islands in the western High Arctic. Its level sediments slope gently into the Beaufort Sea, and the Plain is still being extended as it both receives the products of erosion and experiences isostatic rebound. Nowhere do these unconsolidated, mainly alluvial surfaces rise far above sea level.

Permafrost

Permafrost, a distinctive physical condition of continuously frozen ground, underlies half the area of Canada. It is defined as ground that remains under the temperature of 0°C continuously for more that two years. Usually, however, permafrost persists for hundreds of years, and in some areas of Canada, it may be 100 000 years old. Two zones are recognized: the discontinuous areas mainly under the boreal forests south to latitude 52°N in the east and 58°N in the west; and continuous permafrost on the tundra lands and on the very high parts of the western mountains. As the name implies, continuous permafrost is found everywhere under the ground except where water bodies are large enough to prevent freezing. The southern boundary closely approximates the July 10°C isotherm and the tree line for the area west of Hudson Bay. Thus, in general, the Arctic has continuous permafrost. Permafrost depths are conditioned by the intensity of sub-zero air temperatures and the length of time that freezing persists.

In the western Arctic and on some of the High Arctic islands where conditions were too dry to generate ice caps, the land surface was exposed to below-zero air temperatures for tens of thousands of years. At sites like Prudhoe Bay, Alaska, permafrost is 600 m thick in places. At Resolute it is 400 m, at Inuvik 100 m, at Churchill 30 m, and at Alert it is 1000 m thick. In a few places, permafrost is found under the sea. Air temperatures at all places in the Arctic rise above freezing for some period during summer, and this melts the top layer. Thus, an active layer of ground melts and refreezes in most years, and this zone thickens from north to south. At any one location, however, the depth of the active layer can vary greatly within a short distance, because the amount of summer melt is a response to a range of micro- and macro-physical conditions. Where water was in the soil when

Pingo near
Tuktoyaktuk, NWT

permafrost formed, it will be present as ground ice ranging in form from small crystals to layers many metres thick. Some types of ice, formed after permafrost, lead to cracks of tundra polygons, and ice cores of blisterlike hills called pingos. Ice in permafrost is stable enough as long as it remains frozen, but any circumstance that alters the thermal balance can cause surface collapse, rapid erosion, and damage to human structures. The surface of the ground in permafrost regions frequently displays patterns of hummocks, circles, polygons, nets, steps, or stripes. They are the effect of freeze-thaw, and occur in the active layer.

CLIMATE

The variety and distinctiveness of the Canadian climate begins with the physical nature of Canada itself. The country lies in a northern latitude, extending northward more than 4500 km from the international border with the United States. This causes strong seasonal differences in the climatic element of temperature and the accompanying patterns of daylight and darkness. In winter, when the sun is overhead south of the equator it is low toward the Canadian horizon during daylight, scarcely more than 25° at noon in southern Ontario in mid December, and not visible at all north of the Arctic Circle. The length of day is correspondingly short, and solar energy available to the land and water surfaces and lower atmosphere is not enough to redress the radiant loss of energy from them. So Canada, because of earth-sun relationships, has long cold winters, and its northern latitude is the prime reason for this.

In mid June sun is overhead at the Tropic of Cancer and as high as 70° above the horizon in southern Ontario, where there are sixteen hours of daylight if twilight is included. Daylight is continuous north of the Arctic Circle, although the solar angle is low. Everywhere in the country the net radiation is positive, with more of the sun's energy reaching the surface than leaving it. The pattern of seasonal shift when net radiation[1] becomes positive, and when the positive balance ends has a strongly latitudinal or zonal character. The High Arctic is positive from early in May to early in September. In mid Canada the positive balance lasts from mid March to mid October, and the southern part of the country is positive from February until November. Net radiation is a basic control for the annual temperature changes.

A second physical attribute of Canada that affects climate is the topographic nature of the land surface. The broad band of western mountains effectively protects the interior part of Canada from receiving moist Pacific air in anything but a modified form. East of the mountains, the great open belt of low-lying land that stretches from the mouth of the Mackenzie River all the way south to the Manitoba–Ontario border is like a great open highway for free moving air, allowing it to spread easily across half of the country. At other times it can be a shallow basin holding standing air either to cool intensely in winter or heat greatly under the sunny skies of summer. This large land area produces a continental climate, with the greatest seasonal extremes of temperature anywhere in Canada, and usually the largest range of diurnal temperatures.

Central Canada, the wide area surrounding Hudson Bay, is a higher block of land with an abrupt rim around its southern margin. There are many lakes in this rough surface, and these, along with Hudson Bay and the Great Lakes, have lower air temperatures and a climate that is generally more humid. The lower Great Lakes drain east to the valley of the St. Lawrence River and Gulf of St. Lawrence, and together form a natural trough where weather activity often migrates to the Atlantic Ocean. Southeast of the lower St. Lawrence River, the Atlantic region has several subparallel ridges and low mountains. These cause local airflow effects, and sometimes are high enough to catch extra rain and snow. There are mountains in Labrador and on the eastern fringe of the Arctic archipelago that may not interrupt major circulation flows but can cause local winds and supply snow to nourish glaciers.

A third physical condition that controls Canadian climate is that the country is bounded on three sides by oceans. In the belt of westerlies, the Pacific Ocean is a starting point for much of the air that crosses Canada. But, because of topography, it does not carry far inland without change. On the east coast, cold water flowing south past Labrador, and the moderately warm water of the North Atlantic Drift moving northeast past Nova Scotia, encounter each other over the Grand Banks, and modify air passing overhead. The long and highly indented coastline of Atlantic Canada creates an intimate relationship between eastern weather and its surrounding ocean. The Arctic Ocean has, considering the large number of islands of the archipelago, the longest of the ocean coastlines. It differs too, in that it is ice covered year round in the central basin, and for at least seven months in Hudson Bay. The effect of the ice cover is to make the ocean behave like a land surface with respect to energy exchange, so that winters extend the continentality of the land to establish a cold dry air mass over a large area. When this ocean does melt around its margins in summer, the cold water ensures a cool summer.

Climate and a Circulating Atmosphere

In an average year Canada loses more heat to outer space than it receives from the sun, and yet the air and the surface do not become colder and colder. The whole of the globe's northern latitudes have this negative annual balance because of the geometry of earth's orbit around the sun, and the inclination of its axis of revolution. But the atmosphere and the oceans transfer surplus equatorial energy poleward to ease the high-latitude energy deficits. Conversely, cold polar air and ocean masses move toward the equator to counterbalance the positive energy load that would otherwise build up in low latitudes. Oceans that contribute to this heat energy transfer do so in currents confined to their basins surrounding the continents, and so have specific geographic current patterns. The atmosphere is a more free-floating envelope around the earth, and the heat energy it transfers, although affected by the warm, cold, and high surfaces over which it passes, works to equalize the energy gradients. The overall airflow patterns, however, respond to the spinning globe beneath. Thus, surface observers at mid latitudes see the circumpolar westerlies and storm phenomena as drifts from west to east.

Canada, in a sense, lies in the contending zone where masses of air with different temperature and moisture qualities mix, giving rise to fronts of cyclonic storms that bring the familiar changeable array of weather in the westerly drift. The air masses themselves take on a certain homogeneity that reflects the qualities of the source regions that generate them.

In winter, the North Pacific Ocean modifies cold air from arctic sources and turns it into the cold and moist polar maritime air, which dominates the west coast of Canada. This air brings thick cloud and winter rain in the form of depressions, with a south to southeast flow on to the coast and across the barrier mountains toward the interior of British Columbia. As it moves over mountains, it becomes colder and drier and meets the arctic air that dominates the centre of the country. West central Canada, the Arctic Ocean and Arctic archipelago are covered with snow or floating ice. These ensure a large area of shallow frigid air, which in late winter and spring pours cold air into eastern Canada. This cold air is dry, but the prevailing sunny weather can be interrupted by snow and blowing snow of the prairie blizzard variety.

Farther east, in the Keewatin and Ungava regions, the storm zone of the north Atlantic reaches inland with snow and turbulent cold weather. Large water bodies like Hudson Bay remain open in early winter and supply moisture for heavy snow accumulation in northern Québec. The Great Lakes have snow-shadow areas to the east, where open water helps produce the snowbelts of the Lower Great Lakes and St. Lawrence Valley. In eastern Canada — the Maritimes, Newfoundland and Labrador — the marine influence of the Atlantic Ocean is strong and tends to delay the real cold of winter. The cold air becomes raw with the increase in humidity, and it produces a procession of cyclonic depressions. These track eastward along the polar front at the lower edge of the arctic air mass over Ontario, Québec, and the north and eastern United States. They are then drawn north to the stormy low-pressure area just beyond south Greenland. In the Far North, the western and central Arctic is intensely cold with light storm activity, but the eastern Arctic has more snow and stronger winds. The northlands as a whole are cloaked in the low sunlight of the short winter days and the darkness of prolonged polar nights.

The lengthening days and the sun's warming rays begin to sweep northward as summer approaches. Canadian summer sees the polar front shift poleward to lie across Alaska and northern British Columbia. It continues across the central plains to the Ontario coast of Hudson Bay and across Québec to Labrador. This front is the boundary zone of dynamic energy exchange between cool and cold air masses in the north and warmer air from lower latitudes. The west coast of Canada falls under the summer influence of a planetary high-pressure cell centred on Hawaii that produces a general flow of Pacific air from the northwest parallel to the coast. Near the ocean the summer air is warm but rarely hot, and clear skies add a sparkle to ocean-mountain scenery. The interior of British Columbia, cut off from marine influences, is dry except on the mountain ridges where showers occur, and over its plateaus radiative heating can produce hot and, sometimes, near desert conditions.

In Yukon, conditions are similar if cooler, and weak frontal storms are common toward the north. Once across the Rocky Mountains, the great plains area of central Canada, including the Mackenzie Valley, has a short, warm summer owing to incursions of drier masses of modified polar maritime air with occasional arctic maritime air. The weather changes are often caused by strong local convective heating, which leads to clouds, showers, or thunderstorms. Weather changes are also caused by the cyclonic depressions that move along a weakened polar front across the northern plains and on farther to the Arctic mainland coast. East of the Mackenzie Valley, in the broad sweep of boreal forest surrounding Hudson Bay, polar maritime air gives way to arctic maritime air. With greater areas of open ocean in the east, the cool summer air penetrates farther south, especially on the eastern side. Frontal storms and higher moisture make weather in this area changeable, and it is often cloudy with about as much rain in summer as the equivalent amount of water falling as snow in winter.

Along the lower Great Lakes, in southern Ontario, and in the Ottawa and St. Lawrence River valleys, much warmer summer temperatures are the rule and the warm weather lasts longer than anywhere else in the country. The moisture is high on account of the surrounding lakes, and convective and some frontal activity can bring thunderstorms and late afternoon showers. South of the lakes, in the eastern United States, particularly as one goes farther south, very warm and humid air drawn from the Gulf of Mexico is a prominent summer feature. The high-pressure cell centred near Bermuda moves this air north and northeast, and each summer in eastern Canada two or three periods of hot sticky weather are signs that gulf air has crossed the border.

In the Atlantic provinces and on the island of Newfoundland, the indented coast with its melting sea ice delays the spring and this means a shorter summer. Conditions are pleasantly warm except near the water and distinctly cool where there are cold currents. The air over this region originates above the land surface farther west, and has more of a continental character that a marine. The contrasting land and water temperatures in the nearshore Atlantic, however, produce cyclonic depressions that cause bouts of cloud and rain as they pass over from southwest to northeast. Far to the north, the Arctic summer temperatures are cool, greatly influenced by the cold open water of the Arctic Ocean channels between the islands, in Baffin Bay, and on the Labrador coast. The arctic front lies across the north, and a procession of weaker disturbances brings cloudy weather. Fog is common in areas next to the water.

Temperatures in Canada

A description of air temperature is conventionally depicted in the two seasons of summer and winter. Canada is, overall, a cold country, and a useful way to portray that fact is to examine the January record of mid winter, using the mean daily low temperature as an illustration of the degree of coldness (Map 2.3a). Not surprisingly, the High Arctic is coldest — sometimes as low as -35°C — but Yukon and the rest of the Northwest Territories are also very cold — down to -30°C. Despite the pattern of mean daily minimums, very cold days do occur in the interior of British Columbia, and 50 to 60 very cold days with temperatures below -20°C are experienced on the Prairies. Even Ontario and Québec can have as many as twenty very cold days in any given year. The lowest temperature recorded in Canada was -63°C at Snag near Whitehorse in Yukon in February 1947.

Winter temperatures feel much colder when the wind blows. Wind chill, a physically measured heat loss that varies with wind speed, is frequently quoted in weather forecasts and used in warning people of the severity of the cold. Sometimes it is expressed as an air temperature equivalent. For example, in January near Edmonton and Calgary, the average low temperature of -20°C to -25°C accompanied by the average wind speeds gives wind chill that feels more like -35°C to -40°C. Wind chill can, of course, apply anywhere in the country. Strong windy conditions in winter are associated with the Prairies and the Arctic. Although the Arctic has up to three times the number of days with blowing snow compared with the Prairies, the plains of the west are well known for their blizzards of drifting snow and poor visibility. They constitute a dangerous winter hazard severely limiting ground travel and sometimes trapping people in vehicles. Other open areas in the country also experience blowing snow, bringing traffic to a halt.

Since most Canadians are urban dwellers, it is worth noting that the large cities have their own local climates, heat islands that are a degree or two warmer than surrounding areas. Heating and lighting a concentration of buildings and homes involves escaped energy that raises air temperatures, and this is especially noticeable in the coldest places. An unusual winter feature found mainly in Alberta is the chinook. It is a warm dry wind that is adiabatically heated when subsiding air descends the eastern slopes of the Rocky Mountains. Temperatures can rise dramatically by as much as 15°C, occasionally more, and persist for a few days, causing a thin snow cover to melt and producing an unseasonable spring. Southern Alberta and Calgary can have as many as 25 days of chinooks in a year. Lethbridge, farther south, might have more than 30.

The only part of Canada that does not have a typical Canadian winter is the west coast. The outer coast and the south coast are mild because Pacific water keeps the air above freezing even in January and February. On occasion, outflows of interior Arctic air break out on to the coast, bringing temperatures down to -10°C or so, but these conditions rarely last more than a week or two.

Map 2.3a   Temperature °C – January – Mean Daily Minimum

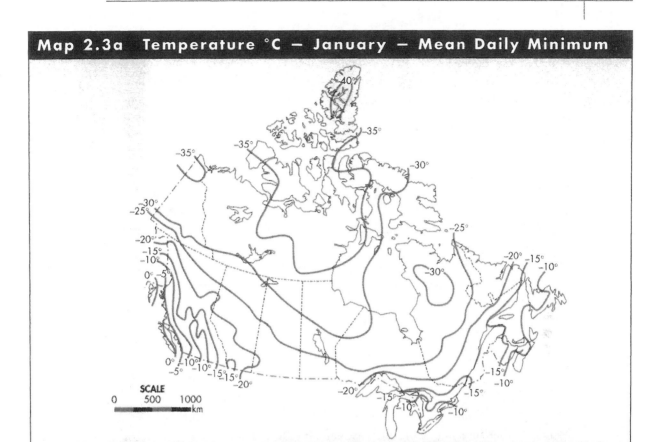

Map 2.3b   Temperature °C – July – Mean Daily Maximum

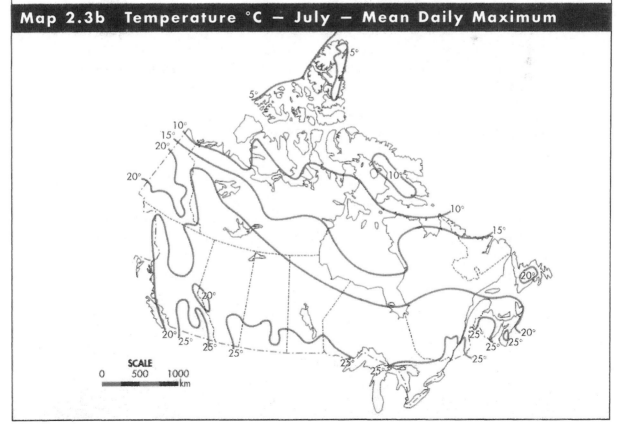

If cold winter is part of Canada's image or reputation, then summer warmth contributes to its prosperity. A typical expression of the Canadian summer is a map showing the average daily high temperatures in the month of July (Map 2.3b). This is what one would expect to experience on the warmest afternoons of mid summer. Most Canadians, living in the southern part of the country, will have July afternoons with a temperature of at least 20°C. One of the warmest parts of Canada is the southern interior of British Columbia, where daily highs in summer are often 30°C. This area can become extremely hot, setting records of 35°C and even 40°C.

It is very warm in the agricultural Prairies, normally at least 25°C and commonly 30°C. The highest air temperature recorded in Canada, 45°C, was measured in July 1937 southeast of Regina. Southern Ontario and southern Québec have mid-summer high temperatures of 25°C but they are not subject to the extremes of the plains, even though temperatures have been as high as 30°C at times. The surrounding Great Lakes modify the air above them, giving bouts of high humidity with hot, sticky weather rather than the usual very warm temperatures in the upper 20°s.

The Atlantic provinces, except perhaps New Brunswick, are affected by air that at some point was near or over the ocean. Thus, summer high temperatures remain close to 20°C, and in inland locations as high as 25°C. Across southern Canada, and even on the cooler coasts, the summer heat is enough to call for widespread use of air conditioning in buildings, homes, and vehicles. Large urban areas with built-up and built-over surfaces are a degree or two warmer than nearby rural landscapes. Exhausts from air conditioning units expel interior heat to the air outside, making a small contribution to the heat island lying over the city. In the north, the valleys and plateaus of the Yukon and the Mackenzie Valley experience 20°C in summer, and there are records of days reaching 30°C. On the Arctic coast, around Hudson Bay, and in the northern islands, the cold open ocean discourages high temperatures in summer. Temperatures of 10°C or even less are common for "warm" days.

Temperature-Related Measurements

There are several climatic indices that are derived from temperature. Summer is the season of plant growth and crop agriculture. Climatic statistics that bear upon crop growth include the frost-free period — the average number of days between the last killing frost of spring and the first one in autumn. The outer coast of British Columbia averages at least 180 days without frost, and the east coast of Vancouver Island and the Lower Mainland east of Vancouver may have 260 frost-free days. This part of Canada has the longest frost-free period. Southern Ontario, especially the interlake peninsula, has 140 days frost free, and at the southern tip of the peninsula near Leamington there are just over 180 days without frost. A different measure, and one very closely related to the growth of the plants, is the growing season. This is the total number of days with a mean temperature above 5.5°C, the temperature above which most plants grow.

The average growing season in Canada is illustrated in Map 2.4. Not unexpectedly, the two best frost-free areas are also where the longest growing seasons occur. The lower coast of British Columbia has on average more than 220 days of growing, and on either side of the Strait of Georgia, the value exceeds 240 days. Similarly, the Ontario peninsula has at least 200 growing days with over 220 days on the southwest tip north of Lake Erie. The cropped part of the Prairies and the Peace River Block are limited to 160 days growing season, and southern Alberta, Saskatchewan, and some of Manitoba may have 180 days or slightly more. The Maritime provinces have more than 180 days of growing season, but Newfoundland Island, at 160 to 170 days, is much more limited. Field agriculture is not restricted to these limitations in Yukon, in the gardens of northern Alberta, and in the upper Mackenzie Valley.

What is not picked up in these indices is the length of day: longer days in the north give northern crops an advantage that compensates for a short growing season. A second energy measure could be recorded, representing the average accumulated degree days of heat.[2] It is a more refined energy index because it measures heat, and is applicable to the subtleties of plant growth. However, apart from being able to compare highs and lows, these values are not as readily understood as the total number of days for plant growth.

Sunshine, of course, has a relationship with temperature but it is also a desirable climatic characteristic by itself. Canada, for the most part, is a sunny country. In the settled south, except for extreme west coast locations and some Maritime sites, places experience sun on 80 per cent of the days of the year. Thus, for a lot of the country, sunshine is as much a feature of winter as it is of summer. The Arctic, with its summer cloudiness and winter darkness, will have sun on fewer than half the days of the year. However, if the measure of sunlight is by total hours instead of by daily appearance, the Arctic looks better: one station, Eureka on Ellesmere Island, has an annual average of 2000 or more hours. This is greater than Vancouver and about the same as Toronto or Montréal.

In general, the northland is less sunny. South Yukon is an exception with over 1800 hours of sunlight. The Prairie cropland area is the sunniest part of the country, with over 320 days of some sun, and 2200 or more hours in a year. Sunny pockets occur in the Okanagan Valley and the Thompson River region of British Columbia as well as in the Interior Plateau. In southern Ontario from the Niagara Peninsula, Hamilton, and the northeast as far as Ottawa there are zones with 2000 hours. The Atlantic region is not very sunny, registering less than 1800 hours, and the Island of Newfoundland measures still fewer hours of sunshine.

Precipitation

The uneven pattern of precipitation in Canada, both areally and throughout the year, may be generalized into four regions: The Pacific Coast, Interior Plains, Eastern Canada, and the North. The northwest–southeast mountains lining up in

Map 2.4   Mean Annual Growing Degree Days

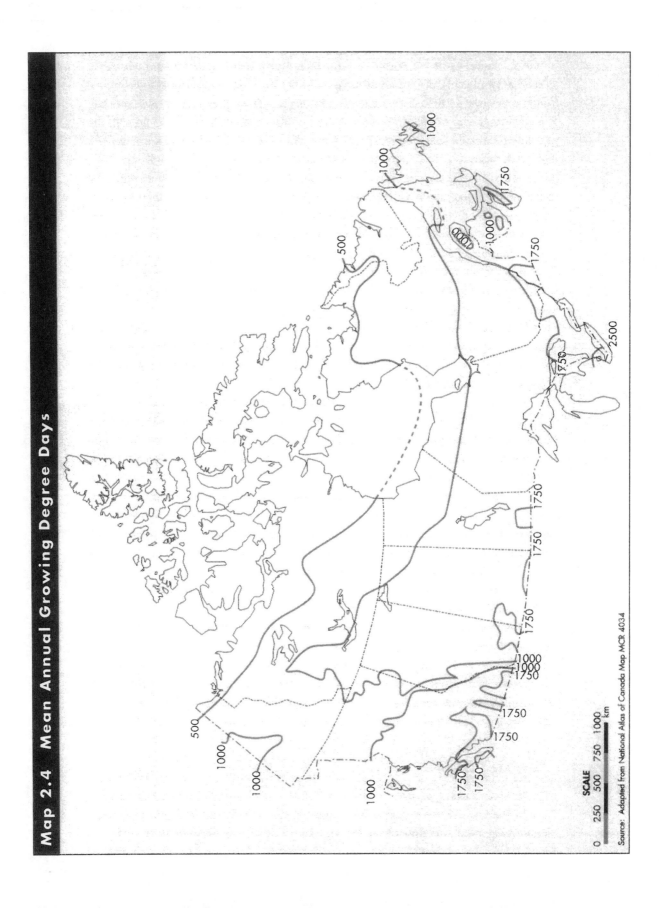

500
1000
1000
1750
1000
1000
1750
2500
1750
1750
1750
1750
500
1000
1000
1000
1000
1750
1750
1750
1750
1750

SCALE
0   250   500   750   1000
km

Source: Adapted from National Atlas of Canada Map MCR 4034

parallel ridges across British Columbia, with the intervening valleys and plateaus, produce practically all the extremes in Canadian precipitation records; they include the largest totals of rain or snow, the least precipitation, and long dry periods. The west coast has a winter wet regime with noticeably dry summers. From October to April or May, moist Pacific air is carried across the province and forced over the grain of the topography to add its orographically cooled condensation to any cyclonic storm activity already there. Thus, more that 3200 mm of rain falls on the outer islands and the exposed parts of the north coast. In winter there are well over 200 days of rain or snow (Map 2.5).

Higher up the mountainsides the proportion of snow is greater, and in the upper basins of the northern Coast Mountains, glacier fields give evidence of the excess of annual snowfall over melt. On the east-facing slopes there are pronounced rainshadow zones and drier conditions. The Thompson-Nicola valleys and the Okanagan are good examples of these zones; with less than 400 mm of precipitation, irrigation is needed for crops. On the west faces of the Columbia Mountains and on the northern Rocky Mountains precipitation levels rise to 800 mm.

Across the Rockies, beyond the lee slopes, there is yet again a rainshadow area that marks the edge of the second precipitation zone, the Interior Plains. This is the centre of Canada in terms of distance from the nearest ocean. This is dry country, far from any major sources of moisture. Southeastern Alberta and southwestern Saskatchewan, the region known as Palliser's Triangle, is very dry with less than 350 mm of rain in a year. Most of the rest of the Plains is only slightly better supplied with moisture with maxima approaching 150 mm. Nevertheless, under these conditions, agriculture is possible because one-third of the precipitation comes as snow and much of it remains as stored moisture until spring. Also, in spring the polar front shifts northward, and cyclonic storm rainfall ensures a June maximum of rain. The snowmelt and spring rain coincide with the growing season for cereal and oilseed crops.

Eastern Canada is the third precipitation zone. From Manitoba eastward moisture-laden air masses, originating from the Gulf of Mexico and the Caribbean, provide precipitation that increases progressively toward the east. Beginning with 500 mm a year at Winnipeg, the rainfall and snowfall totals increase to 800 mm at the Lakehead, and approximately the same in southwestern Ontario. Toronto has 750 mm, and Montréal 900 mm. The proportion of the precipitation that falls as snow is about one-third at Winnipeg, one-fifth at Toronto, and one-third at Montréal.[3] Eastern Canada has snowy winters, and the rainfall and snowfall regime in Ontario and Québec is pretty evenly spread throughout the year. Atlantic Canada is similar, although, with so much ocean surface nearby, total annual precipitation is greater, ranging from 1200 mm to 1400 mm. About 80 per cent of precipitation on the east coast falls as rain, but the 20 per cent of snow is still a substantial amount.

The fourth precipitation region is the North, including the Arctic. This also is a dry region, drier even than the Prairies. This may seen unusual with so much open water in channels between the islands or in Hudson Bay, but these surfaces

Map 2.5   Mean Annual Precipitation in Millimetres

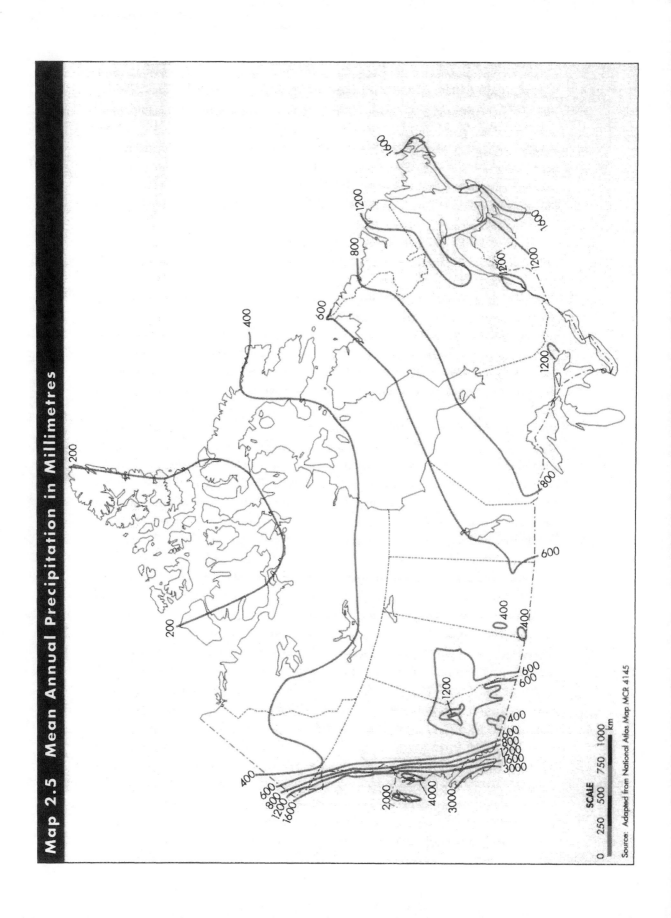

SCALE

0   250   500   750   1000
km

Source: Adapted from National Atlas Map MCR 4145

are ice covered, and the air is so cold that it cannot hold much moisture. Snowfall is very light, but there is enough to make the ground white, and over the tundra the winter winds sweep what little snow there is into hard-packed surfaces and dense drifts. Blowing snow is a more common weather feature than new snowstorms. During late summer when open water returns, the air does not warm up to any extent over the cold water surfaces, so rainfall remains light. Total precipitation for stations on the Queen Elizabeth Islands is so low that the area resembles desert conditions — Resolute, for example, has only 130 mm of precipitation, with half falling as rain and half as snow.

Precipitation-Related Features

The snow that falls on Canada must eventually melt; melting can occur as a result of warming during the winter, or through water loss due to sublimation, evaporation, or plant uptake. But the northward advance of spring weather across the continent brings on the main snow melt. Most of this happens without risk to humans as creeks and rivers swell or lake levels rise. Floods, and flood damage, however, are major threats in years of heavy snow pack, and many rivers overflow their banks. North-flowing rivers like the Red River at Winnipeg and the Mackenizie River are prone to ice-jam floods since they melt first in their headwaters and are plugged downstream by still frozen channels. In mountainous regions like the Fraser River drainage basin in British Columbia, the watersheds have extra heavy snow packs. Serious flooding is a high risk in such areas, especially in streams that occupy long stretches of level land. Rivers that drain from large lakes, like the Great Lakes, have small fluctuating levels because the lake-holding basins can more easily accommodate large quantities of meltwater.

Scarcely a year goes by without floods being reported at one or more locations in Canada. Flash floods, when they happen, are summer phenomena and result from heavy downpours caused by convective rainfall and thunderstorms. Two parts of the country are prone to thunderstorms: the prairie region, especially central Alberta, and southern Saskatchewan and Manitoba. Both areas average 20 to 25 thunderstorms a year. These storms develop from convective heating, a summer characteristic of the continental climate in this part of Canada, but some storms can be triggered by cold front activity. Such storms, with their deep vertical development, quite often produce hail that can destroy swaths of crops, so hail insurance is an established feature of prairie agriculture. Another region of the country with frequent thunderstorms is southern Ontario and the upper St. Lawrence Valley. Both convective and frontal displacement are the reasons for the storms, and southwestern Ontario averages more than thirty storms a year. While not as severe as in the Prairies, hail is a risk in Ontario, and severe damage has occurred to greenhouses in market gardening areas.

Too much water as run-off or flood is most often a feature of short duration, usually subsiding in a few days. Too little water is a long-term affair, both in a single season and in cycles over several years. The consequences are marginal crop agriculture and costly irrigation. The measure of precipitation adequacy for plant growth is based upon potential evapotranspiration, defined as the quantity of water that would evaporate from an open water surface plus the transpiration by plant cover with root access to all the water needed to effect maximum transpiration.[4] These values can be matched on a monthly basis against precipitation in order to relate surplus or deficit to potential need. By taking account of moisture stored in the ground, the accumulating water balance from many stations identifies where water shortage can occur. On average, there is an annual water deficit all across southern Canada except for the north shore of Lake Superior, eastern Québec and the Atlantic provinces. The maximum deficits of more than 300 mm of water are in the semi-arid regions of interior British Columbia, that is, in the Okanagan and Thompson River valleys.

The dry Palliser's Triangle in southeastern Alberta and southwest Saskatchewan has an average deficit of more than 200 mm of water, and the rest of the Prairies over 100 mm. The deficit in Ontario is less than 100 mm. These shortages start to become important in mid summer in the west, and a bit later is the humid east. The very low precipitation of the Far North does not result in deficits because of reduced evaporation and the prolonged period of frozen ground. Finally, the role of precipitation in distributing chemical and particulate pollutants across Canada has to be noted. The global circulation of the atmosphere picks up persistent and potentially dangerous materials from source regions all over world, particularly from industrial belts and places where pesticides have been applied, and sweeps them in the westerly circulation northward to reach all parts of Canada. Acid rain from sulphur and other compounds is endangering the health of lakes in eastern Canada downstream from the North American industrial heartland. When deposited farther north, these manufactured compounds tend to remain because of the lower temperatures. They persist and enter the food chain, creating unknown risks for humans.

VEGETATION

Natural vegetation, whatever form it takes, is an integrator of other physical attributes of the land. The climate of places sets limits for plant growth; the underlying soil also imposes conditions on vegetative response; and both of these controls are affected by topography. Vegetation therefore evolves unique patterns of plant associations and ecological integrity, and their geographic expression is best illustrated in vegetation maps.

Tundra

The tundra is a land without trees, covering a quarter of Canada north of the boreal forest, and like other vegetation zones, responding in its variety to overall climactic conditions. Since this is a treeless area, the boundary with the boreal forest is called the tree line. In fact, it is less of a line and more like a belt of transition. In Arctic

Québec this belt is 400 km wide, with tree islands or patches in protected sites around lakes or along rivers. The transition is more abrupt in the Northwest Territories where, in places such as river valleys, the ground cover of trees stops abruptly and the tundra begins. In much of the mainland tundra, the land is completely covered with grasses, sedges, and heath plants to form a continuous green surface except where there are rock outcrops. Toward the mainland coast and on the southern Arctic islands, lichens and mosses are found on ground and rock surfaces. The Queen Elizabeth Islands and one or two large islands nearby present a barren rock desert surface where plant cover is limited to river flats (Map 2.6).

In the mountain areas the tree line becomes the timber line, and above it, where soil permits, alpine vegetation is found. The timber line at the 49th parallel has a 2000 m elevation, and it declines northward to 1200 m at the Yukon–British Columbia boundary. Farther north the alpine tundra may drop to 900 m above sea level. The vegetation cover there is quite patchy owing to the rocky nature of mountain tops, but alpine meadow grasses and heath plants are found on dry locations and sedges and mosses on wet sites.

Boreal Forest

This is Canada's most extensive forest cover. It reaches right across the country in a belt 1000 km to 1200 km wide, but within it there are so many differences in tree and other plant associations that the whole forms a mosaic of great variety. The trees are mostly conifer or needleleafs. Black and white spruce occur across Canada, with lodgepole pine in the far west and jackpine through central and eastern parts. Alpine fir grows on the western mountains and balsam fir to the east, with tamarack throughout the forest in wetter locations.

Broadleaf trees, mainly aspen, are found both in patches and large pure stands, and paper birch grows where there is good drainage. Thus, in the far northwest it is mainly white spruce and alpine fir, in northern Alberta spruce and aspen are characteristic, and northern Saskatchewan has mainly black spruce and jackpine. Manitoba is similar to Saskatchewan but has more tamarack, while in Ontario and Québec spruce and balsam fir are dominant. Fire is a part of the ecology in areas of inaccessible forest, bringing on broadleafs first in the post-fire succession, then the climax needleleafs. The forest, therefore, is very much a patchwork of tree types in constant evolution.

Three different areas can be identified along the northern part of the Boreal Forest. West of Hudson Bay and east of Hudson Bay are two broad transition zones intermediate between the dense forest and the northern tundra. They are open woodlands with trees widely spaced or in patches along stream courses or lakeshores, and a tundra-like surface with lichens showing through. It is usual to define the southern limit of the Boreal Transition Zones as the southern extent of the lichen woodland. Towards the northern limits the trees are much smaller. Trees over 300 years old may be no more than 10 m high. Between the two transition zones is a wetland in the Hudson Bay Lowlands. The area slopes gently towards the

Map 2.6 Vegetation

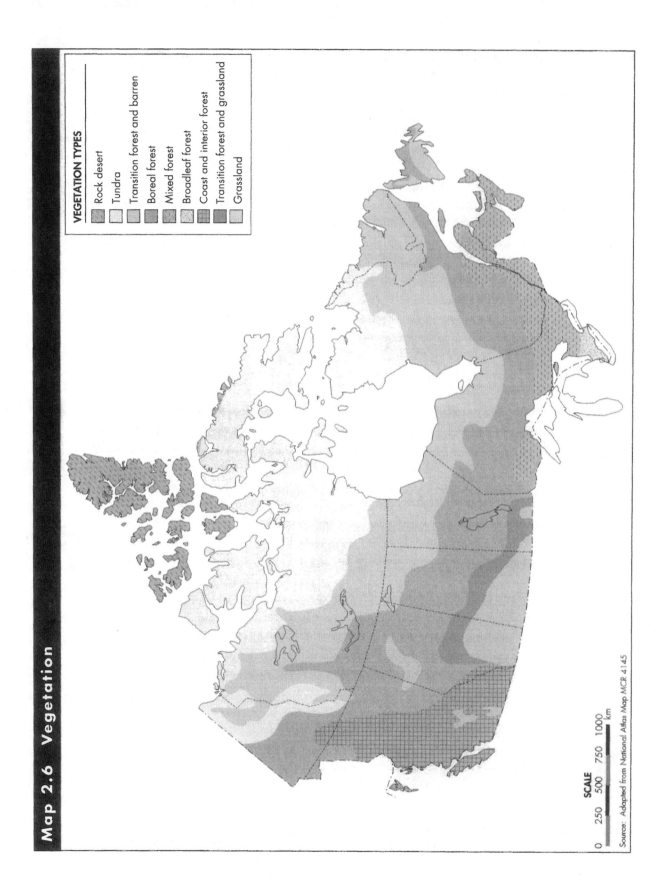

VEGETATION TYPES

- Rock desert
- Tundra
- Transition forest and barren
- Boreal forest
- Mixed forest
- Broadleaf forest
- Coast and interior forest
- Transition forest and grassland
- Grassland

SCALE

0 250 500 750 1000
km

Source: Adapted from National Atlas Map MCR 4145

Bay and is poorly drained. There are frequent peat and string bogs and much organic soil, and the trees are mainly black spruce and tamarack.

Coast and Interior Forest

The wet mild Pacific coast supports some spectacular forests. Mature evergreen trees, centuries old and reaching heights in excess of 80 m, present the most luxurious growth to be found anywhere in Canada. The exposed north coast, opposite Queen Charlotte Sound and including the Queen Charlotte Islands, has a forest cover dominated by western hemlock and western red cedar. Sitka spruce is found near the shorelines. From Vancouver Island south, Douglas fir is the dominant tree. On the extreme south of Vancouver Island, Garry oak reaches its most northerly extent. The best growth of these trees is at lower elevations. With increasing height, the vegetation becomes sub-alpine, and mountain hemlock, amabilis, and subalpine fir dominate. The coast forests have long been recognized for their productivity with high wood-volume per unit area of land, and they have been harvested extensively. There is now, however, increasing concern for the ecological values of forest lands and sustainability in production.

The Columbia Mountains of east central British Columbia experience heavy precipitation and almost match the coast for a wet environment. An interior version of the coast forest is found here with western hemlock as the dominant tree. Western red cedar is also present and on drier sites there are western larch, blue Douglas fir, and lodgepole pine. Broadleaf trees include poplars and paper birch. The subalpine zones are distinguished by the cold climate of high elevation; the vegetation is characterized by conifers that can meet those conditions. Within this forest, Englemann spruce is very common, increasingly so from north to south where it predominates along with subalpine fir. Toward the eastern slopes and in the well-drained sites to the north, lodgepole pine is present.

Montane forest is the natural cover for almost all of the Interior Plateau of British Columbia, and is often called the Interior Douglas fir forest. The southern part surrounds the steppe grassland and is closely related to the semi-arid core. Ponderosa pine, and interior Douglas fir at higher elevations, are the two types of tree most commonly found. The low ground of the woodland is quite open and resembles a parkland transition from grass to close-grown stands. In the central montane, interior Douglas fir is predominant across the central plateaus, but as the land rises eastward to the wet west slopes of the mountains, hemlock takes over. The northern third of this zone has increasing amounts of aspen, lodgepole pine, and white spruce.

Grasslands

The grassland area of British Columbia is located in the Okanagan Valley and along the South Thompson River. It is a very dry region in summer and short grass dominated by bunch grasses is the natural cover. It differs from other Canadian grasslands in that sagebrush persists even when the land is grazed. Agricultural practices have altered the cover; irrigation for commercial cropping, and cattle grazing mean the plant associations as they would have occurred naturally are hard to find. Toward the tops of high hills and upland bluffs there are some trees, and they become more closely spaced as the moisture increases toward the summits.

North Americans usually refer to the grasslands as prairie. From southeast Alberta, through southern Saskatchewan and into southwest Manitoba, and south of Winnipeg the nature of the prairie reflects the rainfall regime. In a series of concentric rings, the precipitation and the type of grass changes progressively from a wet north to a drier south. The Red River Basin south of Winnipeg had a natural tall-grass prairie, representing a luxuriance compatible with dependable rainfall and a dark organic rich soil. Since cultivation, this type of prairie is largely gone, preserved now in ecological museum plots. Most of the grass in Saskatchewan and the northern part of Alberta are mixed grasses, medium in height like cereal crops. A lot of this land has been broken for agriculture, and natural grassland is restricted to areas that resisted the plow. In the southwest and near the international border the grass is a short-grass prairie — plants about 30 cm high, that grow, flower, and seed in a very short season following spring rain and before the summer drought. These grasses are used for grazing but have very low carrying capacities.

Parkland surrounds the grasslands as a narrow strip against the Alberta mountains, but some 200 km wide across the northern edge from Alberta into Manitoba. It is a transition zone to the boreal forest representing progressively better growing moisture conditions toward a cooler north. Near the grass, it begins with bluffs of broadleaf trees in groups, and along stream courses or around sloughs, and eventually becomes more continuous. Aspen and poplar are common trees. Along rivers and their tributaries in Manitoba, the Manitoba maple and cottonwoods grow to be large trees and lend distinctiveness to the landscape.

Southern Broadleaf Forest

A broadleaf or hardwood forest is the natural cover of the southern half of the Ontario peninsula. With an absence of conifers, it is the northern extreme of the Carolinian forest where beech, sugar, and red maple are the most common trees. Other hardwoods are present including elm and ash, oak, black walnut, and sycamore trees. Cultivation has largely eliminated the original cover, but farm woodlots and parks preserve examples of what this forest type once was.

Southeastern Mixed Forest

Both broadleaf and conifer trees coexist in the mixed forest. Mapped as a broad zone from west of Lake Superior right across to the Maritimes, there are regional differences within it. The broadleafs are mainly sugar and red maple, beech, some oak, elm and birch, with rarer examples of basswood. The conifers include the much sought-after white pine, red pine in the east, several spruce varieties including Norway spruce, firs, and hemlocks, and in wetter sites, some cedars. The northern boundary of this forest loosely coincides with the northern limit of white pine, the great commercial tree of early lumbering in Ontario and Québec. The northern side of the mixed forests takes on more of the character of the boreal forest. Some of this natural cover has been sacrificed to cultivation, and long periods of commercial forestry have made changes in the natural succession. The sugar maple is the basis of a significant industry in Ontario and Québec, producing maple syrup and other sugar products.

CONCLUSION

Landforms, climate, and vegetation as they affect, or are influenced by, human occupance are the essential elements of Canada's physical geography. At the same time, from the perspective of population distribution, soils and waterways are the key to present-day concentrations of settlement. These two resources determined where early settlers from Europe took up residence. Rivers like the St. Lawrence were the only mode of transportation for bulk materials in those early days, and the rich soils on or near its banks ensured a livelihood from farming.

Access by early European settlers to the interior of what is now Canada was quite limited for a long time after the establishment of New France on the lower reaches of the St. Lawrence, and there were no prospects for permanent settlement west of the Great Lakes. Hence, as people arrived from Europe, they settled where others were already rooted. As political events shifted control of the land westward, the next area to be successively occupied was southern Ontario — a part of the country better endowed with soils and waterways than anywhere else. Again, people arrived and joined those already there.

Even in later years, in the first decade of the twentieth century, with Canada now a single country, as immigration from Europe peaked and large tracts of the Western Interior were settled, areas around the St. Lawrence River retained their dominance as the main concentrations of settlement. Today in Canada, as in so many other places around the world, these early settlement patterns, conditioned as they were by the physical environment, explain the long-term distribution of population and, in Canada's case, define the nation's heartland.

N O T E S

[1] Net radiation is complex and takes account of energy in air turbulence, evaporation and condensation, and photosynthetic processes, and in heating and cooling of the surface below.

[2] A degree day, measured in °C, is the positive temperature difference between 5.5°C and the mean air temperature for that day. That value is added to the same difference on the next day and so on until all the degree days are accumulated for the year.

[3] Snow is measured in centimetres and converted to precipitable water at a 10:1 ratio. Thus, if one-third of Montréal's 900 mm of precipitation falls as snow, there would be 300 cm of snow, certainly enough for a snowy winter.

[4] Potential evapotranspiration (PET) is greatly dependent on air temperature as a measure of water-holding capacity of the air and energy available for evaporation. The summer maximum air temperature map (Map 2.3) gives a good indication of areas of high evapotranspiration. PET is also adjusted for the length of day since plant transpiration is conditioned by daylight, so the longer daylight of northern latitudes somewhat counteracts the lower temperature.

chapter 3

Donald Kerr**

Emergence of the Industrial Heartland, c. 1760–1960*

The remarkable paradox of central Canada is that, although it can claim status as Canada's heartland, it has been and continues to be sharply divided along the Ottawa River into two distinctly different linguistic and cultural regions. At Confederation there was a marked contrast between the deeply rooted, French-speaking Catholics living in Canada East and the strongly British, English-speaking Protestants of Canada West. These two contrasting jurisdictions remain joined in an uneasy political alliance.

If central Canada has little or no cultural or social unity, what justification is there for demarcating a region beyond some vague historical identity? The answer lies in the economic realm. Squeezed between the Shield and the American border on a relatively small lowland, central Canada has attracted innumerable factories, trading enterprises, and financial institutions, both foreign and domestically owned. As the centre of industry, trade, and finance, it controls the Canadian economy. Although there has been little social connection between individuals or communities in Québec with those in Ontario, there has been a strong and deeply rooted economic interaction and interdependence.

* From 1987 to 1993, three volumes of the *Historical Atlas of Canada* (H.A.C.) were published by the University of Toronto Press (see notes 4, 5, and 18). Maps, diagrams, graphs, and text from these volumes provide an excellent background source for readers of this chapter.

** I would like to acknowledge with thanks the help of two colleagues, professors Thomas McIlwraith and Gunter Gad, Department of Geography, University of Toronto, in the revision of this chapter.

The purpose of this chapter, then, is to provide a geographical study of the Industrial Heartland's changing economic structure and its dominant position in the Canadian space economy by 1960. The chapter emphasizes historical economic geography and aims to interpret geographic change through the examination of four interrelated themes: resources and economic development; interregional trade and growing interdependence between Ontario and Québec; the Industrial Heartland's role in economic development and control *vis-à-vis* the rest of Canada; and the impact of growth on the spatial organization of central Canada's economy.

Although the relationship between resource endowment and economic growth is complex, staple theory provides an appropriate framework for analysis.[1] Put simply, the export of furs and, later, timber products provided revenue to support the beginnings of a domestic economy. In the nineteenth century the agricultural resources of the region proved to be the critical catalyst for growth. In fact, it might be argued that had the direction of geological events been somewhat different, causing the ancient boundary of the Canadian Shield to lie anywhere from 100 to 150 km further south of its present position, the course of Canadian history might have changed. If the lowland had been compressed more tightly between the Precambrian mass to the north and the Appalachians to the south, thereby reducing the amount of arable land significantly, a viable agricultural society could not have been supported and Canada as we know it may not have survived. Be that as it may, there is no question that in the nineteenth century wheat became the main staple of the agricultural lowland. Producers found overseas markets, and the rising optimism stimulated land clearing and immigration, subsequently encouraging domestic manufacturing, railway building, and an expanding urban network.[2]

A succession of staples — furs, timber and then wheat — were shipped through the Great Lakes–St. Lawrence system across the North Atlantic to markets in western Europe. From an early date the merchants of Montréal took charge, directing the flow of goods in and out of the "Commercial Empire of the St. Lawrence" by exchanging foreign manufactured goods for staples. They thus developed the infrastructure of transportation systems, commercial contacts, and financial mechanisms necessary for the flow of goods and information between Upper and Lower Canada. Although closed by ice to ships during the winter and interrupted by rapids and waterfalls, the St. Lawrence–Great Lakes system played a decisive role in the pattern of trade, and not surprisingly the most progressive of the early towns were ports. Later, canals, railways, highways, and telegraph and telephone lines were built to facilitate trade, and new manufacturing towns sprang up away from the St. Lawrence and lower lakes. The roots of modern interregional trade had been established.

By the middle of the nineteenth century, changes in the economic structure of the emerging core *vis-à-vis* the rest of Canada could be detected — the region was showing signs of national importance. The growth of manufacturing industries and financial institutions and the strengthening of wholesale and retail trading houses

Early view of Montréal from Saint Helens Island

all coincided with the development of railways, and gave strength and diversity to the region's economic base.

Thus, at the time of Confederation, the central Canadian economy had reached a level of maturity that allowed it to take an active role in the development of the hinterland. Aided by the projectionist policies of the federal government, especially those embodied in the National Policy of the late 1870s, centrally based companies expanded into the periphery; through corporate policies of centralization, they reduced or eliminated the ability of many hinterland firms to compete. The imposition of high tariffs on most agricultural machinery made it almost impossible for farmers in western Canada to buy machinery from anywhere else but southern Ontario. The Bank Act of 1871 fostered the rise of national banks that, by 1920, through merger and financial acquisition, made the financial houses of Montréal and Toronto the undisputed masters of accounts from the Atlantic to the Pacific. Low postal rates created opportunities for large catalogue distributors, such as Eaton's of Toronto, to challenge local retailers and regional wholesalers. In similar fashion, the weak corporate trust laws before 1929 encouraged central Canadian institutions to take over sectors of the economy in the Maritimes.

Despite their consequences, such measures integrated the growing national economy. The Industrial Heartland controlled wholesale and retail prices, interest and insurance rates, and trading and investment policies, and in doing so, it created a system of dominance that has persisted to the present. It would be incorrect, however, to conclude that control was absolute, because some sectors of regional economies, such as the British Columbia forest industry and the Alberta oil industry, have prospered independently of central Canada. Ironically, with the increasing penetration of direct American investment in Canada in the twentieth century, many critical decisions have been made in New York, Detroit, and other American cities, bypassing the Industrial Heartland completely. Nonetheless, the extraordinary geographical concentration of economic power in the central region is a distinct characteristic of Canada.

What impact, then, did this economic growth have on the internal geographical organization of the heartland? While some towns grew into large, multifunctional centres, others remained unchanged, part of a stagnating local economy. The group of towns and cities stretching from Québec to Windsor came to form the economic core of central Canada, often described as "Main Street." Railways and highways integrated this bustling corridor, which stood in sharp contrast to the stable rural economies of eastern Ontario, the Chaudière Valley of Québec, and the Huron Uplands of Ontario. Yet, within the corridor itself, growth was differential; some once prosperous old ports such as Cobourg having declined.

In contrast, by 1960 the two metropolitan centres of Montréal and Toronto grew to account for over 20 per cent of the population in Canada, 35 per cent of the manufacturing, and most of the critical corporate and financial decision making, thus creating a strikingly bipolar spatial pattern. In fact, while these prosperous metropolitan centres gave the heartland national pre-eminence, they also created, at the regional-provincial scale, an internal core–periphery structure: Montréal within Québec and Toronto within Ontario.

All of these geographical changes have taken place gradually but consistently over the course of two hundred years, beginning in the mid-eighteenth century and reaching maturity by the early 1960s. During this *longue durée*[3] — the historical geography of a long time span — strong economic and political forces have made the Industrial Heartland the dominant region of Canada.

STAPLES AND EARLY DEVELOPMENT

In 1760, the area north of the Great Lakes was taken over by Britain from France, the formal transfer of which was defined by the Treaty of Paris in 1763. Within less than fifteen years, the British colonies to the south were in revolt, the successful completion of which brought the United States into being. Early on, the Americans urged the old French colony through invitation and by invasion (1775–76) to join the revolution and make Québec the fourteenth colony (state). The loyalty to Britain of the French elite (church leaders and seigneurs), the neutrality of the peasants, and the superiority of British troops in crushing the siege of the city of Québec, thwarted the revolutionary goal of driving the British from the continent. Thus, early Canada survived with its predominantly French population along the banks of the St. Lawrence and a very small and scattered population in Nova Scotia and Newfoundland.

Understandably, the geographic framework of Canada was well in place at the time of the Conquest, the product of a settlement history of at least 150 years. Québec and Montréal, which by 1760 had grown to house populations of 8000 and 5200 respectively, dominated the economic and social life of the colony. Standing guard over the estuary of the St. Lawrence, the city of Québec maintained a superb location where the French had centred their military operations and administered the colony. Deep water, without significant currents or tidal variations, provided an ideal harbour for sailing ships, and Québec functioned as the terminus for virtually all Atlantic shipping. Montréal, situated at the Lachine Rapids and commanding the Ottawa, St. Lawrence, and Richelieu valleys, existed primarily as a transshipment centre and controlled much of the fur trade. Although farming provided the livelihood for the bulk of the population, the fur trade created some wealth for the colony, encouraged exploration of the back country, and brought the French into contract with the Native peoples of the area. At the time of the Conquest, the British acquired an established colony of some 60 000 French-speaking Roman Catholics who had cleared considerable land for agriculture within a few kilometres of the St. Lawrence River, and who had built two cities in which virtually all commercial, political, and religious activities were centred.[4]

In the period immediately following the British Conquest, the economic geography of Lower Canada changed very little. The fur trade not only persisted but expanded, as it was increasingly controlled by British traders who eventually merged many small operations into the continent-wide North West Company. Although French Canadians lost control of much of the fur trade, they continued to provide most of the work force. Important, too, was the immediate movement of French Canadian farmers onto empty seigneurial lands, where they produced a modest surplus of wheat for export.[5] Linear villages ran through the countryside, but most remained commercially insignificant. In contrast, Loyalist immigrants, refugees from the American Revolution, took up land in the Eastern Townships within the framework of a rectangular survey. In Montréal and Québec, the number of Scottish and American entrepreneurs and poor English and Irish immigrants increased markedly, reducing, by the 1830s, the native French population to less than half of the total.

Around the turn of the nineteenth century, rich pine, spruce, and maple resources of the central lowlands and neighbouring Shield began to be exploited. Newly formed companies appropriated large tracts of land and cut timber along the Ottawa, Saguenay, and other valleys of central Canada. Squared timber and lumber were exported to Britain when, during the Napoleonic Wars, traditional Baltic supplies were cut off. Subsequently, tariff legislation gave preference to Canadian suppliers. For the most part, the timber trade was organized by merchants in Montréal and Québec.[6]

The territory to the west, partitioned off as Upper Canada by the Constitutional Act of 1791, and opened for European settlement, was predominately forest country and sparsely populated by Native peoples, who had been reduced in number by war-

Lumber ship,
Québec, 1872

fare and disease. At first, refugees from the American Revolution (Loyalists) and, later, post-revolutionary Americans accounted for most of the early settlement, to be joined in the decades following the War of 1812 by large numbers of immigrants from the British Isles. The population of Upper Canada grew rapidly, and in a period of nine years — from 1842 to 1851 — it more than doubled from 450 000 to 952 000 people.[7] Nearly all of these settlers practised mixed farming. Wheat became the most important cash crop and, along with timber, was exported to Britain.[8]

The success of wheat in Upper Canada may be explained, at least in part, by the region's superior physical resources. In terms of present climatic conditions, southwestern Ontario has a growing season that is longer, warmer, and less susceptible to unseasonable frost than southern Québec. Windsor, which has an energy supply of almost 2400 growing degree days (Celsius) and an average frost-free period of 170 days (and thus is similar to areas of the American Midwest), stands to contrast to the city of Québec, which maintains an average of just less than 1600 growing degree days (Celsius) and a frost-free period of 132 days (conditions resembling those of northern New England). For the most part, the soils of southern Ontario were more productive than those of Québec, and although the soils of both regions deteriorated in the nineteenth century because of improper cultivation, the acid soils of Québec suffered more. Wheat grown there also became susceptible to insect infestation and disease at an earlier date than in Ontario, reducing yields accordingly.

The "failure" of the wheat economy in Lower Canada, combined with rapid population growth in relation to a shortage of agricultural land, initiated an out-migration of French Canadians to the United States in the 1840s. With the decline in available arable land in Québec, and the persisting high birth rate, French-Canadian families continued to migrate to textile towns in New England throughout the nineteenth century. In many cities they established large French Catholic parishes.[9]

The differential impact of the wheat trade on the early economic landscape can be described in a number of ways.[10] In Canada West, a varied trading network — composed of lake ports, inland towns, villages, and rudimentary roads — took shape and facilitated the export of wheat. There was no comparable development in Canada East where, apart from Montréal and Québec, the urban hierarchy remained poorly developed. For example, by 1850, some thirty-eight urban centres housing over 1000 people could be identified in Canada West. The populations of Kingston, Hamilton, London, and Ottawa ranged between 5000 and 25 000, and Toronto stood at 30 800. In contrast, there were only sixteen centres in Canada East that exceeded the urban threshold and fourteen of these had fewer than 5000 people. The two largest centres — Montréal (57 700) and Québec (42 000) — were exceptional indeed (Map 3.1). In short, these different urban hierarchies, whose basic form still exists, have their origins early in the nineteenth century and are apparently related to commercial agriculture more than to any other factor.[11]

Capital accumulated from the sale of wheat and timber, and the injection of funds by the British for defence and administration supported the rise of manufacturing in urban centres early in the nineteenth century. Invariably, small manufacturing enterprises — including grist mills, sawmills, tanneries, and distilleries — grew up at settlements with access to water power, and these activities were soon augmented in some places by industries producing consumer goods such as farm

Church of the 1850s in rural Ontario

Map 3.1   The Industrial Heartland in the Mid-Nineteenth Century

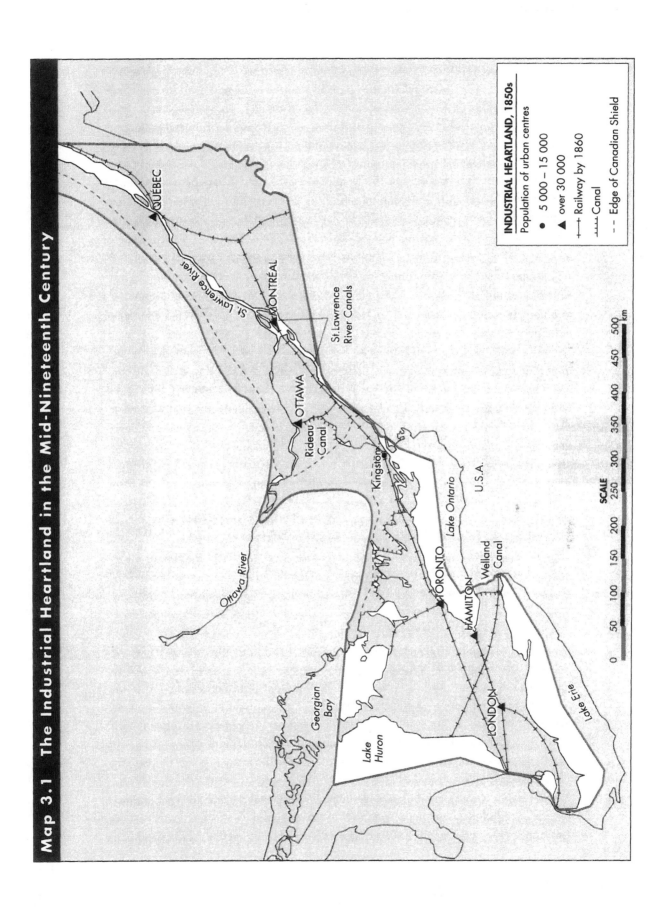

INDUSTRIAL HEARTLAND, 1850s

Population of urban centres

- ● 5 000 – 15 000
- ▲ over 30 000
- ┼┼ Railway by 1860
- ⌒⌒ Canal
- – – Edge of Canadian Shield

QUÉBEC

MONTRÉAL

St. Lawrence River

St. Lawrence
River Canals

OTTAWA

Ottawa River

Rideau
Canal

Kingston

Lake Ontario

TORONTO

U.S.A.

HAMILTON

Welland
Canal

LONDON

Georgian
Bay

Lake
Huron

Lake Erie

SCALE

0   50   100   150   200   250   300   350   400   450   500
km

implements and furniture. In the parlance of staple theory, the sale of resource commodities abroad created capital to purchase necessities and even luxuries, an increasing number of which were produced locally. Later, on a larger and urban scale, first in the 1840s in Montréal and then in the 1850s in Toronto, manufacturing expanded significantly in response to the growing market of Canada West.[12]

The economy of Montréal continued to prosper in the face of a depressed local agriculture. This juxtaposition of rural depression and urban buoyancy can be explained by the stimuli of immigration and the wheat and timber trade. The merchants of Montréal organized trade by pressing the Legislature of Lower Canada for funds to build, widen, and deepen canals for the improvement of transportation along the St. Lawrence system. The most serious impediment to ocean-going vessels, apart from the ice of winter, was the shallow and treacherous water at the western end of Lac St. Pierre. Through the deepening, widening, and straightening of channels, unimpeded access to Montréal was gained in 1848. An increasing number of vessels then sailed directly inland, bypassing Québec.

As a result of these developments, between 1850 and 1899, the value of the import and export trade at Montréal increased from $8.9 million to $129 million, while over the same period at Québec it remained virtually unchanged ($7.2 million to $9.9 million). Even as canal building increased upstream from Montréal, plans for railroad construction were well under way, and Montréal businessmen pursued this activity as well.[13] In fact, a short line had been built to bypass the rapids on the Richelieu River along the water route from Montréal to New York as early as 1836, and in 1847 the Champlain Railway was opened around the Lachine Rapids. Of far more importance was the completion in 1853 of a railroad from Montréal to Portland on Maine's Atlantic coast. This line extended westward as the Grand Trunk to Toronto in 1856 and to Sarnia by 1859 (Map. 3.1).[14]

Complementing the canal and rail developments of the 1840s, the flow of information also improved. In 1848 telegraphic communications were opened between Montréal and Toronto, and in 1856, following the regular scheduling of rail services, the time required to deliver a letter between the two cities had dropped from about ten days to less than twenty-four hours.

As transportation and communication improved, and as the economy became more concentrated geographically, some cities grew at the expense of others. The city of Québec found itself in an increasingly marginal position as new rail routes converged on Montréal. At Confederation, when its population of 50 000 was only one-half the size of Montréal, it was still engaged in a desperate struggle to gain access to Montréal and Toronto through the building of the North Shore Railway.[15] The eventual bankruptcy and sale of this railway further reduced the economic role of Québec as a rival to Montréal.

In Upper Canada a similar struggle for urban leadership eventually gave primacy to Toronto. Although its small but well-sheltered harbour provided some natural advantages, Toronto's rise must be explained by its role as political capital and by its merchants' staunch belief that it would gain control over the Northwest. In fact, the notion of Toronto exerting control over trade in the Northwest began early in the writings and policies of Governor Simcoe. By the late 1840s, George Brown was writing frequently in the newly formed *Globe* about the importance of western trade and the need to acquire western lands. Subsequently, in the late 1850s the Canadian government awarded a consortium of Toronto entrepreneurs a contract to develop roads and railways in the Thunder Bay district through the new Northwest Transport Company.

At the same time, increasing amounts of wheat were being shipped through the port of Toronto and, although the merchants of Hamilton and London competed aggressively, those in Toronto dominated. The extension of the Great Western Railway in the mid 1850s diverted substantial trade from Hamilton, previously the foremost wholesaling centre. Kingston, equal in size and importance to Toronto in the early nineteenth century, possessed a restricted agricultural hinterland and failed to keep pace. Thus, the increasing concentration of merchants, financiers, and some manufacturers in Toronto provided the basis for regional control, and eventually Toronto mounted a sustained challenge against the hegemony of Montréal. In fact, through the 1850s and 1860s, Toronto entrepreneurs partially freed themselves from Montréal's grasp, strengthening ties with New York via the Hudson–Mohawk route and developing a stronger industrial base.[16]

By Confederation, then, the Industrial Heartland had emerged as an integrated and distinctive economic region, bound together by a network of transportation and communication facilities. The region's principal sources of capital were derived from the sale abroad of wheat and timber, from the expenditures of the British government on local defence and administration, and increasingly from the transfer to Canada of undetermined amounts of capital by British immigrants. The large-scale benefits of the staple trade in wheat and timber accrued not only to the merchants of Montréal, but also to their rivals in Toronto. There is little doubt that at the time of Confederation, many merchants, manufacturers, and financiers of these two cities were ready to participate in the exploitation of both the new western territories and the old colonies of Nova Scotia, New Brunswick, and Prince Edward Island.

THE INDUSTRIAL HEARTLAND A CENTURY AGO

By the 1880s, the spatial arrangement of settlement and transportation facilities in the Industrial Heartland had crystallized to the extent that the outline of today's patterns of regional differentiation were well in place (Maps 3.2 and 3.3). Agricultural settlement had reached its outer limits in southern Ontario where, in fact, some rural depopulation had already occurred. Migration to the cities, especially Toronto, as well as movement to the United States, accelerated dramatically. Québec farmers, occupying all but the poorly drained parts of the St. Lawrence Lowlands, had invaded, with varying degrees of success, portions of the

Appalachian roughlands and even the terrain of the Shield. Almost 3500 km of railroads provided connections to most of the region. Urbanization proceeded apace with approximately one of every four people living in towns or cities by 1881. The agricultural economy was giving way to industrialization.

Within Québec, regional patterns distinguished the St. Lawrence Lowlands from both the Eastern Townships and the fringe of the Shield, and further emphasized the importance of Montréal and, to a lesser extent, the city of Québec. Rural French Canadian society was strongly entrenched in the Lowlands where mixed farming was prominent. Although modest progress had been made in commercial farming in Québec, especially in dairy production, population pressure on agricultural lands increased. For this reason above all others, out-migration from the Lowlands continued to New England, to the crowded tenements of Montréal, to the clay lowlands of the Lac St. Jean area, along the south shore toward Gaspé, and into the Eastern Townships. A few towns of the Lowlands, such as St. Hyacinthe, had expanded in response to both the increasing commercialization of farming and the expansion of railways, and provided another outlet for the surplus rural population.

In the more rolling country of the Eastern Townships, French and British settlers interspersed in almost equal numbers. The French arrived late, pre-empting poorer lands bypassed earlier by the British and purchasing farms the British had since vacated. Sherbrooke, with good rail connections, functioned as the regional capital and, in addition to providing basic services, housed several industrial establishments. To the north of the St. Lawrence, on the fringes of the Shield, with its thin, acid, stony soils, attempts at settlement proved to be difficult.[17]

In sharp contrast to rural Québec stood Montréal, the Canadian metropolis, which in the early 1880s had a population of almost 200 000. Its strong and diversified economy relied on trade and related manufacturing, a fact reflected in the large number of wholesale establishments, commission agents, financial institutions, and entrepôt manufacturers. Water-borne traffic had increased in the second half of the nineteenth century to make Montréal Canada's most important seaport. By the early 1880s manufacturing already boasted over 32 000 workers employed in a variety of enterprises making ships, foundry products, marine engines, railway engines and equipment, shoes, tobacco, sugar, and flour. Led by the energies of Scottish, English, and some American entrepreneurs, manufacturing in Montréal thus accounted for approximately half the provincial total, a ratio that did not change significantly.

Fronting the north side of the St. Lawrence River, Montréal in the late nineteenth century extended west beyond Atwater Avenue and well into St. Henri, east beyond the rue d'lberville into Hochelaga, and north beyond Mount Royal Avenue. French Canadians, increasing in number, comprised nearly 60 per cent of the population and were found mainly to the east and northeast of the business core. In the core, British institutions dominated the economic life of the city. Their control stretched from the docks and wharves along the river, through the nearby wholesaling districts, to the financial district on St. James Street, and on to the new retail

Map 3.2 Distribution of Population in the Industrial Heartland, 1881

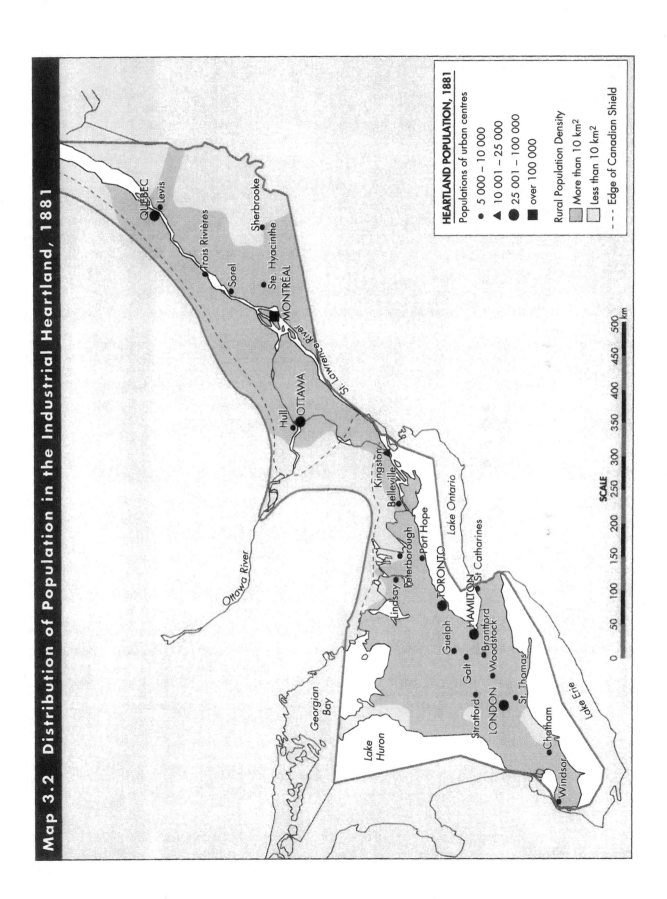

HEARTLAND POPULATION, 1881

Populations of urban centres

- · 5 000 – 10 000
- ▲ 10 001 – 25 000
- ● 25 001 – 100 000
- ■ over 100 000

Rural Population Density

- More than 10 km²
- Less than 10 km²
- - - - Edge of Canadian Shield

QUÉBEC
Lévis
Trois Rivières
Sherbrooke
Sorel
Ste Hyacinthe
MONTRÉAL
St. Lawrence River
Hull
OTTAWA
Ottawa River
Kingston
Belleville
Port Hope
Peterborough
Lindsay
Lake Ontario
St Catharines
TORONTO
HAMILTON
Guelph
Brantford
Galt
Woodstock
Stratford
LONDON
St. Thomas
Chatham
Windsor
Lake Erie
Lake Huron
Georgian Bay

SCALE

0   50   100   150   200   250   300   350   400   450   500
km

Map 3.3 Urban-centred Manufacturing in the Industrial Heartland, 1881

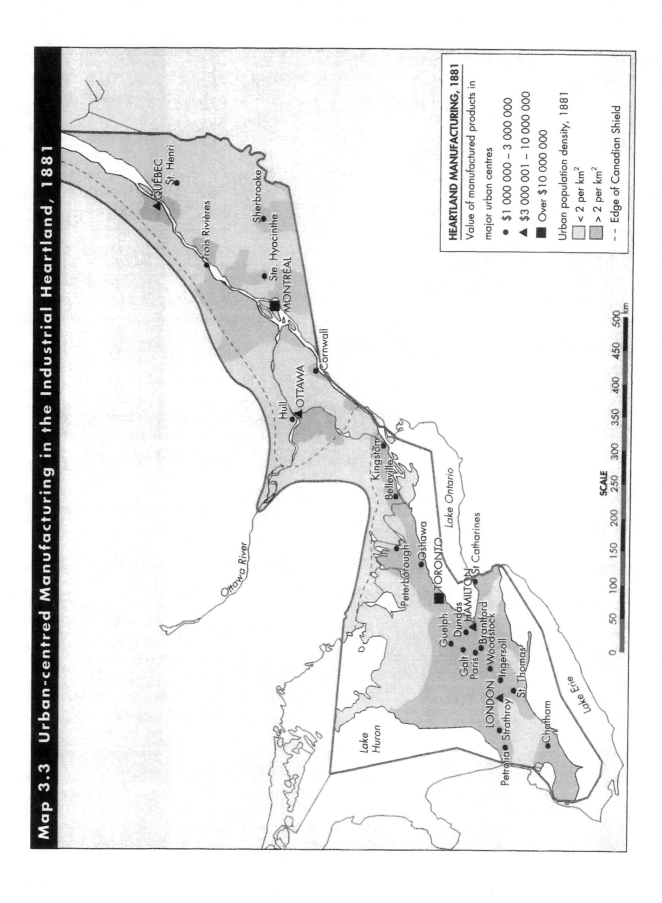

HEARTLAND MANUFACTURING, 1881

Value of manufactured products in major urban centres

- • $1 000 000 – 3 000 000
- ▲ $3 000 001 – 10 000 000
- ■ Over $10 000 000

Urban population density, 1881

< 2 per km²
> 2 per km²

– – Edge of Canadian Shield

SCALE

0   50   100 150 200 250 300 350 400 450 500 km

QUÉBEC
St. Henri
Trois Rivières
Sherbrooke
Ste. Hyacinthe
MONTRÉAL
Cornwall
Hull
OTTAWA
Kingston
Belleville
Peterborough
Oshawa
TORONTO
St Catharines
Guelph
Dundas
HAMILTON
Galt
Paris
Brantford
Woodstock
Ingersoll
LONDON
Strathroy
St. Thomas
Petrolia
Chatham

Ottawa River
Lake Ontario
Lake Huron
Lake Erie

district of Notre Dame Street built in the 1860s. Along Dorchester, St. Catherine, and Sherbrooke streets, middle-and upper-class housing extended to Westmount, where a new street plan had just been drawn and a few houses built. French Canadian economic institutions, fewer in number but by no means insignificant, had taken up positions mainly within and just to the northeast of the downtown area.

The growing industrialization of the nineteenth century had created distinctive landscapes, especially in the suburbs around the Lachine Canal. The development of hydraulic power at the Mill Street and St. Gabriel locks in the late 1840s gave rise by the 1870s to rolling mills, foundries, a sugar refinery, flour mills, and sash and door, tool, and textile factories. In response, working-class houses were built close by, creating distinctive Irish and French Canadian neighbourhoods. Further west, the old village of St. Henri was being integrated into the industrial fabric of the city, with the building of the Montréal Street Railway car barns, a large sewing machine plant, an abattoir in the 1860s and 1870s, and a spate of industrial development in the later decades. Further south, closer to the river, the Grand Trunk Railway had built a complex of shops in the 1850s that provided employment for hundreds of workers, many of whom had been brought directly from England. Point St. Charles, located close by, had become a distinctive English working-class district. A similar pattern of industrialization extended to the northeast of the downtown area, from Molson's Brewery, built in 1790 at the foot of rue Papineau, to include, by the 1870s, sugar, tobacco, and rubber factories, a gas works, a cotton mill, and the shops of the Montréal Street Railway Company and of the Québec, Montréal, Ottawa, and Occidental Railway. Houses were built for French Canadians at St. Jacques and Ste. Marie in the East end. These were the forerunners of the innumerable francophone districts, such as Hochelaga and Maisonneuve, that were constructed in the late nineteenth and early twentieth centuries.[18]

Downriver at Québec, where the lumber trade had previously stagnated, shipbuilding was in decline; and the exodus of the English, which began at mid century, had accelerated. As a result, the city was growing very slowly, if at all. Nevertheless, by emphasizing its role as a provincial capital after Confederation, maintaining important functions in church administration, supporting a few labour-intensive industries (such as shoemaking), and serving a moderately prosperous but rather small agricultural area, Québec managed to withstand serious economic decline. The die had been cast, however, and Québec would thereafter remain less than one-quarter the size of Montréal.

By 1880 in southern Ontario the general aura was one of well-being. Wheat still accounted for much of the improved farm acreage, but its decline had already set in, and the change to more modern specialized farming was well underway.[19] A combination of low wheat prices, increasing incidences of blight, and declining soil productivity forced farmers to seek new options, but fortuitously, important structural changes in agriculture made possible both alternative strategies and the maintenance of attained living standards. New drainage techniques encouraged the

Ogilvie flour mill, Lachine Canal, 1896

use of rich lands otherwise too wet for cultivation, including, in particular, areas of level land in Essex and Kent counties. A relatively long and warm growing season facilitated the successful cultivation of corn and related crops.

Other sectors of the agricultural industry prospered as well. Improved strains of dairy cattle, mechanical inventions such as the cream separator, and expanding markets for butter and especially cheese in Britain all supported a remarkable expansion of dairy production. Although the dairy industry was widespread, concentration was apparent even in the 1880s in Oxford County in the southwest and in Dundas County in the east. Elsewhere, the urban market for fruits and vegetables expanded, and because of general improvements in farming methods, horticulture became increasingly specialized in the Niagara Peninsula and along the shores of the Great Lakes. Here, risks of unseasonable spring and late summer frosts and winter kill were minimal. On the eve of electrification and the introduction of the automobile, thousands of urban horses still pulled wagons and carriages. The large market for hay provided a supplementary yet significant cash crop for farmers.

Throughout rural and small-town Ontario, the late Victorian landscape bore a striking uniformity. Red or buff brick houses, either solid or veneered over wood, predominated to a degree not seen elsewhere in Eastern North America. The typical rectangular farm of 40 hectares had a one-and-a-half storey brick house, a wooden barn with a gable roof, fields of grain, hay, and pasture, and a woodlot. An increasing number of wooden silos showed that the Ontario Agricultural College, founded in Guelph in 1874, was exerting some influence. The only differences from these British patterns were the farms of French Canadians in Eastern Ontario, those of Mennonites in Waterloo County, and those of Americans who had settled along the shore of Lake Erie.

Standing out sharply from this rural countryside were the industrial cities of southwestern Ontario, including Toronto, Hamilton, Brantford, Galt, and Guelph (Map 3.2 and Map 3.3). By the early 1880s, Toronto possessed a population of almost 90 000, making it twice the size of Hamilton, its nearest rival. Building on its advantageous location as a focal point of rail and road routes, Toronto had steadily increased its share of the production and distribution of goods in southern Ontario and had contributed significant capital to its own activities and to those of other parts of Ontario and western Canada. In fact, Toronto exhibited, on a somewhat more modest scale, the same multifunctional characteristics as Montréal. The city housed various types of wholesaling, including dry goods, food products, and medicines; manufacturing, particularly industries producing food and beverages, clothing, boots and shoes, machinery, furniture, and metal products; and financial institutions, which varied from insurance and trust companies to banks and a stock exchange.[20]

These activities brought change to the city. By the early 1880s, Toronto had spread across much of the level terrain of the old Lake Ontario Plain. It ran a northward course along Yonge Street and showed more growth to the northwest than to the northeast. To the regret of many, the railways had pre-empted the lakeshore area, thereby reducing access to what were once attractive recreational grounds. Industries followed the railways and wholesalers strengthened their concentration nearby, especially on Front Street. Predominantly British and Protestant and divided into distinctive classes, the Toronto townscape revealed the stately mansions of the wealthy on Jarvis Street, the tenements of the poor to the east and west of the business core, and everywhere red brick Presbyterian and Methodist churches.

Urban change also prevailed elsewhere. Hamilton was in a state of transition — manufacturing having expanded to offset declines in wholesaling and financial activity. Strong entrepreneurial and municipal leadership partially explain the rise of textile and metal fabricating to augment well-established clothing and hardware industries. Much the same kind of development characterized Brantford, where the manufacture of farm machinery and engines dominated, and also Guelph, which produced sewing machines and musical instruments. The growth of manufacturing in these and other nearby places fostered the growth of population in the Toronto-centred region, creating a legacy of concentration that has continued to influence the geographical structure of southern Ontario.

But other parts of Ontario also experienced significant urban and industrial growth. The discovery of petroleum in the extreme southwest, at Petrolia, provided the basis for modest industrial development and paved the way for significant growth in oil refining at nearby Sarnia a half-century later. In anticipation of the western Canadian grain trade, flour milling was established at the Georgian Bay port of Collingwood. Along the Grand Trunk corridor, on the north shore of Lake

Ontario, railway cars were built at the Cobourg Car Works, and just over 150 km to the east, at Kingston, a short-lived industrial revival produced textiles, locomotives, and biscuits. Ottawa, despite its status as the capital of Canada, was dominated by the lumber industry that employed many and provided the financial basis for a local bank. In some other communities in eastern Ontario, such as Cornwall and Almonte, Montréal interests had built textile mills.

THE DRIVE TO INDUSTRIAL MATURITY

Toward the end of the nineteenth century, the Canadian economy began to grow rapidly. From 1891 to 1921 the gross domestic product almost quadrupled from $858 million to $3.2 billion, and rose to $13.6 billion in 1961 (constant 1913 dollars).[21] Leading this surge was central Canada, where the population increased from just over 3 million in 1881, to 5 million in 1921, and then to almost 10 million in 1961. Employment in all sectors except agriculture experienced a similar upward spiral. In fact, by 1913 Canada ranked third in the world after the United States and Britain in the output of manufactured goods per capita.[22] The extent to which this expansion was due to the acquisition of new markets in western and eastern Canada, and just how much was the result of continued regional growth, is, of course, impossible to measure. But it is clear that many manufacturing firms, trading companies, and financial institutions of the heartland — aided by federal policies of high tariffs and favourable rail rates, and employing considerable entrepreneurial initiative to reduce and eliminate competitors in the quest for markets — soon came to control much of the Canadian economy.

The increasing number of mergers in the banking industry illustrates this geographical and economic concentration.[23] By the early 1880s, there were at least forty four chartered banks in Canada operating a small network of about three hundred branches. The largest maintained head offices in either Montréal or Toronto and accounted for 71 per cent of all assets, but there were also at least twenty eight others scattered in sixteen centres, including Hamilton, Ottawa, Winnipeg, Halifax, Québec, and Victoria. Through merger and acquisition, or simply because of insolvency, the number of banks had shrunk to eleven by the mid 1920s, but the banking network itself totalled over four thousand branches from coast to coast. National control from either Montréal or Toronto had eliminated all the regional banks but one, a small enterprise based in Weyburn, Saskatchewan. Particularly serious for hinterland interests was the disappearance of the Winnipeg banks, which on the eve of World War I managed almost three hundred branches scattered throughout western Canada, as well as the disappearance of those in Halifax as they supported local industry.

This take-over pattern, reflecting the merger movements so characteristic of Canada in the early twentieth century, had its root in the policies of the Bank Act of 1871. Unlike its American counterpart, the Act did not restrict the establishment of branches beyond political (provincial) boundaries. Furthermore, by setting special requirements for the establishment of any new bank, it made entry into the system difficult for both domestic aspirants and foreign banks alike. Whether the large national banks such as the Royal Bank and the Bank of Commerce were as sensitive to the needs of peripheral areas as regional banks might have been is debatable, but what is important is that the image of metropolitan control of banking — in fact, of most aspects of finance — was strengthened in the minds of all Canadians, particularly those in the peripheral regions. Indeed, the intrusion of metropolitan financial institutions into the economy of the Martimes in the late nineteenth and early twentieth centuries was widespread, and has long been a source of regional protest and discontent. Regional banking was eliminated when the Bank of Nova Scotia transferred its major management facilities to Toronto (1900), the Merchants Bank of Halifax was transformed into the Royal Bank based in Montréal (1904), and the Halifax Banking Company was absorbed by the Bank of Commerce (1903), the People's Bank by the Bank of Montréal (1905), and the Union Bank by the Royal Bank (1910).

Although all sectors of the Canadian economy grew rapidly, the most apparent trend was the transformation of the heartland into an integrated, strong, but internally differentiated industrial region. By the 1920s, all segments of a diversified industrial economy were well in place, including primary iron and steel, automobiles, electrical goods, industrial machinery, chemicals, food and beverages, textiles, clothing, and shoes. Not only did employment in manufacturing double from about 200 000 in 1880 to 400 000 by 1920, and then more than double to just over one million by 1960, but its share of the total labour force in the heartland rose to just over 25 per cent. Because manufacturing was largely urban-oriented, it became the catalyst for growth for many towns and cities.[24] Peterborough, an old industrial centre, grew in population from 6800 in 1881 to more than 45 000 in 1951 when 55 per cent of its labour force was employed in manufacturing.

Despite Montréal's early industrial revolution, the towns and cities of southern Ontario matched its pace of development. By 1880, manufacturing employment in Ontario exceeded that in Québec by a ratio of four to three. Since then the gap between the two provinces has continued to widen in Ontario's favour. Significant, too, are differences in the structure of provincial manufacturing.[25] In 1880, Québec led in the production of leather and tobacco products; Ontario's strength rested on foods, beverages, and metalworking. Beyond these difference, considerable structural similarity prevailed, but by the early 1900s Québec supported a slightly higher share of clothing, shoes, and textiles. The rise first of the electrical apparatus industry and later of primary iron and steel and automobiles gave Ontario undisputed leadership in these important categories by the 1920s. Other changes gradually took hold in the postwar period. The major differences and similarities in structure are outlined in Table 3.1. These differences in industrial structure do not negate the principle of a unified economic region; rather, they foster complementary activities and functional integration.

At this point, two interrelated questions need to be raised. First, what factors explain the concentration of manufacturing in the Industrial Heartland? Second, why did manufacturing have a differential impact on the industrial structures of Ontario and Québec? Several factors provide the basis for at least a partial explanation: the character of the resource and energy base; human initiative and entrepreneurship; the role of government policy; foreign direct investment; and corporate mergers and monopolistic control.

TABLE 3.1   Overrepresentation and Underrepresentation in Manufacturing in Québec, 1951

Type[a]	Group	Value Added in Québec as Percentage of the Same Group in Ontario	Employment in Québec as Percentage of the Same Group in Ontario	Percentage of Overall Under-representation in Employment[b]
A	Tobacco	602	414	+ 495
	Clothing	173	171	+ 146
	Textiles	137	156	+ 123
	Leather	110	123	+ 77
	Paper	109	102	+ 46
B	Petroleum and coal	96	43	- 39
	Wood products	79	90	+ 29
C	Foods and beverages	59	61	- 12
	Chemicals	58	81	+ 16
	Non-metallic minerals	53	56	- 26
D	Printing	49	52	- 26
	Miscellaneous	40	46	- 34
	Iron and steel and non-ferrous	37	40	- 42
	Electrical operators	36	39	- 44
	Transportation equipment	22	44	- 37
	Rubber	22	45	- 39
All Manufacturing		58	70	

[a] Type A: absolutely greater than in Ontario, and relatively more important in Québec. Type B: absolutely less than in Ontario, but relatively more important in Québec. Type C: absolutely less than in Ontario, but of approximately same relative importance as in Ontario. Type D: absolutely less than in Ontario, and relatively less important than in Québec.

[b] Employment in Québec was 70 per cent of employment in Ontario. This was used as a base level in expressing over- or underrepresentation in industrial groups. Employment more or less than the employment required to make Québec employment equivalent to 70 per cent of employment in Ontario was expressed as a percentage above (overrepresentation) or below (underrepresentation) the employment required to produce 70 per cent.

Source: R. Keith Semple, "Quarternary Places in Canada," in John N.H. Britton ed. *Canada and the Global Economy* (Montréal: McGill-Queen's University Press, 1996), pp. 356 and 357.

Any discussion of the relationship of resources and energy to industrial development emphasizes the complexity of the issue. In the nineteenth century, energy for the industrial economy was provided initially by wood and water power and later by coal. Because wood was plentiful everywhere, it had little impact on the regional variation of economic growth. Coal was imported in increasing amounts after 1860 from the American Appalachians, giving the lake ports of western Ontario — particularly Hamilton and Toronto — a definite advantage of accessibility.[26] Even when coal shipped from Nova Scotia became available in the last quarter of the nineteenth century to St. Lawrence ports, dealers in Québec and eastern Ontario still had to pay a higher price than their competitors to the west. Industrial progress in southwestern Ontario's metalmaking and metalworking sectors was due at least in part to the accessibility of cheaper American coal, made even cheaper early in the 1900s by subsidies from the Ontario government.

In the pre-industrial economy, many industrial entrepreneurs explored the possibilities of using water power. Streams were harnessed by building small dams at waterfalls or rapids, creating reservoirs that channeled water to turn the wheels that powered grist mills and sawmills. On a larger and more elaborate scale, the hydraulic power scheme associated with the development of the Lachine Canal in Montréal in the 1840s attracted moderately large industries such as shoe factories and machine and engine works.

Toward the end of the century new technological developments harnessed large flows of water to generate electricity, and the huge reserve of central Canada's "white coal" became the catalyst for considerable industrialization.[27] The modern electrical age in Canada was born in the late 1890s, when a hydroelectric station with a small generating capacity was opened at Niagara Falls. At first there was widespread apprehension that any locally produced electricity would be exported to the United States or monopolized by Toronto, but the government of Ontario responded to these fears by establishing an industrial inquiry in 1905. As a direct outcome, the Ontario Hydro Electric Power Commission was established to regulate the production and transmission of electricity by private utilities.[28] Under the aggressive leadership of Adam Beck, Hydro (as it became known) expanded its role not only in regulating the private utilities but in producing its own power. So extensive was its growth that by the 1920s most of the private companies had been expropriated, and Ontario Hydro was in firm control. It was the policy of the Commission to disperse electricity as widely as the technology of transmission allowed. Thus factories at Guelph, London, and other towns and cities within the transmission radius of Niagara Falls were assured a supply. The Commission also equalized rates whenever feasible, regardless of distance from the generating site. This policy enhanced the industrial development of towns and cities throughout southern Ontario. Furthermore, Beck offered domestic customers relatively low rates, thus encouraging the consumption of electricity for all sorts of household appliances and, in turn, stimulating the growth of the electrical appliance industry in southern Ontario.

In Québec, by contrast, private utilities, fully supported by the provincial government gained control of the development and distribution of hydroelectric power.[29] One of the most important was the Shawinigan Water and Power Company, comprised of a consortium of American and English Canadian entrepreneurs who in 1899 decided to build a plant at Shawinigan Falls, located about 40 km north of Trois-Rivières on the St. Maurice River. The company immediately adopted a policy of selling low-cost electricity to heavy industries on a long-term basis. This policy led the emergence, as early as 1900, of both an aluminum plant and a pulp mill at the site of the first generating station. In 1903, the company acquired another large market and an international reputation for long-distance transmission by selling power to the Montréal Light and Power Company, some 145 km away. Successive developments along the St. Maurice River (obtained through leases from the provincial government) and the acquisition of other facilities at Grand Mère and La Tuque gave the company control of *La Mauricie* by 1931. At each stage of development, surplus power was created that led the company to maintain and expand its policy of persuading industries in need of cheap electric power to located in the valley. As a result, electrochemical and electrometallurgical industries appeared at Shawinigan, and pulp and paper and other resource industries were established at Trois-Rivières, Grand Mère, and La Tuque. Similar developments soon took place on the Saguenay River, where a large aluminum mill was built at Arvida in 1926. Other power-consuming industries located at Chicoutimi.

In contrast, the much smaller Southern Canada Power Company of the Eastern Townships inaugurated a policy of industrialization and market expansion to sell power to light industries. The Montréal Light and Power Company formulated no industrial policy whatsoever and launched no campaigns to persuade industry to locate in the Montréal area, even though it held a firm monopoly on the distribution of power. The provision of hydroelectric power in Québec thus had the same effect on the spatial distribution of manufacturing as it did in southern Ontario — dispersion — but for entirely different reasons.

The extent to which industrial growth in central Canada may be explained by the rise of commercial agriculture is debatable.[30] It can be argued that in the nineteenth century Ontario farmers consistently produced a surplus of agricultural products for sale to local food processors and foreign markets, using profits from this trade to purchase a variety of products including new and technically improved farm machinery. Clearly, their demand for farm equipment stimulated industrialization, but it cannot be said, for example, that Hamilton's iron and steel complex owes its expansion only to the material needs of farm implement manufacturers.

In Québec, commercial agriculture was less developed. Although dairying had become moderately widespread by the 1880s, the sale of products was confined mainly to the Québec market and returns were correspondingly low. As a result, the purchasing power of Québec farmers was weak, and incentives for industrialization stemming from agriculture were minimal. Québec agriculture did, however, create an indirect impact on industrial development. The combination of a shortage of agricultural land and rapid rural population growth, especially in the last quarter of the nineteenth century, created a situation of serious population pressure, forcing many people off the land and into the growing industrial centres of Québec and the United States. Because of their large numbers, these migrants provided cheap labour for labour-intensive industries such as tobacco, shoe, clothing, and textile manufacturing. Women and children formed a relatively large segment of the work force, and working conditions in these Anglo-dominated industries were often deplorable. Although some improvements were made by World War II, wages remained low.

Initiative and entrepreneurship also played a vital role in industrialization. Much of the heartland's industrialization can be traced to entrepreneurs of Scottish and American origin who were prepared to experiment with new techniques, work hard, pay low wages, and reinvest their profits in expansion. In Montréal, and throughout Upper Canada, artisans, engineers, inventors, and entrepreneurs of American and British origin teamed up with merchants to build shipyards, foundries, flour mills, and chemical plants.

The role of French Canadian entrepreneurs is more controversial. It has long been held that their participation in the industrial process was minimal, explained by the *mentalité* that placed a higher value on careers in the professions and in the Roman Catholic church than on those in industry. Evidence now shows that the participation of Québécois businessmen in shipbuilding, shoe manufacturing, textiles, transportation, trade, and banking was greater than previously acknowledged.[31] The fact that their contribution was not more impressive must be explained by shortages of investment capital, limited access to English-controlled industrial and financial institutions, and strong language and social barriers rather than by any absence of capitalistic values or lack of entrepreneurial skills. In fact, entrepreneurial behaviour *per se* has little, if any, relevance in explaining the differences in the industrial structures of Québec and Ontario.

Nor should traditional views of the conservative role of the Roman Catholic church remain unchallenged, for the church was not entirely opposed to industrialization. At times, it acted boldly as an entrepreneur. In the 1870s, for example, the Grey Nuns financed the building of a large warehouse in downtown Montréal for lease to commercial enterprises. During the period of feverish railway promotion in the nineteenth century, the church participated by investing directly or by giving tacit approval to projects. Later, as *La Mauricie* became a development focus, the church supported and in fact encouraged industrialization by working with managers and investors.[32]

What role did government policy play in industrialization? It has long been stated that the real basis of industrialization in the heartland was tariff protection from foreign competition. The roots of industrial protection can be found in scattered policies of the mid-nineteenth century, but the most important and all-embracing legislation was enacted in 1879 as a prime element of the National Policy.[33] Although the debate on the precise effects of these tariffs continues, it is apparent that manufacturers in Ontario and Québec used this protective shield to build a nationwide market. Thus, loggers in British Columbia and fishermen in Nova Scotia wore boots made in Montréal; farmers in Saskatchewan and Prince Edward Island purchased agricultural machinery from Toronto; and lawyers in Edmonton and Halifax drove automobiles made in Windsor. Manufacturing in the Industrial Heartland was given protected access to a captive national market. The largest part of this market was in Ontario and Québec, of course, but the markets in peripheral areas were often crucial in maintaining the economic viability of companies.

The National Policy also encouraged the branch plant character of manufacturing in the heartland.[34] To win Canadian markets, American direct investment jumped the tariff barrier to build plants throughout Canada, especially in Ontario. Westinghouse, Gillette, and Singer Sewing Machine, to name some well-known American firms, were among the approximately one hundred companies operating in Canada at the turn of the century. They were soon joined by hundreds more and, by 1960, American corporations accounted for at least half of central Canada's manufacturing activity. American industry favoured Ontario over Québec largely because of Ontario's higher market potential and the proximity of southwestern Ontario to the head offices of parent organizations in the United States manufacturing belt.

Closely related to the expansion of American branch plants was the increasing dependence of Canadian industry on foreign technology.[35] Much of the machinery that provided the infrastructure for Canadian industry was either being imported

Early 1900s multi-storied factories, southern Ontario

from the United States or made in Canada under licence from American enterprises. Throughout the early and mid-twentieth century, the rapidly changing technology of American industries became irresistible to most Canadian manufacturers, who eventually succumbed to its accessibility. Many promising Canadian industries of the early 1900s, such as automobiles, had disappeared by the late 1920s, swallowed up in the rapidly changing and increasingly concentrated American industry. By contrast, the steel industry of Hamilton, brewing and distilling, and furniture manufacturing were among those that remained under Canadian ownership.

As the pace of manufacturing growth accelerated at the turn of the century, so did the move toward large corporations. Large size often meant cheaper production costs, and savings in production could be applied to the costs of transporting goods to distant markets. In the increasingly competitive national market, small companies located across Canada were frequently absorbed by heartland interests. The textile industry of the Maritimes is but one example of an industry that all but disappeared because of industrial concentration.[36] Geographically, the main effects of the merger movement were to centralize operations in southern Ontario and Québec, especially in the vicinity of Toronto and Montréal, and to close down peripherally situated plants. Thus, central Canadian corporations came to control many sectors of Canada's manufacturing structure, particularly those industries producing consumer goods.

THE INDUSTRIAL HEARTLAND IN THE LATE 1950s

How did sustained economic development during the course of the *longue durée* affect the spatial organization of the heartland? The main effect was urban concentration (Map 3.4). Between 1881 and 1961, while the region's population increased from just over three million to almost ten million, the proportion living in towns and cities rose from 25 to 70 per cent. Montréal (2.1 million) and Toronto (1.8 million) dominated all other urban places. Although most cities in the Windsor–Quebec City corridor grew at rates exceeding the national average, the populations of some hamlets, villages, and even towns, especially those located off the beaten track, actually decreased.

By the 1950s, Montréal and Toronto together exerted a profound influence on Canadian society. From 1871 to 1961, Montréal's share of Québec's population rose from 9 to almost 40 per cent; Toronto's share of Ontario's increased from 3.5 to just over 25 per cent. Together, they housed over 20 per cent of Canadians in 1961, and their 490 000 manufacturing workers comprised more than one-third of the Canadian total. This total exceeded that of the four Atlantic provinces (90 000), four western provinces (230 000), and northern fringes (55 000) combined. Of equal importance was the striking concentration of corporate and financial power, as the tertiary sector added its force to urban growth. Mergers had reduced the number of Canadian banks to nine, all of which had head offices in either Toronto or Montréal. Based on corporate assets, 75 per cent of all insurance companies and 80 per cent of all trust and loan companies had head offices in these two cities. Fully 95 per cent of all stock market transactions and practically all dealings in bonds were made on the Toronto and Montréal markets. Of all the cheques written in Canada in the postwar decade, approximately 37 per cent passed through Toronto clearing houses. Montréal cleared another 25 per cent. Finally, at least three out of every four Canadian corporations were based in Toronto and Montréal.[37] The internal organization of both cities reflected the ascendancy of metropolitan functions — corporate and financial districts emerged in their downtown cores, and the demand for industrial and residential land created significant suburban expansion.

Other regional changes occurred to the southwest of Toronto in peninsular Ontario, where industrialization continued to have a strong urban impact. The output of automobiles and chemicals in Windsor and Sarnia, of farm machinery in Brantford, of food and beverage products in London, and of electrical appliances and steel in Kitchener and Hamilton continued to increase, though over half of this output was produced by American branch plants. Agriculture responded to technological and industrial change, as people left farms for cities and new specializations developed. Based on a warm and long growing season (by Canadian standards), cash crops such as corn, soybeans, and sugar beets prospered in the extreme southwest. Tobacco became concentrated in Norfolk County, and grapes, peaches, and other fruits were grown in the Niagara Peninsula below the escarpment. Dairying remained widespread, easily meeting the increased urban demand for fluid milk and other dairy products. Southwestern Ontario's primary, secondary, and tertiary sectors had built a strong and diverse economic base, whose influence extended across Canada, and whose geographic pattern has remained stable in the postwar decades.

Those areas embracing the Huron Uplands, fringing the Shield, and running eastward through central and eastern Ontario experienced limited or no urban and industrial growth. They created a hinterland within a heartland. Three of the more promising centres of the late nineteenth century, Collingwood, Brockville, and Cobourg stagnated. Peterborough, using hydroelectric power to build an industrial base, and Ottawa, the national capital, were exceptions. Agricultural in the area was based on cattle production in the Huron Uplands and mixed farming and dairying elsewhere. In rural Ontario as a whole, farm depopulation was endemic by the late 1950s, but society remained strongly conservative and for the most part firmly British and Protestant.

The St. Lawrence Lowlands between Montréal and Québec remained much as it had since the Conquest — the rural heartland of French Canadian society. Although spotted by urban and industrial activity, this area retained a high density of rural population ratio; dramatic depopulation did not come until the 1970s. Farmers of the Lowlands practiced mixed farming with an emphasis on dairying for the urban market. There were a few small to medium-sized cities, such as St. Hyacinthe in which

Map 3.4 Distribution of Population in the Industrial Heartland, 1951

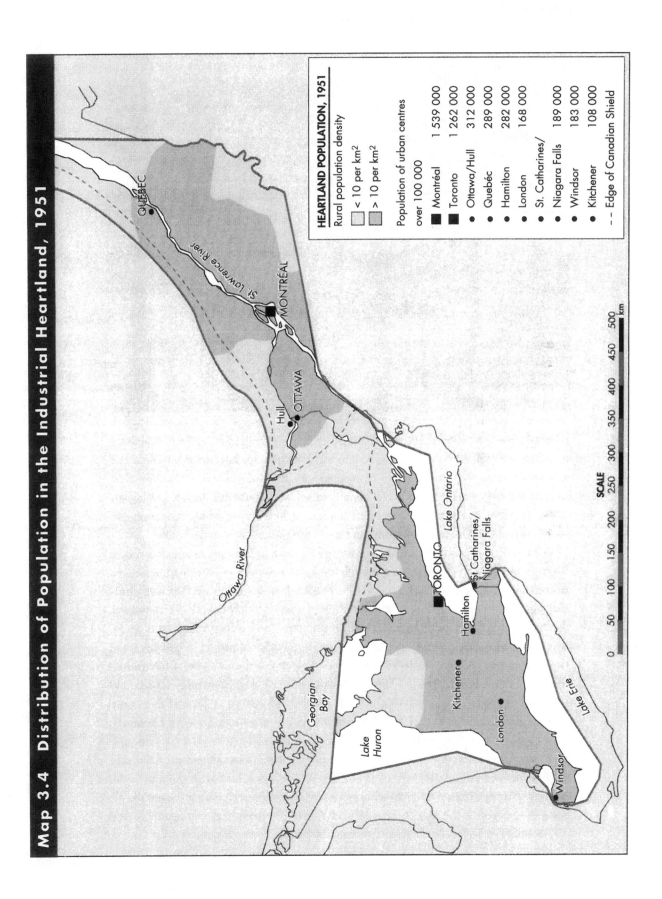

HEARTLAND POPULATION, 1951

Rural population density

  □ < 10 per km²

  ▨ > 10 per km²

Population of urban centres
over 100 000

■ Montréal	1 539 000
■ Toronto	1 262 000
● Ottawa/Hull	312 000
● Québec	289 000
● Hamilton	282 000
● London	168 000
● St. Catharines/	
Niagara Falls	189 000
● Windsor	183 000
● Kitchener	108 000

-- Edge of Canadian Shield

QUÉBEC

St. Lawrence River

MONTRÉAL

Hull

OTTAWA

Ottawa River

Lake Ontario

TORONTO

St Catharines/
Niagara Falls

Hamilton

Georgian
Bay

Lake
Huron

Kitchener

London

Lake Erie

Windsor

SCALE

0   50   100   150   200   250   300   350   400   450   500
km

General Motors
Co. of Canada
plant in Oshawa

labour-intensive textile, shoe, and furniture factories were found. Québec, housing a population of 375 000 in 1961, was sustained by its historic roles in government and religious affairs, but functioned increasingly as a focal point for the lower St. Lawrence Valley and as a centre for the manufacture of pulp and paper. During the 1960s and 1970s, the city of Québec became the focus of powerful economic and social changes in Québécois society, and more than doubled its population.

Flanking the Lowlands on the south, the Eastern Townships retained a mixed English and French population, although the proportion of the latter had increased in some districts to over 80 per cent. Sherbrooke continued to thrive as a regional and industrial centre, reaching a population of some 50 000 people. To the north, on the edge of the Shield, engineers, entrepreneurs, and industrialists succeeded where farmers failed, and the harnessing of the St. Maurice, Saguenay, and Gatineau rivers by large-scale hydroelectric projects continued to stimulate industrialization and urban growth. The combined population of Shawinigan Falls and Grand Mère, which stood at 6300 in 1901, reached 19 500 in 1921, and then more than doubled to 50 000 in 1960. Over 8000 people were employed in chemical, textile, pulp and paper, and several other minor industries. In the Lac St. Jean district, the Chicoutimi-Jonquière-Arvida complex, which took shape after the mid 1920s, counted almost 100 000 in 1960. Many were associated with the aluminum industry that dominated the industrial structure. In retrospect, energy-based developments proved to be the forerunners of Québec's industrial expansion of the 1970s and 1980s, led by the Manicouagan and giant James Bay projects.

METROPOLITAN DOMINANCE

It has been emphasized that through the *longue durée* Montréal and Toronto grew to metropolitan status to dominate not only the central region but also all of Canada beyond. Not only did they become centres of wealth and power diversified in function, but provided a wide range of services in specialized retailing, medicine, and education, as well as offering the cultural attractions of the theatre, art galleries, and museums. In particular, the head offices of corporations and financial institutions congregated in the metropolis where policies were formulated and decisions made to be diffused to all corners of the hinterland. In reverse, funds, materials, and people moved from the hinterland of Montréal or Toronto, either directly or through regional cities. Here indeed was dominance and dependency — the most extreme form of the heartland–hinterland structure.

As the tentacles of corporate capitalism spread out through the country, much growth and prosperity followed. On the other hand, negative decisions resulting in the removal of local companies through acquisition or merger, widespread foreclosures on mortgages, or denial of funds for investment in enterprises brought forth bitter resentment, and reactions from politicians and the press were legion. Among the many that appeared, the one in the June 1913 issue of the *Grain Growers Guide of Winnipeg*[38] was particularly colourful and vitriolic. Lashing out at the banks, railways, and manufacturers of agricultural implements for causing high interest rates, high rail rates, and the high cost of farm machinery, the guide identified fifty-two plutocrats who in its opinion controlled the Canadian economy. Not surprisingly, all were of British origin and nearly all lived in either Toronto or Montréal.[39]

Although Montréal and Toronto shared power, their relative positions changed through time. Montréal was by far the larger and more important until Confederation, after which the gap narrowed. Although Toronto's population did not exceed Montréal's until the late 1970s, its status as a corporate financial centre grew to almost equal that of Montréal as early as the 1880s. After Montréal reached a final high point in the late 1920s, Toronto in the 1930s began a very modest ascendancy that became larger and more significant thereafter. What was the nature of this changing status and how can it be explained?

As Montréal's wholesale trade expanded early in the nineteenth century, so, too, did the banks, insurance companies, and other financial institutions that served trans-Atlantic and interregional trade. Although Montréal merchants were in control of most of the economy, Toronto traders formed local financial institutions to facilitate the rapid growth of the regional economy, especially in the trade in wheat. By the late 1840s, Toronto interests had also achieved some success in breaking Montréal's monopoly by exporting goods over the Erie Canal system through New York. On the eve of Confederation, Toronto had developed a trading and financial infrastructure remarkably similar to that of Montréal. Both cities were headquarters for chartered banks and insurance and trust companies, both housed small but active stock and bond markets, both supported numerous wholesale establishments, and both had spawned an increasingly strong and diversified industrial structure. Although miniscule compared to world cities such as London and New York, they operated within separate but interrelated decision-making systems, and soon rivalled each other for the control of investment and trading in the new Dominion.

For two decades after Confederation, growth remained sluggish, out-migration of people mainly to the United States exceeding in-migration by far. In the mid 1890s the economy began to grow very rapidly, and from 1896 to 1929, during what has been called "The Great Transformation," the Canadian population more than doubled and gross domestic product quadrupled. Assets of financial institutions, most head offices of which were located in either Toronto or Montréal, grew from $500 million in 1896 to almost $7 billion by 1929. At the same time innumerable enterprises in manufacturing, wholesaling, trade, and electrical energy grew into complex corporations through expansion or merger to establish metropolitan head offices. Thus, old corporations in Montréal such as the Canadian Pacific Railway were joined by Shawinigan Water and Power and the Aluminum Company of Canada amongst many, while in Toronto the venerable Massey Harris Corporation began to share listing on the stock exchange with Imperial Oil, Loblaws, and many others.

During and after the completion of the railway to Vancouver, both cities participated vigorously in the opening of the Prairie West through the wholesale trade and by establishing branch banks.[40] After 1920, challenges from Winnipeg diminished and Toronto exerted a somewhat greater influence. Traders and entrepreneurs from Montréal and Toronto laid claim to much regional business in the Maritimes as they reached out to conquer the entire Canadian hinterland.[41] As Canada's leading seaport and through its control of the national rail network, Montréal strengthened its transportation function. In contrast, in the last 1890s Toronto financiers responded more enthusiastically than their Montréal counterparts to mineral discoveries on the Shield. Through banks and investment houses prepared to invest in exploration for and development of mineral deposits and, with speculation on the Standard Stock and Mining exchange (1899), Toronto became increasingly an important mining centre. By the late 1920s, of the 50 major mining companies in Canada, 25 or half had head offices in Toronto, and only 4 in Montréal.

With the crash of 1929, the rapid expansion in the Canadian economy was interrupted for a decade. During the Depression of the 1930s, plant closures, bankruptcies, and unemployment affected both Montréal and Toronto. Although the functioning of the corporate-financial sector was severely dislocated, much of its infrastructure remained intact. In particular, trade on the stock markets continued at a very much reduced scale. For the first time, in the early 1930s, the value and volume of transactions on the Toronto market (mainly mining shares) exceeded those on the Montréal stock exchange. A rise in the price of gold in the midst of a very depressed market aroused interest, thus simulating the development of new mines on the Shield. The Toronto financial community benefitted accordingly. A merger of the old Standard Stock and Mining exchange with the Toronto Stock Exchange in 1934, and the introduction of more stringent regulations on listings and speculation gave much greater confidence to foreign and domestic investors.[42] With the general recovery of the economy, more corporate and governmental financing joined that of mining to increase the strength of the Toronto market. More specifically, by the late 1930s the Toronto Stock Exchange accounted for 57 per cent of the trade in Canada, Montréal 41 per cent, with the remaining 2 per cent divided between Winnipeg, Vancouver, and Calgary. All of this led the *Financial Post* to claim, in 1939, that "evidence is not lacking to show that Toronto has passed Montréal as a centre of finance" and that "several head offices have left Montréal to locate in Toronto."[43]

The pace of trade in the Toronto market accelerated in the late 1940s and 1950s, with the widescale development of resource-based industries. Much of the development of the Alberta oil industry and of new uranium and copper mines was heavily financed from Bay Street. Montréal's share of transactions on the stock market continued to shrink to one-quarter in 1961 while Toronto's increased to two-thirds.

Through the post-World War II period other segments of the corporate financial structure favoured Toronto over Montréal. In particular, the insurance industry became increasingly concentrated in Toronto through the migration of companies from Montréal, the establishment of new firms, and the choice of a Toronto location by branches of American firms. Whereas Montréal accounted for a larger number of companies (mainly British) through the 1920s, Toronto had surpassed its rival by 1950 and a decade later the gap had widened.[44]

Although the evidence points to an increasing differential growth between the two metropolitan centres, it must be emphasized that Montréal in 1960 remained a dynamic and robust centre. Its share of total corporate financial power as measured by assets was 40 per cent (Toronto 41 per cent), and of the 70 leading companies in Canada 23 had head offices in Montréal and 26 in Toronto. But in insurance and advertising Toronto had forged ahead, and in mining it was certainly dominant with over two-thirds of all head offices and the residence of over half of all directors of major mining companies (Montréal 11 per cent).

The shifting status of Montréal and Toronto has a deep historical basis, due to multicasual processes. Of great importance has been the growth of a strong and diverse economic region in southwestern Ontario, comprising productive agricultural land and a number of highly industrialized cities such as Hamilton, Kitchener, London, and Windsor.[45] These cities accounted for a somewhat larger percentage of high-value manufacturing that their counterparts in Québec. As transportation facilities and other communications improved, these cities formed the basis of a functionally integrated region. As its leading industrial, financial, and trade centre, Toronto became the focal point for the region, providing an increasing number of specialized services and growing accordingly.

Although Montréal had developed connections with other Québec cities, they were generally weak and insignificant, there being nothing comparable to the economic strength of southwestern Ontario in its Québec hinterland.[46] Furthermore, its anglo-dominated financial community turned its back on investments in French Canadian enterprises in the Québec hinterland, favouring support for large anglo or American industries such as electric power at Shawinigan or pulp and paper at Trois-Rivières.

An increasingly powerful force was American enterprise, which sought out banks and brokerage houses in Toronto, more so than in Montréal, for investment in Canada. More specifically, American manufacturing branch plants favoured a location in southwestern Ontario and especially Toronto by far. Of all the American-controlled manufacturing firms that could be identified in Canada in 1960, over 60 per cent were located in the Toronto region, only 10 per cent in Montréal.[47]

Regardless of whether or not the economic dislocation of the Depression of the early 1930s triggered the differential growth of Toronto over Montréal, the fact is that given the forces at work it was inevitable that Toronto would acquire a large share of the nation's corporate financial business by mid century. In the decades that have followed, other forces (especially those within the framework of Québec nationalism) have added another dimension to a process that had already been set in motion.

PERSPECTIVE ON THE *LONGUE DURÉE*

Over the course of the *longue durée* — as the heartland developed from a frontier region to become an industrialized, urbanized, and integrated economic centre — it functioned as the centre of Canadian development. A moderately rich resource base fostered initial growth, and the production and export of wheat and timber were critical in generating revenue for reinvestment and economic diversification. Indeed, the early staple economy and the linkages it spawned explain much nineteenth-century growth.

There is, however, another advantage associated with the resource base. After the partitioning of North American territories in the late eighteenth century, Britain retained a relatively small and narrow lowland along the southern flank of the Shield. In Lower Canada, the arable lands in the St. Lawrence Lowlands proved to be critical to the survival of the rural population; in Upper Canada, almost four times as much arable land, combined with a more favourable climate, supported a much stronger agricultural economy. Good land is important to the market economies of all societies, and it was crucial to the development of central Canada. At a later stage the development of hydroelectric power resources played a role in shaping the region's economic character.

In spite of linguistic and cultural differences, the economies of Québec and Ontario became more intertwined during the *longue durée*. From the late eighteenth century, important trading lines grew up along the St. Lawrence–Great Lakes waterway. Staple products were exchanged for British manufactured goods at Montréal, thus establishing the basis of an ever-growing exchange system. The early infrastructure of canals, waterways, telegraph lines, and rudimentary roads eventually gave way to modern systems of transportation (rail, road, air) and communication (telephone, telex, fax, computer). These, in turn, carried an increasingly large and more diverse volume of goods, messages, and information, interregional trade between southern Québec and southern Ontario being by far the largest in Canada, airline travel between Montréal and Toronto exceeding that of any other pair of cities, and long-distance business telephone calls between the two cities ranking first by far.

Using aggressive strategies of acquisition and merger, facilitated by government policies, central Canadian institutions reduced or eliminated considerable competition in the periphery. By the close of the *longue durée* in the 1950s, the Industrial Heartland, and in particular Montréal and Toronto, remained the unchallenged base of metropolitan power in Canada.

NOTES

Since 1986, when the second edition of *Heartland and Hinterland* was written, several important studies have been published, a small selection of which follows:

In the field of economic and social history:

Ian Drummond, *Progress Without Planning — Economic History of Ontario, 1939–1967* (Toronto: University of Toronto Press, 1987).

Paul-André Linteau, René Durocher, and Jean-Claude Robert, *Histoire du Québec contemporain: le Québec dupuis 1930* (Montréal: Boréal Express, 1986).

Douglas McCalla, *Planting The Province: Economic History of Upper Canada, 1784–1870* (Toronto: University of Toronto Press, 1993).

Brian Young and John Dickinson, *A Short History of Québec: A Socio-Economic Perspective* (Toronto: Copp Clark Pitman, 1988).

In urban history and geography:

James Lemon, *Toronto Since 1918* (Toronto: Lorimer Press, 1985).

Paul-André Linteau, *Histoire de Montréal Depuis La Confédération* (Montréal: Boréal Express, 1992).

Other studies:

Thomas McIlwraith, *Looking For Old Ontario* (Toronto: University of Toronto Press, 1997).

John Warkentin, *Canada — A Regional Geography* (Toronto: Prentice Hall, 1996), chapters 8 and 9.

[1] Staple theory was developed by the distinguished Canadian scholar, Harold Innis, in his studies of the fur trade and cod fisheries in Canada. For a succinct summary, see Richard Pomfret, *The Economic Development of Canada* (Toronto: Methuen, 1981), pp. 33-38. See also Roy I. Wolfe, "Economic Development" in John Warkentin, *Canada — A Geographical Interpretation* (Toronto: Methuen, 1968), chapter 8, for an application of staple theory in Canada.

[2] John McCallum, *Unequal Beginnings: Agriculture and Economic Development in Québec and Ontario until 1870* (Toronto: University of Toronto Press, 1980).

[3] On the nature and meaning of the *longe durée* in historical analysis, see Fernand Braudel, *On History*, trans. Sarah Matthews (Chicago: University of Chicago Press, 1980), pp. 27-34.

[4] Society and economy in pre-Conquest Québec are well described in R. Cole Harris and John Warkentin, *Canada Before Confederation* (Toronto: Oxford University Press, 1972), chapter 2. See also plates 45–56 in *H.A.C.*, Volume I, *From the Beginning to 1800*, R. Cole Harris (ed.), 1987.

[5] For a superb description and discussion of agriculture in Québec in the nineteenth century, see plates 13 and 40 in *H.A.C.*, Volume II, *The Land Transformed*, R. Louis Gentilcore (ed.), 1993.

[6] The timber trade is discussed in William Marr and Donald Paterson, *Canada: An Economic History* (Toronto: Macmillion, 1980), pp. 61-73. See also plate 11 in *H.A.C.*, Volume II.

[7] Louis Gentilcore, "Settlement," in Ontario, ed. Louis Trotier, *Studies in Canadian Geography* (Toronto: University of Toronto Press, 1972), pp. 23–44.

[8] Douglas McCalla, *Planting the Province: Economic History of Upper Canada, 1784–1870* (Toronto: University of Toronto Press, 1993), chapter 5. See also plates 14 and 41 in *H.A.C.*, Volume II.

[9] On the theme of out-migration, see "Le Québec et l'Amérique Française: Le Canada, La Nouvelle-Angleterre et le Midwest," Numéro Spécial, *Cahiers de Géographie du Québec*, 23 (1979) and plate 31 in *H.A.C.*, Volume II.

[10] The best summary of this relationship is McCallum, *Unequal Beginnings*.

[11] See plate 10 in *H.A.C.*, Volume II.

[12] This pattern of development is discussed fully in James G. Gilmour, *Spatial Evolution of Manufacturing of South Ontario, 1851–1891*, Department of Geography Publication Series, No. 10 (Toronto: University of Toronto Press, 1972).

[13] Gerald Tulchinsky, *The River Barons* (Toronto: University of Toronto Press, 1977).

[14] See plate 20 in *H.A.C.*, Volume II.

[15] Brian Young, *Promoters and Politicians: North Shore Railways in the History of Québec, 1854–1885* (Toronto: University of Toronto Press, 1978).

[16] On the economic growth of Toronto, Hamilton, and Kingston, see Jacob Spelt, *Urban Development in South Central Ontario*, Carleton Library Series, No. 57 (Toronto: McClelland and Stewart, 1972); Douglas McCalla, "The Decline of Hamilton as a Wholesale Centre," Ontario History, 65 (1973), 247–54; and Brian S. Osborne, *Kingston — Building on the Past* (Westport, Ontario: Butternut Press, 1988).

[17] R. Cole Harris, "Of Poverty and Helplessness in Petite-Nation," *Canadian Historical Review*, 52 (1971), 23–50.

[18] The discussion of the Montréal landscape in the 1880s draws upon personal communication with David Hanna and the work of David B. Hanna and Frank W. Remiggi, *Les Quartiers de Montréal: Un Guide d'Excursion* (Montréal: L'Association Canadienne des Géographes, 1980) and the superb studies of Montréal by Paul-André Linteau on plate 14 in *H.A.C.*, Volume III, and

by Sherry Olson and David Hanna on plate 30 of *H.A.C.*, Volume III, *Addressing the Twentieth Century*, Donald Kerr and Deryck W. Holdsworth (eds.), 1990.

[19] R. M. McInnis, "Perspectives on Ontario Agriculture," *Canadian Papers on Rural History*, 1992, and plate 11 in *H.A.C.*, Volume III.

[20] Gregory S. Kealey, *Toronto Workers Respond to Industrial Capitalism, 1867–1892* (Toronto: University of Toronto Press, 1980), chapter 2; and Jacob Spelt, *Toronto* (Don Mills: Collier-Macmillan, 1973), chapter 2 and 5. See also plate 50 in *H.A.C.*, Volume II.

[21] See plate 3 in *H.A.C.*, Volume III.

[22] W. Arthur Lewis, *Growth and Fluctuations, 1870–1913* (London: George Allen and Unwin, 1978), p. 163.

[23] The trends in banking in Canada at this time are discussed on plate 9 in *H.A.C.*, Volume III.

[24] See plate 13 in *H.A.C.*, Volume III.

[25] James Gilmour and Kenneth Murricane, "Structural Divergence in Canada's Manufacturing Belt," *The Canadian Geographer*, 17 (1973), 1–18.

[26] David Walker, "Energy and Industrial Location in Southern Ontario, 1871–1921," in *Industrial Development in Southern Ontario*, eds. David Walker and James H. Bater, Department of Geography Publication Series, No. 3 (Waterloo: University of Waterloo, 1974), pp. 41–68, and plate 46 in *H.A.C.*, Volume II.

[27] See plate 12 in *H.A.C.*, Volume III.

[28] Viv Nelles, *Politics of Development* (Toronto: Macmillan, 1974).

[29] John Dales, *Hydro-Electricity and Industrialization in Québec, 1989–1940* (Cambridge, Mass.: Harvard University Press, 1957).

[30] McCallum, *Unequal Beginnings*, pp. 83–114.

[31] Paul-André Lineau, René Durocher and Jean-Claude Robert, *Histoire du Québec contemporain* (Montréal: Boreal Express, 1979), and Paul-André Linteau, *Maisonneuve, Comment des Promoteurs Fabriquent Une Ville* (Montréal: Boréal Express, 1981); Ronald Rudin, *Banking en français: The French Banks of Québec, 1855–1925* (Toronto: University of Toronto Press, 1985); and Paul-André Linteau, *Histoire de Montréal depuis la Confédération* (Montréal: Boréal Express, 1992).

[32] William Ryan, *The Clergy and Economic Growth in Québec, 1896–1914* (Québec: Laval University Press, 1966); and Brian Young, *In Its Corporate Capacity: The Seminary of Montréal as a Business Institution, 1816–1876* (Montréal: McGill-Queen's University Press, 1986).

[33] The National Policy is the focus of a special issue of *The Journal of Canadian Studies*, 14 (1979).

[34] Glen Williams, "The National Policy Tariffs: Industrial Underdevelopment through Import Substitution," *Canadian Journal of Political Science*, 12 (1979), 333–68.

[35] Glenn Williams, *Not for Export* (Toronto: McLelland and Stewart, 1983).

[36] See plate 9 in *H.A.C.*, Volume III.

[37] Donald Kerr, "Metropolitan Dominance in Canada," in *Canada: A Geographical Interpretation*, ed. John Warkentin (Toronto: Methuen, 1968), pp. 531–55.

[38] "Who Owns Canada?" *Grain Grower's Guide Winnipeg* (25 June 1913), pp. 11–34.

[39] For a detailed discussion of Toronto's plutocrats, see plate 15 in *H.A.C.*, Volume III.

[40] Donald Kerr, "Wholesale Trade on the Canadian Plains in the Late Nineteenth Century: Winnipeg and its Competition," in *The Settlement of the West*, ed. Howard Palmer (Calgary: University of Calgary Press, 1977), pp. 130–52.

[41] L.D. McCann, "Metropolitanism and Branch Businesses in the Maritimes, 1881–1931," *Acadiensis*, 12 (1983), 111–125.

[42] Ian Drummond, *Progress Without Planning — Economic History of Ontario, 1867–1939* (Toronto: University of Toronto Press, 1987), page 337.

[43] *Financial Post* (Toronto), January 7, 1939, p. 1.

[44] See plate 55 in *H.A.C.*, Volume III.

[45] See plate 51 in *H.A.C.*, Volume III.

[46] See the seminal paper by Albert Faucher and Maurice LaMontagne, "History of Industrial Development," in *French Canadian Society*, eds. Marcel Rioux and Yves Martin, Carleton Library Series, No. 18 (Toronto: McClelland and Stewart, 1964), pp. 257–71.

[47] See plate 51 in *H.A.C.*, Volume III.

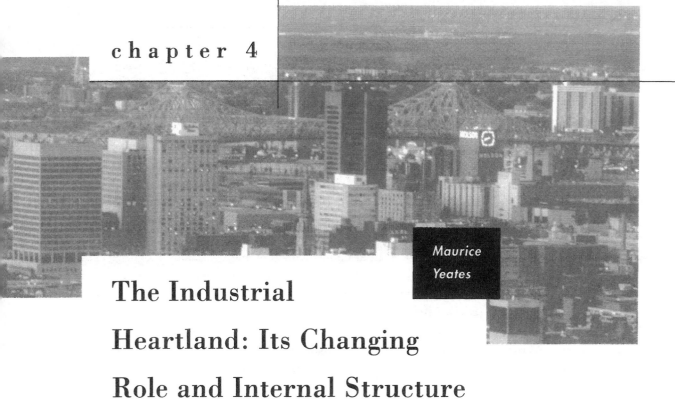

c h a p t e r   4

Maurice
Yeates

The Industrial Heartland: Its Changing Role and Internal Structure

By the 1950s, the Industrial Heartland functioned as the undisputed core of Canadian economic development. In all spheres of economic activity — trade, transportation, manufacturing, and finance — the heartland predominated. This position was consolidated in the immediate postwar years, and culminated in many ways with the signing of the Automotive Trades Agreement (Auto Pact) between Canada and the United States in 1965 for the auto/transport industry had come to underlie much of the economic strength of the heartland.[1] Since 1970, fluctuations in the economic strength of the heartland have made the heartland–hinterland picture much more complex.

In the early 1980s and early 1990s the country experienced major cyclical downturns in the economy, slowing the growth of the heartland much more than the western hinterland, which was becoming influenced by the burgeoning economies of nations around the Pacific Rim. However, by the mid 1990s the economic fortunes of parts of the heartland had strengthened consequent to its overwhelming enmeshment, reinforced by the Canada–United States Free Trade Agreement (1989), with restructuring economic activities in the Midwest and northeastern United States.[2]

Demographic changes have also had an enormous impact on heartland–hinterland relationships. The rapid switch from "baby boom" in the 1950s and 1960s to "baby bust" in the 1970s and 1980s, followed by a much diminished "echo" in the 1990s, has had a profound effect on the rate of urban growth, and on the general orientation of society.[3] There has also been a high rate of immigration, on the average accounting for one-fifth of annual population growth since 1959. The vast majority of new immigrants have settled in the major metropolitan areas of Canada. Before 1970 most immigrants were of European origin; after 1970 most immigrants are from south and southeast Asia. This post-1970 immigration has fueled much of the population growth of Vancouver, Toronto, and Montréal, turning these metropolitan areas into highly multicultural societies.

Like its counterpart in the northeastern United States, the traditional core of Canadian power and innovation is reflecting post-industrial change, the transition to an information based economy and society, and increasing globalization. People in the heartland are being forced to contend with a number of issues — polarization of job opportunities, governmental deficit reduction and constrained funding for educational and social services, and a need to provide equal access to employment opportunities for extremely diverse multi-ethnic/cultural populations. In short, the Industrial Heartland is in a state of transition to a new economy and a different type of society from that existing in the 1970s. Our purpose in this chapter is to assess the changing social and economic character of this urban-industrial region — the "Main Street" of Canada — in the post-World War II period.

Before we turn to major indicators of change, however, it is important to establish a clear definition of the Industrial Heartland. The region's physical, economic, and social characteristics are integral to understanding the forces of change and the heartland's emerging role in the country as a whole.

THE HEARTLAND:
Physical Definition

The geographical extent of the Industrial Heartland is indicated in Map 4.1.[4] The heartland extends some 1000 km from Windsor to Québec City and averages about 300 km in width. Although it is but a small part of the country as a whole, and also a small part of the two provinces that comprise the region, it is large when compared with other political and geographical entities in the world. It is larger than England and Wales combined, for example, and is about two-thirds the size of France. If the Canadian heartland were transported to Europe as an independent political entity, it would be one of the largest countries in that part of the world.

The area has been defined on the basis of population densities, major spheres of urban influence, and physical features.[5] In general, the heartland is highly urbanized, so that population densities are much higher than elsewhere in the country. More than half of Canada's metropolitan areas and urban agglomerations are concentrated in this area, and together they comprise an urban system that integrates and defines the region. The heartland is not, however, a uniform physiographic region. It encompasses two sections of the St. Lawrence Lowlands (termed the West and the Central St. Lawrence Lowlands by the National Atlas of Canada) in which

Map 4.1 The Geographical Extent of the Industrial Heartland

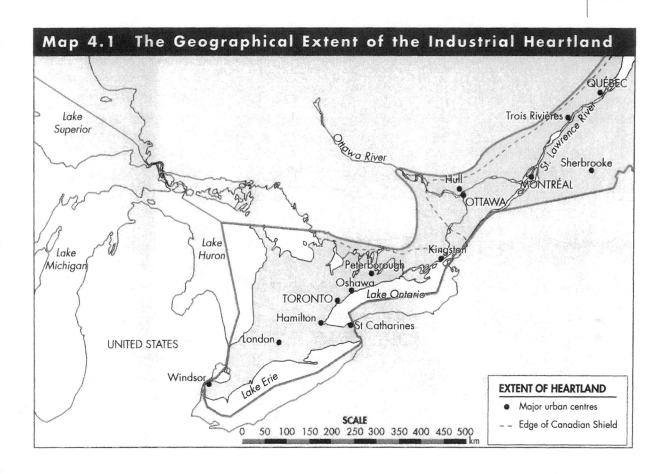

EXTENT OF HEARTLAND
- Major urban centres
- – – Edge of Canadian Shield

SCALE
0 50 100 150 200 250 300 350 400 450 500 km

most of the population and cities are located; and part of two upland and generally barren physiographic units (the Canadian Shield and the Appalachians).

The West St. Lawrence Lowland has been affected by several physical processes, most recently (some 15 000 years ago) by retreating glacial action (Map 4.2). In some areas, deposits have produced fertile soils (Map 4.3), but in others the surface debris, including drumlin fields, makes for poor farmland. A major feature of this Lowland region is the Niagara Escarpment, which extends from the Bruce Peninsula to the Niagara Peninsula and then into New York state. It is noted for recreation (hiking and downhill skiing) and spectacular scenery.

The small portion of the Canadian Shield within the heartland has the appearance of a dissected plateau, varying in height from 945 m at Mont Tremblant to only a few hundred metres in eastern Ontario. Although this area has little good topsoil and is basically inimical to settlement, it is the location of much recreational development, including ski resorts (in the Laurentians and Gatineau Hills) and second homes (such as in Muskoka) for the urban population. The prong of the Shield that leads into the Thousand Islands separates the West and Central Lowlands.

The area of the St. Lawrence Lowlands extending east of the Thousand Islands to near the city of Québec consists of broad areas of clay interspersed with pockets

Map 4.2   Physiographic Regions of the Industrial Heartland

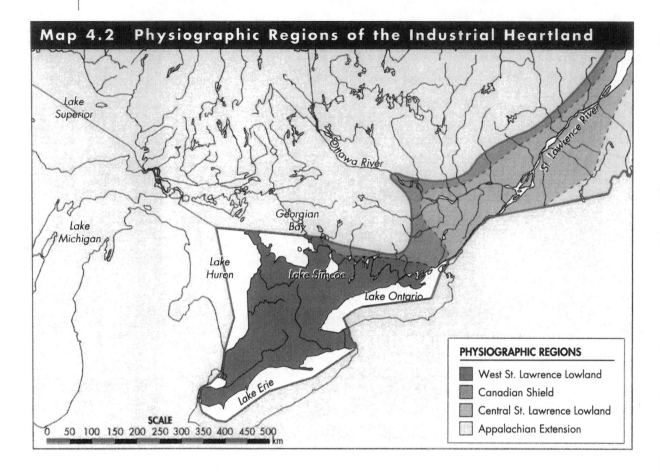

PHYSIOGRAPHIC REGIONS

- West St. Lawrence Lowland
- Canadian Shield
- Central St. Lawrence Lowland
- Appalachian Extension

of sandy soil. The flat plain, which is excellent for agriculture (Map 4.3), is relieved of its monotony by the eleven Monteregion Hills, three of which are outliers of the Canadian Shield. The other eight are made of hard volcanic rocks, remnants from the erosion of the softer material around them. One of these, Mount Royal in Montréal, is a prized site for recreational and residential use.

The southeastern portion of the heartland is part of the Appalachian physiographic region, which extends from the United States, through southern Québec, and into the Maritimes and Newfoundland. In the heartland, this well-worn mountainous system is not particularly high, most peaks being between 600 m and 1200 m above sea level. The upland areas are covered with forests or rough scrub. They support forest activity, but abandoned farms are evidence of the poor soils and difficult slopes. The river valleys, full of deposits and glacial infill, have attracted considerable farm settlement. By comparison to the Lowlands, the area is sparsely populated.

The most favourable climatic and soil conditions for agriculture and settlement are found in southern Ontario. Although the whole heartland is located in a humid continental, short summer climatic zone, the warmest summer climate, greatest heat accumulation, and shortest winters are in Essex County around Windsor, on

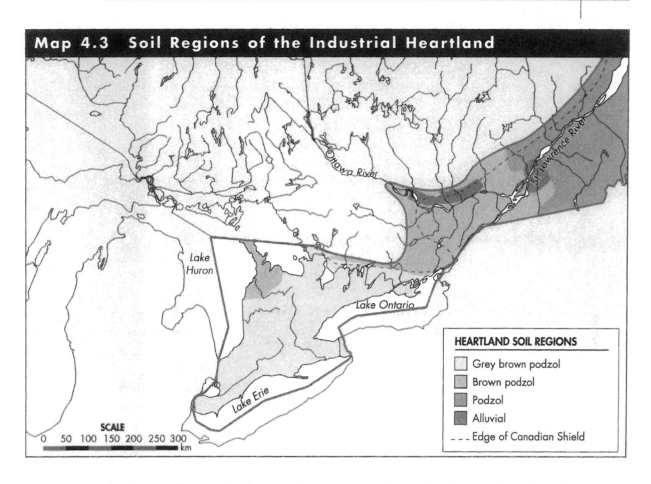

Map 4.3   Soil Regions of the Industrial Heartland

Ottawa River

St. Lawrence River

Lake Huron

Lake Ontario

Lake Erie

SCALE

0   50   100   150   200   250   300
km

HEARTLAND SOIL REGIONS

☐ Grey brown podzol

☐ Brown podzol

☐ Podzol

☐ Alluvial

--- Edge of Canadian Shield

the Niagara Peninsula (the region's prime area for tender fruit production), and along the coastal regions of Lake Erie and Ontario. Elsewhere, beyond southwestern Ontario, the winters are longer and colder, the summers shorter and cooler.

THE HEARTLAND:
Economic Definition

The role of southern Ontario and southern Québec as the core of the Canadian economy can be illustrated in three ways. First, with particular indicators comparing heartland and hinterland, we can demonstrate recent changes in the importance of the core *vis-à-vis* the rest of the nation. Second, we can show how the heartland and hinterland are interconnected, and the pivotal role that the heartland plays in national interaction. Third, we can compare heartland and hinterland with respect to international trade. The first approach describes the heartland and hinterland system; the second shows how interaction ties this system together in a pattern of flows; and the third demonstrates the external general pattern of the trade that results from the heartland–hinterland arrangement.

Heartland–Hinterland Indicators

Table 4.1 lists specific indicators that describe the heartland–hinterland system. Although there are some fluctuations, the information emphasizes the continued dominant role of the heartland. The area of the heartland is small compared to the hinterland, even when the comparison is restricted to that of the "occupied" territory or "ecumene." But, over 54 per cent of Canada's population is located in the heartland, a concentration greater than that in the core of any other developed nation, and this population is, on the average, 6 per cent wealthier (on a per capita basis) than that in the hinterland. The highly concentrated nature of the heartland "market" as compared with the much more dispersed nature of the hinterland "market" is emphasized in the population distribution map of Canada in Map 1.1 and Map 1.2.

In economic terms, the major distinguishing feature of the heartland is its dominance in manufacturing. This position is apparently strengthening relative to the hinterland, even though the rate of growth in manufacturing employment between 1971 and 1990 was comparatively low. The greatest single source of employment, and the fastest growing in terms of jobs created, is, however, the tertiary of service sector (excluding construction, transport, and communication). Since tertiary employment is distributed roughly in accordance with the population, the heartland does not dominate in this sector of economic activity in the same way as it does in manufacturing.

Farm cash receipts and hectares in farmland provide indicators of the relative importance of agricultural staples in the economy of the hinterland as compared with the core. Only 12 per cent of the farmland in the country is located in the heartland, and this proportion has been declining rapidly in recent years. Nevertheless, on this limited amount of farmland is generated production that yields over one-third of the total cash receipts received by the nation's farmers. Clearly, the agricultural activities in the area are generally intensive, involving high revenue per

Agricultural area of southern Ontario

TABLE 4.1   Some Indicators of the Changing Role of the Windsor-Québec City Core Area in Canada, 1951-1991

Indicators	Core	Periphery	% in Core
Area (000s km²)			
Total area	176	9 799	1.8
Occupied area	176	1 099	14.0
Population			
1951 (000s)	7 272	6 737	51.9
1961	9 745	8 583	53.2
1971	11 920	9 648	55.3
1981	13 194	11 154	54.2
1991	14 684	12 320	54.4
Per Capita Income			
1961 ($ millions)	1 315	1 010	
1970	2 600	2 300	
1981	9 550	9 685	
1991	17 890	16 480	
Manufacturing Employment			
1971 (000s)	1 342	500	72.9
1991	1 534	550	73.6
Service Employment[a]			
1971 (000s)	2 376	1 821	56.6
1991	4 739	3 590	56.9
Farm Cash Receipts			
1966 ($ millions)	1 533	2 741	35.9
1971	1 747	2 766	38.7
1981	6 356	12 325	34.0
Hectares in Farmland			
1966 (000s)	10 023	60 444	14.2
1971	8 950	59 714	13.0
1981	8 075	57 815	12.0
Canadian Patents/Trademarks			
1951 (per cent)			77.7
1961			82.5
1971			82.6
1981			71.2
1986			77.1

Source: Patent and trademark data from S. L. B. Ceh and A. Hecht, "A Spatial Examination of Inventive Activity in Canada: An Urban and Regional Analysis Between 1881 and 1986" *Ontario Geography* 35, 1990, 1424; rest of data compiled by the author from Statistics Canada information.

[a]In order to compare 1991 experienced labour force data with that for 1971, the 1971 employment data has been increased on a pro rata basis to remove the "industry unspecified or undefined" category in the 1971 census, as this category does not exist in the 1991 census.

hectare of economic activities. The high levels of productivity are due to two main factors: the excellent nature of much of the farmland; and the accessibility of urban markets, which generate a large local demand for perishable but high-value agricultural commodities such as dairy products, meats, vegetables, and fruit.

The patent/trademark information emphasizes the nature of this core region as the dynamic heartland of the country. During this post–World War II era, almost 80 per cent of patents and trademarks have been filed by people and businesses located within the heartland region. Though this high percentage does not necessarily imply that the work leading to the patent or trademark was done by a resident of the heartland, these indicators do demonstrate that it is through heartland-located institutions and business-headquartered enterprises that much of the innovative activity of the country is protected and controlled.

Indicators relating to the heartland, therefore, reflect many of the theoretical characteristics that typify core regions. It contains the majority of the country's population, concentrated in a relatively small part of its territory. Within the region is the vast majority of the nation's high-valued added industries, particularly manufacturing, and a significant share of its agriculture — as measured by farm cash receipts. The region is also the locus of innovation, or, perhaps more precisely, the locus of control of innovation. These attributes all translate into generally greater individual wealth — the only period during which the heartland–hinterland wealth differential disappeared was during the 1970s, when oil and gas prices reached high levels and force-fed much high-wage economic growth in the west.

Heartland–Hinterland Patterns of Interaction

The way in which urban areas of the heartland and hinterland are connected is demonstrated with airline traffic outflow data in Figure 4.1 for 1971 and 1990. The arrows represent the direction of primary and secondary airline traffic outflows (domestic and U.S. flights) from metropolitan areas in percentages of the total outflow from each Census Metropolitan Area (CMA). Thus, in 1990, the primary outflow from Winnipeg, representing 25.4 per cent of its outflow traffic, was to Toronto. The second largest outflow from Winnipeg, representing 11.6 per cent of its outflow traffic, was to Vancouver. The primary outflows, therefore, delineate the first-rank connectives, while the secondary outflows present a pattern of second-rank interactions.

The basic pattern emphasized in the first-ranked outflows in 1971 is the link between most large urban areas in the periphery to the heartland, primarily Toronto. Montréal and Toronto form a reciprocal pair, each with their highest outflow to the other — providing a dynamic illustration of the level of interconnection between Canada's two largest metropolises. Regional centres in the heartland, particularly Vancouver, Winnipeg, and Halifax, serve as intermediary nodes linking the periphery to the core. The importance, in 1971, of Montréal as a hub for the eastern part of the hinterland is emphasized in the pattern of second-rank outflows.

FIGURE 4.1 Heartland and Hinterland Interaction, 1971 and 1990, as Indicated by Primary and Secondary Airline Traffic Flows

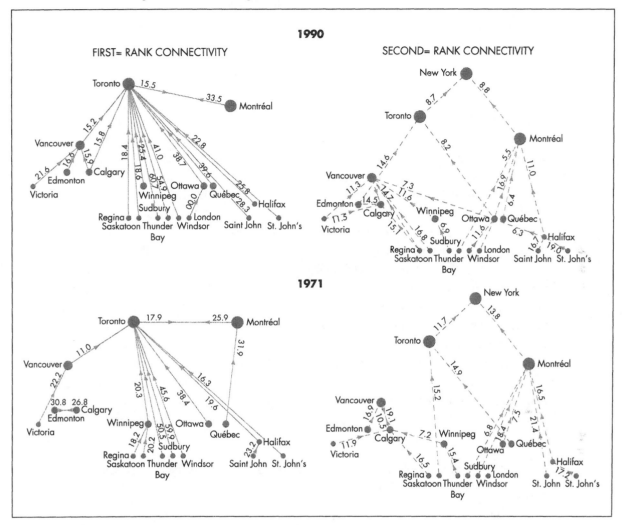

The most interesting feature of the pattern of the 1971 second-rank outflows is that New York is the focus of the second largest volume of flows from both Montréal and Toronto. The primary links, therefore, identify the interlinked nature of the core and the ties between the heartland and the hinterland; the second-rank patterns reveal the relationship between Montréal and the eastern hinterland, and the strength of the external relationships between Montréal and Toronto to the focus of North American business and finance.

The pattern of primary and secondary flows in 1990 exhibits a number of differences from that for 1971. The first-rank pattern of 1990 connectivities indicates that Toronto has reinforced its position as the leading focus of travel from all metropolitan areas in the country. In general, links up the urban size hierarchy have strengthened

(Montréal to Toronto, Vancouver to Toronto, Toronto to New York), while links down the size hierarchy have weakened. Furthermore, both first- and second-rank connectivities indicate the growing importance of Vancouver as hub in the west hinterland. This, as will be discussed in ensuing sections, reflects fundamental changes in the location of economic activity, and, therefore, heartland– hinterland relationships.

Heartland–Hinterland Patterns of External Trade

The external trade patterns of the core and the periphery relate directly to the different economic characteristics of the two regions. Unfortunately, as external trade information is not available for parts of provinces, the heartland in this section will be regarded as Ontario and Québec, and the periphery as the rest of Canada. This division means that the primary-producing areas of northern Ontario and Québec — involving mainly mining and partial refining (Sudbury, Chicoutimi-Jonquière, Rouyn-Noranda), forestry, and hydroelectric power — are included within the core. The data are therefore weighted more to primary production than is actually the case.

During the early 1990s, Canada's balance of trade has been positive — that is, Canada exports more by total value than it imports (Table 4.2). This favourable balance is, however, somewhat misleading because the balance of payments — which includes financial transfers relating to loans, interest and dividend payments on foreign capital, international capital transfers, and foreign travel — has been negative. However, the trade data in Table 4.2 indicate that Canada basically imports manufactured products and pays for these with the export of crude materials, food products, partially fabricated materials, and some manufactured end products. Furthermore, the data indicate that between 1982 and 1993 the share of manufactured end products in total imports and

TABLE 4.2 Canada's Imports and Exports by Commodity Grouping, 1982 and 1993

Commodity Group	Percent of Exports		Percent of Imports	
	1982	1993	1982	1993
Crude materials, inedible	18.1	12.2	12.9	5.3
Food, feed, beverages, tobacco	12.1	6.9	7.1	5.9
Fabricated materials, inedible	34.2	32.1	17.5	18.8
Manufactured end products, inedible	34.8	46.9	60.8	67.3
Other	0.8	1.9	1.7	2.7
Total	100	100	100	100
Total volume ($ millions)	81 464	176 756	67 355	169 560

Source: Statistics Canada "Summary of External Trade, 1982"; and "Summary of International Trade, 1993."

exports has increased, while that of crude raw materials has decreased.[6] This increased share of manufacturing in the country's international trade is related in large part to the even greater integration of the North American automobile industry[7] during this period.

The different roles of heartland and hinterland in trade are illustrated in Figure 4.2. The periphery dominates the export trade (80 per cent) in crude materials (particularly oil, natural gas, and forest products), most of which go to the United States (73 per cent) and the rest to other foreign areas (27 per cent). Moving down the diagram to groups of products involving higher levels of fabrication, the contribution of the hinterland becomes smaller and that of the core larger. Crude materials are often partially fabricated, usually in the core, and then exported to the United States. On the other hand, food products involve both heartland and hinterland almost equally (51 per cent and 49 per cent), and three-quarters of the exports go to the rest of the world. Manufactured end products are produced virtually entirely in Ontario and Québec and exported almost exclusively to the United States (88 per cent).

This overwhelming link with the economy of the United States, and the role that the core plays in this link, is emphasized with respect to imports. Since Canada is a staple-producing country, imports of food and crude materials (usually subtropical and fresh vegetables and fruits, and oil to eastern Canada) are few compared with fabricated materials and end products. The greatest value of imports by far involves a whole variety of end products that come from the United States and go to the core. A large share of the end product trade with the United States (imports and exports) involves automobiles and parts, for since the institution of the Canada–United States Auto Pact in 1965, the auto industry in North America has been integrated by the major companies, and there is a considerable flow across the border of these high-value products.[8] This integration has been renewed in the Canada–United States Free Trade Agreement (1989) and extended to include all traded items, including services and transport.[9]

Thus, the heartland–hinterland arrangement of the Canadian economy is manifest to a large extent in the general pattern of international trade. Staples and food products form a large share of the exports, and these emanate to a large extent from the periphery. The Industrial Heartland dominates both the export and import of fabricated products. The ties to the United States are strong and reciprocal, for not only does the United States dominate Canada's trade pattern, but Canada is also the largest single trading partner of its southern neighbour.

CONCENTRATION AND CHANGE IN THE HEARTLAND

Although the proportion of Canada's population residing in the heartland increased steadily to 1971, a small relative decline has since set in (see Table 4.1). The period of greatest heartland population growth occurred between 1951 and 1971,

FIGURE 4.2   The Import-Export Trade of Canada's Heartland–Hinterland Space Economy

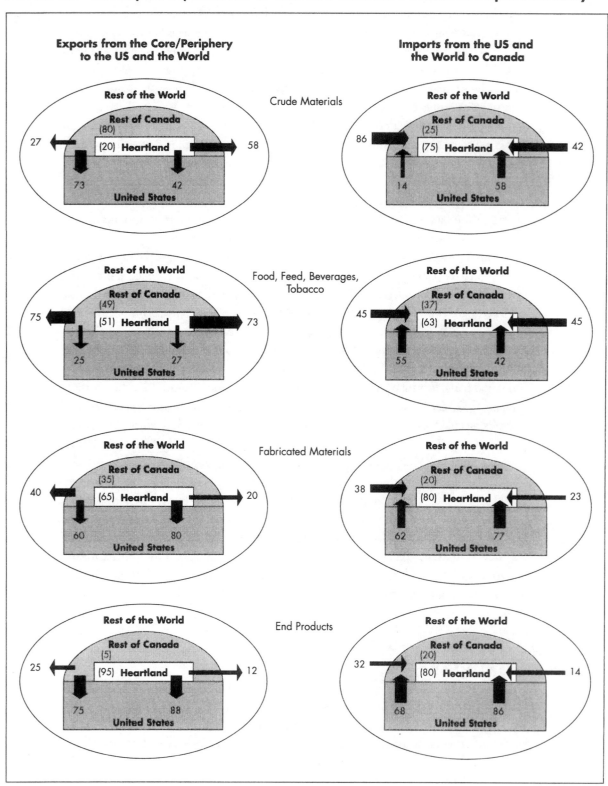

when the area's share of the nation's population increased from 51.9 per cent to 55.3 per cent. This period of concentration was followed by a decade of deconcentration between 1971 and 1981, and stabilization between 1981 and 1991. During the 1990s it appears that this stabilization of the heartland's share of the national population is continuing, with particularly high rates of growth being experienced in and around metropolitan Toronto and Ottawa–Hull.[10]

The factors that maintain the comparative strength of the heartland are: its proximity to markets and capital in the United States; its relatively large local domestic market; agglomeration economies derived from a relatively well-educated and large urban population; the productivity of agriculture; and the highly integrated urban system of southern Ontario and southern Québec. Although investment in large resource-based projects in the hinterland (such as the Hibernia oil field off Newfoundland) periodically attract capital away from the heartland, the long-run effects of such developments are to boost economic activity in the heartland because most of the manufacturing and organization of industry in Canada is focussed in this part of the nation.

Urban Concentration

The concentration of the nation's manufacturing into the industrial core is matched by a concentration of the population into a few large urban areas. Map 1.2 (page 7) illustrates quite vividly the nature of this concentration — one large concentration focusses on Toronto and other on Montréal; two intermediate-size concentrations occur around Ottawa–Hull and Québec; and a number of small concentrations occur around Windsor, London, Kitchener–Waterloo, St. Catharines–Niagara, Oshawa–Whitby, Kingston, Sherbrooke, and Trois Rivières. The CMA[11] populations relating to these places accounted for 76.3 per cent of the heartland population in 1991 (Table 4.3).

Three general features of this urban distribution should be emphasized: (1) the Census Metropolitan Areas of Toronto and Montréal, each with a population of over 3 million, clearly dominate over all other urban areas; (2) there are many more large urban places in southwestern Ontario than there are in the Québec portion of the heartland; and (3) there has been a wide variation in 1971–1991 CMA growth rates. Oshawa–Whitby, Ottawa–Hull, Kitchener–Waterloo, and Toronto each experienced twenty-year growth rates at least 50 per cent above the national average (26.6 per cent). On the other hand, Windsor, Montréal, and Hamilton experienced growth rates lower than the national average.

Linguistic Concentration and Multiculturalism

In addition to this urban concentration, the heartland is pivotal because it is the centre around which the multicultural nation has been forged. The heartland consists of the densely populated areas of two provinces in which the two linguistic

TABLE 4.3   Population of Census Metropolitan Areas (CMAs) and Large Urban Agglomerations (UAs), 1971 and 1991

Region or Province	Metropolitan Area	1971 CMA/UA Population in thousands	1991 CMA/UA Population in thousands	1971-1991 Percentage Change
Heartland		11 920	14 684	23.2
Ontario	Toronto	2 628	3 893	**48.14**
	Ottawa-Hull	602	920	**52.82**
	Hamilton	498	600	20.48
	London	282	382	35.46
	St. Catharines-Niagara	286	365	27.62
	Kitchener-Waterloo	239	356	**48.95**
	Windsor	249	262	5.22
	Oshawa-Whitby	120	240	**100.01**
	Kingston	106	136	28.31
Québec				
	Montréal	2 743	3 127	13.99
	Québec City	481	646	34.31
	Sherbrooke	106	139	31.13
	Trois-Rivières	104	136	30.76
Hinterland		9 648	12 320	27.7
Maritimes				
	Halifax (N.S.)	223	321	**43.95**
	St. John's (Nfld)	132	172	30.31
	Saint John (N.B.)	107	125	16.82
Prairies & Shield				
	Winnipeg (Man.)	540	652	20.74
	Saskatoon (Sask.)	126	210	**66.67**
	Regina (Sask.)	141	191	35.46
	Chicoutimi-Jonquière (Que.)		161	
	Sudbury (Ont.)		158	
	Thunder Bay (Ont.)		124	
West				
	Vancouver (B.C.)	1 082	1 603	**48.15**
	Edmonton (Alb.)	496	840	**69.34**
	Calgary (Alb.)	403	754	**87.09**
	Victoria (B.C.)	196	289	**47.45**
CANADA		21 568	27 297	26.6

Note: Percentages in bold indicate 20-year growth rates at least 50 per cent above the national growth rate.

groups that founded the nation in post-Contact times are concentrated. These linguistic groups are focussed in distinct parts of the heartland, reflecting a cultural polarity that on the one hand has sparked social and political conflict, but on the other has encouraged a dynamic complementarity of social and economic values.

Nearly 80 per cent of the population in the Québec portion of the heartland is French-speaking, and the only two parts of the province in which the proportion drops below 80 per cent are along the Ontario and United States borders, and on the Île de Montréal (Map 4.4). A small part of Ontario, along the Québec border around Hawkesbury, is also predominantly French-speaking. The location of the English-speaking population is virtually a mirror image of the French (Map 4.4 and 4.5). Almost the whole portion of Ontario located in the heartland is English-speaking, the only exceptions being along the Québec border. Of particular note is the English-language concentration in the western part of Île de Montréal.

Since 1951, foreign immigration has been providing a large share of annual population growth. During the period of high domestic birth rates up to 1963, net immigration contributed about 10 per cent of national net annual population

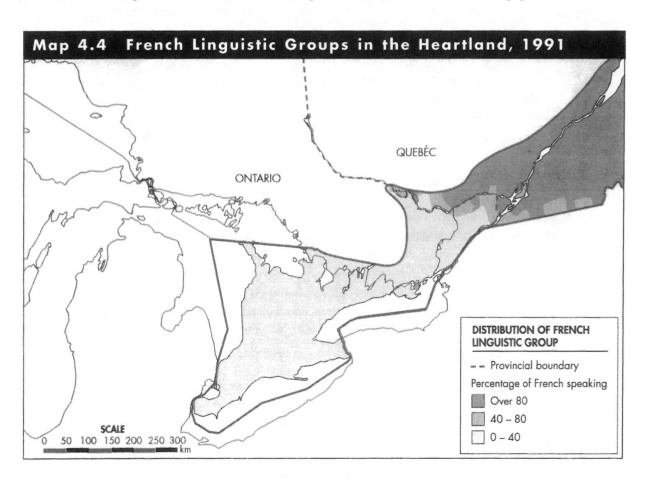

Map 4.4 French Linguistic Groups in the Heartland, 1991

QUEBÉC

ONTARIO

DISTRIBUTION OF FRENCH LINGUISTIC GROUP

– – Provincial boundary

Percentage of French speaking

- Over 80
- 40 – 80
- 0 – 40

SCALE

0 50 100 150 200 250 300
km

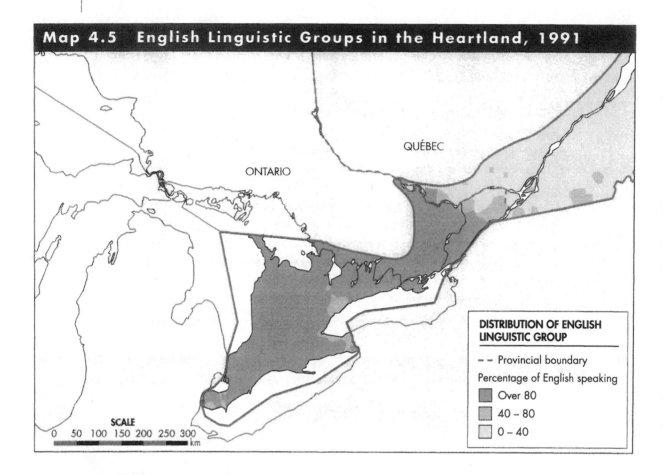

Map 4.5   English Linguistic Groups in the Heartland, 1991

QUÉBEC

ONTARIO

DISTRIBUTION OF ENGLISH LINGUISTIC GROUP

– – Provincial boundary

Percentage of English speaking

Over 80

40 – 80

0 – 40

SCALE

0   50 100 150 200 250 300
km

increase. With decreasing domestic birth rates immigration accounted for an increasing share, rising from 20 per cent in the early 1970s to almost 30 per cent in the 1980s, and above 30 per cent in the 1990s. Prior to 1971, much of the immigration originated from Europe. Since 1971, the chief sources of immigration have been from non-European areas. Throughout the postwar period, the vast majority of immigrants settled in Canada's three major metropolises. The result is that Toronto and Montréal are now even more multicultural places than they have ever been, with populations of many faiths, races, and languages. With this concentration of immigration into the largest urban areas, the metropolises of Toronto and Montréal have become culturally far more heterogeneous than most other urban and rural areas within the heartland.[12]

The high level of population heterogeneity existing in these two major metropolises is illustrated in Table 4.4 with respect to the Toronto CMA in 1991. Nearly 41 per cent of the population in the metropolis is born outside of Canada. Of this foreign-born population, 46 per cent is from Europe, primarily Italy, the United Kingdom, and Portugal. This group represents the population that immigrated prior to 1971. The post-1971 immigrant population originates mainly from southeast Asia, the Caribbean

TABLE 4.4   Place of Birth of the Population in the Toronto CMA, 1991

Area	Population	Per Cent
Canada	2 296 380	59.4
United States of America	36 470	2.33
Central and South America	99 120	6.33
Caribbean and Bermuda	128 720	8.22
United Kingdom	182 770	11.67
Italy (est.)	289 520	18.40
Portugal (est.)	125 480	8.00
Other European countries	542 660	34.64
Africa	59 190	3.78
India	70 450	4.50
China	220 535	14.10
Other Asian countries	342 085	21.83
Oceania and other	7 165	0.46
Non-permanent resident	98 105	6.26
Immigrant Population	*1 566 735*	*40.6*
Total	3 863 115	100.00

Source: Statistics Canada, Cat. 95-354, Table 2

and Bermuda, Central and South America, India, and various African countries. As well, the influx of Native peoples has added another component to the mix, one that creates its own stresses in and around major urban areas (see Table 4.5).

TABLE 4.5   Populations of 13 000 or More with Aboriginal Origins in Heartland Cities and Provinces, 1991

City	Aboriginal Population
Montréal	45 000
Toronto	41 000
Ottawa–Hull	31 000

Province	Aboriginal Population
Québec	69 000
Ontario	143 000

Source: Royal Commission on Aboriginal Peoples

The Concentration of Manufacturing

As noted earlier, although manufacturing employment in Canada increased little between 1971 and 1991, the heartland has strengthened its position as the manufacturing centre of the nation. Between 1971 and 1991 manufacturing employment in the core increased 14.3 per cent as compared with 10 per cent in the hinterland. These rates of increase were much less than those experienced in the 1951-1997 period when manufacturing employment in Canada as a whole increased by 30 per cent.[13]

Manufacturing industries are found in a large number of heartland cities, but the bulk of the employment is in major metropolitan areas large enough to generate local industrial complexes. The metropolises listed in Table 4.3 account for about three-quarters of the manufacturing labour force in the heartland, with the CMAs of Toronto and Montréal providing 45 per cent of the total in 1991. Hamilton has about 5 per cent of all core manufacturing jobs; Kitchener–Waterloo 3.5 per cent; St. Catherines–Niagara, Windsor, London, Québec, Oshawa–Whitby, and Ottawa–Hull between 2 and 3 per cent of the total; and the remaining places about one per cent each.

All types of manufacturing are represented in the core (Map 4.6), but over 50 per cent of the labour force is concentrated in transport equipment, food and beverage processing, metal fabricating, and electrical products. The various industrial types can be divided into two groups on the basis of industry-wide estimates of value-added per worker in the manufacturing process. Low value-added industries tend to have low capital/labour ratios and pay relatively low wage rates, whereas high value-added industries tend to have high capital/labour ratios and pay relatively high wage rates.

Stelco Hamilton works

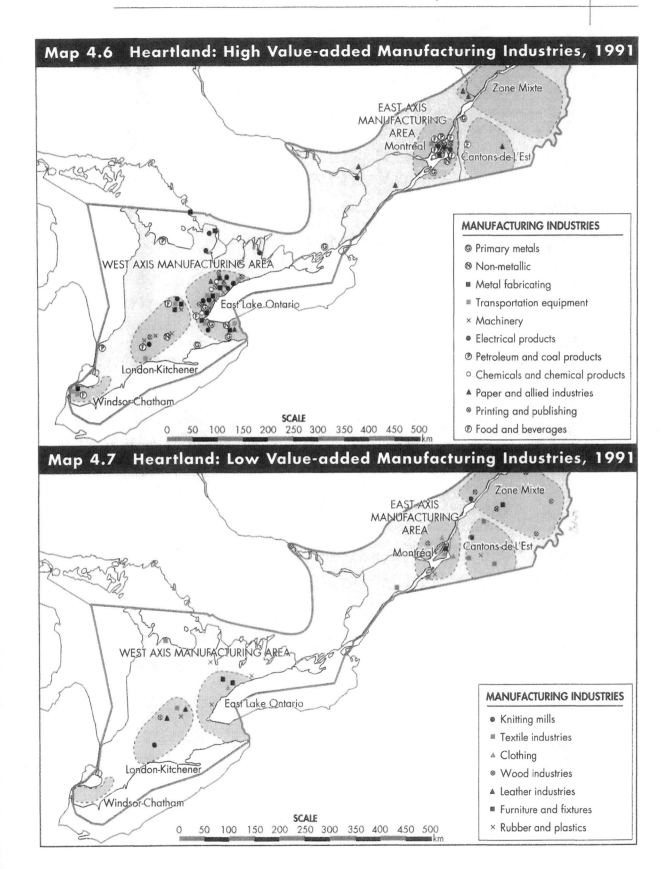

Map 4.6   Heartland: High Value-added Manufacturing Industries, 1991

MANUFACTURING INDUSTRIES

ⓖ Primary metals
Ⓝ Non-metallic
■ Metal fabricating
✳ Transportation equipment
✕ Machinery
● Electrical products
ⓟ Petroleum and coal products
○ Chemicals and chemical products
▲ Paper and allied industries
⊛ Printing and publishing
ⓕ Food and beverages

EAST AXIS MANUFACTURING AREA
Montréal
Zone Mixte
Cantons-de-L'Est

WEST AXIS MANUFACTURING AREA
East Lake Ontario
London-Kitchener
Windsor-Chatham

SCALE
0 50 100 150 200 250 300 350 400 450 500 km

Map 4.7   Heartland: Low Value-added Manufacturing Industries, 1991

MANUFACTURING INDUSTRIES

● Knitting mills
▓ Textile industries
▲ Clothing
⊛ Wood industries
▲ Leather industries
■ Furniture and fixtures
✕ Rubber and plastics

EAST AXIS MANUFACTURING AREA
Montréal
Zone Mixte
Cantons-de-L'Est

WEST AXIS MANUFACTURING AREA
East Lake Ontario
London-Kitchener
Windsor-Chatham

SCALE
0 50 100 150 200 250 300 350 400 450 500 km

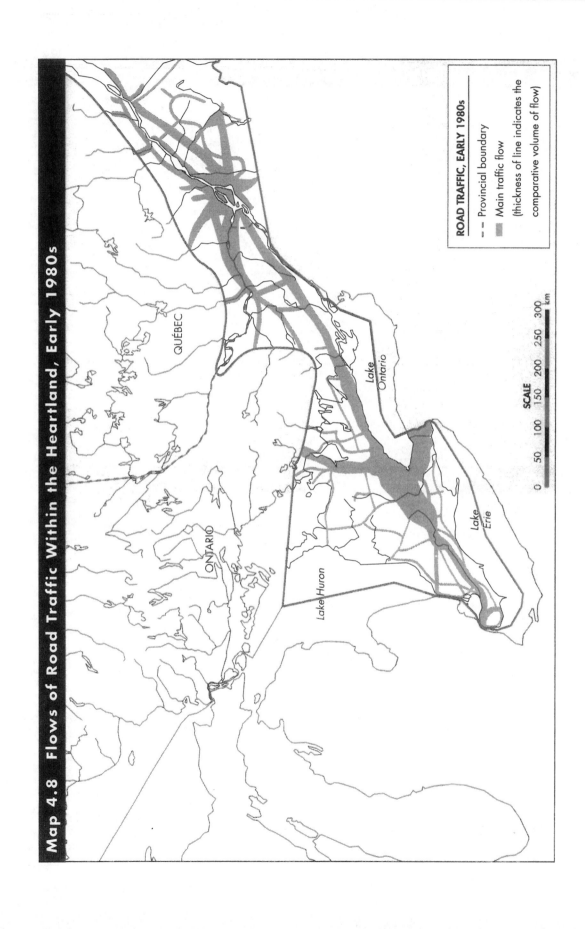

Map 4.8 Flows of Road Traffic Within the Heartland, Early 1980s

ROAD TRAFFIC, EARLY 1980s

- - - Provincial boundary

Main traffic flow
(thickness of line indicates the
comparative volume of flow)

QUÉBEC

ONTARIO

Lake Huron

Lake
Ontario

Lake
Erie

SCALE

0   50   100   150   200   250   300
km

The contrasting distributions of these two types of industry are illustrated in Map 4.6 and Map 4.7. The high value-added industries (Map 4.6) tend to be clustered in the largest metropolitan areas: Toronto and Montréal. But the concentration is particularly evident in the area around eastern Lake Ontario from Oshawa–Whitby to Hamilton, and extending westwards along the major limited access highway (401) and rail routes to Kitchener–Waterloo, London, and Windsor–Chatham.

The low value-added industries (Map 4.7), on the other hand, appear to be quite dispersed. Some (such as clothing and furniture and fixtures) are present in the two largest metropolises, but most are found in the smaller towns beyond the major urban centres. Prior to 1971 there was a particular pattern to these industries in Québec, with the textile industry being concentrated in the townships immediately to the east of Montréal, an area defined somewhat more broadly than usual as Cantons-de-l'est. Since 1971, however, this industry has been decimated, and many factories have closed. Wood and paper industries are centred in the small towns somewhat farther from Montréal and in Québec, in a territory described as "zone mixte" because it supports a variety of economic activities.

This emergence of high and low value-added geographic concentrations of industry may be related to the differing availabilities of two factors of production — capital and labour — in the two provinces over the past few decades. As capital has been relatively more abundant than labour in southern Ontario, capital-intensive (and hence high value-added) industries have tended to concentrate in that part of the heartland. In Québec, labour has been relatively more abundant than capital, leading to more labour-intensive (and hence low value-added) industries. This structural difference has been a major cause of the variation in rates of growth in manufacturing employment between these two parts of the heartland during the 1951-1991 period. Thus, in summary, the major regions of industrial specialization are:

1. *The East Axis Manufacturing Area* consisting of

   (a) The Montréal Region: with mainly high value-added industries (transport equipment, electrical products, chemical products, printing, publishing), and some low value-added industries such as clothing, the region had about 20 per cent of Canada's manufacturing labour force in 1991.

   (b) Cantons-de-l'est including Magog, Sherbrooke, Drummondville, and Cowansville: this region has historically been a textile area interlinked with the clothing industry of Montréal.[14]

   (c) "Zone mixte": in which are found mainly industries based on wood in cities and towns such as Trois Rivières, Shawinigan, and Arthabaska. The city of Québec stands out as an anomaly with a slightly wider manufacturing base.

2. *The West Axis Manufacturing Area* consisting of

(a) East Lake Ontario: focussing primarily on Oshawa–Whitby, Toronto, and Hamilton. With 32 per cent of the country's manufacturing employment in 1991, it has the largest concentration of manufacturing employment in Canada. Most of the industrial activities involve high value-added manufacturing. The dominant industrial types are transport equipment (particularly automobile assembly, trucks, farm equipment), metal fabricating, primary metals, electrical products, and newer high-technology industries such as computer equipment.

(b) The London–Kitchener/Waterloo corridor: involving high value-added high-technology industries in Kitchener–Waterloo, automobile manufacture in St. Thomas, and metal fabricating in Kitchener and Guelph.

(c) Windsor–Chatham: focussing almost entirely on high value-added automobile and metal fabricating industries.

There are also two areas of minor manufacturing concentration that should be mentioned. One is the limited, but growing, area of manufacturing employment in Ottawa–Hull. The chief industry was once pulp and paper manufacturing, but the more rapidly growing modern industry involves computer and communications hardware and software production, which is subsumed partly under the "electrical products" heading. The second area of minor concentration is the petrochemical industry in Sarnia, which is based largely on the resources of Alberta and reflects the locational requirements of chemical and plastic manufacturers for accessibility to the central Canadian market.

Manufacturing Restructuring

Manufacturing in the heartland, particularly the automobile industry, has been greatly affected by the restructuring of industrial activity that has been occurring in North America since 1971. This restructuring has been in response to greater competition from German, Japanese, and southeast Asian producers of manufactured goods (see Table 4.6). This greater competition has been with respect to both the price and quality of many manufactured goods.[15] Lower prices and better-quality products were being achieved by some exporters in these countries through new product and process technologies.

For example, during the 1970s North American automobile production remained in a "Fordist" mode of industrial organization epitomized by assembly-line production with long product lines to achieve economies of scale. In this type of production, inventories are large, and quality control procedures permit a certain (very small) percentage of faults. The newer product and process technologies, pioneered in Japan, involving quality-circle modes of production, heavy use of robots in certain parts of the manufacturing process, and just-in-time delivery systems, decreased costs and increased quality.[16] More effectively, however, in terms of costs, this competition undermined the good benefits, wage rates, and types of job-protection procedures that had been negotiated between production workers and management.[17] With widespread introduction of some of these newer product and process technologies, production has been maintained, but costs reduced through lay-offs and reorganization. Much of this reorganization has been achieved through the contracting out of parts of production to lower-wage companies.

This restructuring has greatly affected manufacturing activity in the heartland. It has led to severe employment downsizing in the automobile industry, and wiped out much of the textile industry in Québec as textile and clothes production has shifted to China and other lower cost areas in the world. Iron and steel production has also been greatly affected as the increased demand for specialized steels has been met by mini-mills, while the demand for general purpose steel, such as that used in rails or girders produced in large plants, has decreased. On the other hand, the higher value-added industries in electronics, process control equipment, and other "high-technologies" have flourished. Ontario has therefore suffered less than Québec in the restructuring, for its manufacturing base was already less labour intensive than that in Québec.

TABLE 4.6   Passenger Cars: Places of Manufacture and Total Canadian Sales, 1981 and 1991

	Total Number of Sales	Place of Manufacture (%)		
		North America	Japan	Other
1981	904 000	71.6	23.0	5.4
1991	873 000	65.6	27.3	7.1

Source: Statistics Canada 1992

THE CHANGING NATURE OF THE HEARTLAND ECONOMY AND DIFFERENTIAL URBAN GROWTH

Since 1951, the Canadian economy has grown considerably, and also undergone great sectoral transition. This is illustrated with respect to the experienced labour force in Figure 4.3 in which the sizes of the circles are proportionate to the magnitude of total employment in the year indicated.[18] The national economy has undergone a highly significant change from that in 1951, when primary and manufacturing activities provided the bulk of employment, and services little more than one-third.[19] By 1991 almost two-thirds of the experienced labour force is in service activities, with primary

FIGURE 4.3   Increasing Importance of Service Employment in the Canadian Economy

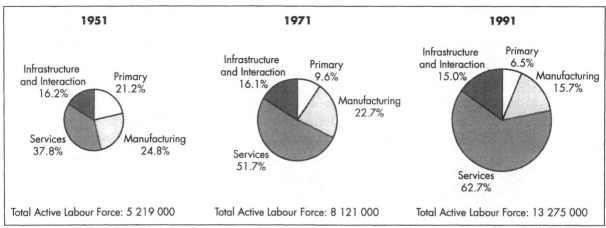

and manufacturing activities contributing little more than one-fifth. The great change, as far as the heartland is concerned, occurred with respect to manufacturing and service activities after 1971. This shift has also been accompanied, since 1971, by a large increase in the number of women in the paid labour force — by 1991, 53 per cent of women were in the paid labour force as compared with 41 per cent in 1971 — and their employment concentrates heavily in the service sector.

The service sector consists functionally of three components: consumer services, producer services, and government/public services. This three-fold distinction is important because the three components incorporate a range in types of end uses, different financial bases, and are associated with different theoretical explanations concerning their location and allocation among urban areas in Canada.[20] There is considerable debate in some cases concerning which particular subgroups of activities should be included within these three components.[21]

Consumer services are those involving distributive activities, where the sale of the product achieves, either directly or indirectly, its end-use of final demand. These activities include retail trade, which involves the sale of manufactured products to final consumers. They also include accommodation, entertainment and recreation, and personal services to private households (such as home decorating) and individuals (such as barbershops and restaurants). As consumer services involve end consumption, their distribution is directly related to the location of the market — that is, the distribution of the population and its purchasing capacity.

Producer services are those that mainly provide inputs to, or are linked with, other economic activities. Included are business services,[22] finance and accounting, insurance, real estate development, and wholesale trade.[23] Because they are linked with other economic activities, producer service firms often chose locations that maximize the possibility of forward and backward linkages with other firms and offices.[24] Thus, whereas consumer service activities locate to maximize access to a highly spread final demand market of individuals and household, producer services locate to maximize access to concentrations of other economic activities (and each other). This implies concentration of location within large metropolitan areas, for it is in these places that firms and offices are clustered.

Governmental and public services involve education, health, and governmental services. Education, especially at the primary and secondary levels, is tied directly to the locations of school-age children. Employment associated with education at these levels is, therefore, related to the distribution of families with children. Post-secondary education facilities may be less demand fixed because their clientele is more mobile. General health-care facilities are also distributed according to the distribution of demand, but less frequently needed specialized services tend to be clustered in metropolitan areas, which have the population base to support such facilities. Governmental services tend to be concentrated in those places which have been selected as capitals, such as Ottawa–Hull, Québec, and Toronto.[25] Apart from these anomalies, there is generally a strong production/consumption link between governmental and public service employment and the distribution of population.

These components of service activity has increased in employment considerably in Canada since 1971 (Table 4.7) — so much so that each alone provides more jobs than the manufacturing sector. Whereas the experienced labour force in manufacturing increased 13.2 per cent between 1971 and 1991, the labour force in the consumer and producer subsectors more than doubled, and the public and governmental labour force increased by 81 per cent. Given the strong purchasing capacity/provision link in the service component, and the equally strong (apart from a few anomalies) production/consumption link with respect to the labour force in governmental and public services, the major locational issue focusses on the concentrating tendencies of the producer service subsector.

Toronto Stock
Exchange

TABLE 4.7   Change in Experienced Labour Force, 1971-1991

Economic Group	1971		1991		1971-1991
	Labour Force (in thousands)	% Change	Labour Force (in thousands)	% Change	% Change
Primary[a]	776.9	9.0	868.0	6.1	11.7
Manufacturing	1 841.8	21.3	2 084.1	14.7	13.2
Services					
Consumer[b]	1 350.9	15.7	2 741.1	19.3	102.9
Producer[c]	988.0	11.4	2 227.3	15.7	125.4
Government & Public[d]	1 858.0	21.5	3 361.2	23.6	80.9
Infrastructure & Interaction[e]	1 304.7	15.1	1 994.4	14.0	52.9
Other	506.6	6.0	948.1	6.6	87.1
Total	**8 626.9**	**100.0**	**14 220.2**	**100.0**	**64.8**

[a]Primary: agriculture, fishing, mining, forestry.
[b]Consumer: retail, and accommodation, restaurants, and personal services, etc.
[c]Producer: FIRE (finance, insurance, real estate), business services, wholesale activities.
[d]Government & Public: government, education, health.
[e]Infrastructure & Interaction: construction, transport, communications.

Source: Data from Coffey (1994), Table 2.1, modified and retabulated by author.

TABLE 4.8   Concentration of the Producer Service Experienced Labour Force in Heartland and Western Hinterland CMAs, 1971 and 1991

Metropolitan (CMA)	1971			1991		
	Population Share %	Labour Force Share %	Concentration Index	Population Share %	Labour Force Share %	Concentration Index
Heartland						
Toronto	12.19	22.21	1.82	14.26	23.54	1.65
Montréal	12.72	16.53	1.30	11.46	14.19	1.24
Ottawa–Hull	2.79	2.83	1.01	3.37	3.88	1.15
Québec	2.23	2.01	0.90	2.37	2.47	1.04
Hamilton	2.31	2.33	1.01	2.21	2.31	1.05
Total	**32.24**	**45.91**	1.42	**33.67**	**46.39**	1.38
Western Hinterland						
Vancouver	5.02	8.28	1.65	5.87	8.51	1.45
Edmonton	2.31	2.83	1.23	3.08	3.53	1.15
Calgary	1.87	2.99	1.60	2.77	4.12	1.49
Winnipeg	2.51	3.68	1.47	2.39	2.53	1.06
Total	**11.71**	**17.78**	1.52	**14.11**	**18.69**	1.32
Rest of Country	56.05	36.31	0.65	52.22	34.92	0.67

Note: Included in this listing are only those CMAs that have 2% or more of the 1991 national producer service experienced labour force.

THE CONCENTRATION OF PRODUCER
SERVICES WITHIN THE HEARTLAND

The five largest CMAs in the heartland increased, by a small amount, their share of the national producer service labour force between 1971 and 1991. These metropolises, listed in Table 4.8, had 33.7 per cent of the country's population in 1991, but 46.4 per cent of producer service jobs. The concentration indices for these five metropolises with respect to the producer service labour force are all above 1.0, indicating that they had a greater share of the national labour force in this category than would be expected based on their population size alone. These indices are, however, clearly greatest for Toronto, Montréal, and Ottawa–Hull. The Toronto CMA has been increasing its dominant national position in producer service employment, while Montréal's share, though remaining large, has been decreasing.

The concentrated nature of the producer service labour force in large urban agglomerations is related in large part to economic and behavioural forces that induce clustering. Economic forces include requirements for high levels of access to financial and securities markets and information flows associated with them; a need for single-point access to national, continental, and global transportation; the synergistic effect of a wide array of possible local forward and backward linkages; and the availability of a high-energy, well-trained, and educated labour force. Behavioural forces include a need by different levels of management for face-to-face-contact for business negotiations, and a general predilection for places with an array of services and amenities.

Information in Table 4.9 provides examples of this clustering as it emphasizes the role of Toronto and Montréal as the headquarters locations of large companies (with annual revenues of more than $30 million) in Canada. Given that the concentration of the headquarter offices of Canadian banks and other financial companies in Toronto and Montréal is well known,[26] the concentration measures in Table 4.9 relating to non-financial industrial subsectors is extremely interesting.[27] In general, headquarter firms located in Toronto and Montréal administer 64 per cent of the total revenue of the non-financial subsectors in the country. In ten subsectors, firms headquartered in these two metropolises administer more than 70 per cent of revenues, and in only two subsectors do firms in these two places administer less than 50 per cent.

THE CHANGING PATTERN OF URBAN
DEVELOPMENT IN THE HEARTLAND

The spread of the population in the heartland is directly related to the concentrating tendencies of economic activities. At the macro heartland–hinterland scale, the concentration of manufacturing and producer services within the heartland underlies the continuing concentration of the bulk of the country's population within the area. Within the heartland, the concentration of manufacturing, producer services, and employment in governmental and public services in a few places underlies the tendency of the population to cluster within four extensive urban regions. Changes in the pattern of clustering people into these four extensive urban regions, therefore, reflects the changing space-economy of the heartland.

Table 4.10 shows the changing concentration of population within the four major urban regions for two twenty-year time periods: 1951–1971, and 1971–1991. The increased clustering tendencies of the heartland economy are illustrated by the dramatic increase in the proportion of the heartland population located within these four regions — from 57.2 per cent of the population in 1951 to 70.6 per cent in 1991. This overall population growth coupled with the increase in concentration means that the population within these four urban regions now exceeds ten million persons.

TABLE 4.9   The Headquarters Location of Companies in the Non-financial Subsector, 1990 (percentages)

Sector	Subsector	Toronto, Montréal	Vancouver, Calgary, Edmonton, Winnipeg
Resource	Mining	64.7	29.3
	Petroleum	33.3	61.3
Manufacturing	Chemical	55.3	28.1
	Electrical	85.9	0.8
	Food	62.9	5.0
	Forestry	56.7	23.8
	Machinery	84.1	—
	Metal	70.7	—
	Miscellaneous	51.1	34.6
	Pharmaceutical	97.4	—
	Transport Equipment	52.8	3.9
	Textiles	93.1	—
Service	Business Services	81.8	12.0
	Communication	77.5	10.6
	Construction	48.6	35.2
	Cooperatives	32.1	32.5
	Miscellaneous	71.8	15.1
	Retail	78.6	13.7
	Transportation	73.7	20.7
	Wholesale	53.3	35.6

Source: Rice (1996), p. 71

TABLE 4.10   The Concentration of Heartland Population Within the Vicinity of Four Major Urban Regions, 1951, 1971, and 1991

Urban Region Concentration	Area (km²)	1951 Number (000)	1951 % of Core	1971 Number (000)	1971 % of Core	1991 Number (000)	1991 % of Core
Toronto (within 65 km radius)	7 563	1 696	23.3	3 413	28.6	5 133	34.9
Montréal (within 65 km radius)	11 938	1 840	25.3	3 174	26.6	3 656	24.9
Ottawa–Hull (within 25 km radius)	1 965	305	4.2	588	4.9	900	6.1
Québec City (within 25 km radius)	1 847	321	4.4	507	4.3	685	4.7
Total, four urban concentrations		4 162		7 682		10 374	
Heartland population		7 272		11 920		14 684	
% within four major concentrations			57.2		64.4		70.6

These time periods reflect two quite different periods of growth in the heartland economy.[28] The first period, from 1945 to 1971, was one of almost continuous economic growth and rapid population increase, when real per capita incomes increased at an unprecedented average rate of 3.1 per cent per annum. This, as can be seen in Table 4.1, was a period of consolidation, during which time the population of the country concentrated into the heartland, and, as can be noted from Table 4.10, also concentrated within the heartland to the four large urban regions. Between 1971 and 1991, when real per capita incomes increased on average at a rate one-half (1.5 per cent per annum) of that of the period of consolidation, there were three economic recessions that affected the heartland quite considerably. As a consequence, the level of concentration of the country's population into the heartland decreased slightly, and, more importantly, there occurred a refocussing of the population within the area to the region around Toronto (and, secondarily, Ottawa–Hull).

The Period of Consolidation

Between 1951 and 1971 the population doubled in areas within the heartland defined as urban. Six geographical patterns were associated with this rapid urban development:

1. By 1971 there had emerged an almost contiguous macro-urban strip from Windsor to Québec. Not only have urban areas around Montréal and Toronto coalesced, but the whole area has become one elongated urban corridor.

2. The integrated nature of this corridor is reflected not only through physical contiguity, but also by flows of goods, people, and information. These flows were facilitated by significant public infrastructure developments — such as the Macdonald-Cartier Freeway connecting Windsor to Québec via Toronto and Montréal. This is illustrated in Map 4.8, which demonstrates the twin-hub nature of the organization of the heartland around Toronto and Montréal. The information in this illustration not only identifies the arrangement of interaction around these hubs, but also the major routes that link the core economy to that of the United States.

3. As counterpoint to the concentration of population into the macro-urban strip, rural depopulation within the heartland continued, with large-scale farm abandonment from marginal lands in the sparsely populated Shield and Appalachian regions, and declining populations in small rural villages and towns located some distance from expanding urban areas.

4. The greatest surge in metropolitan growth occurred around Toronto and into southwestern Ontario, for while the broader Montréal region increased its share of the core population, the 65 km-radius region around Toronto increased its share even more. This shift in population balance toward the Toronto region is related in part to the growth of high value-added manufacturing, particularly

St. Lawrence
Seaway near
Cornwall

FIGURE 4.4 Rapid Suburbanization: Suburbanization in Laprairie County, South of Île de Montréal

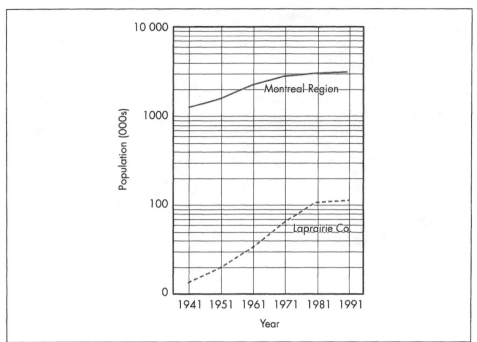

auto parts and assembly, in the area, and a gradual drift of financial and office activities from Montréal to Toronto during the 1960s.[29] This office-function drift reflected the increasing strength of the Toronto financial community.

5. At the same time as the population concentrated into metropolitan areas, there was also extensive suburbanization of population and some economic activities within these regions. The first great wave of middle-income suburbanization in heartland urban areas occurred between 1951 and 1971, associated with the housing needs of baby-boom families.[30] This is illustrated for one suburban area, Laprairie County, south of Montréal across the St. Lawrence River, in Figure 4.4. Whereas the rate of population growth for the Montréal region as a whole slowed down considerably during the 1961–71 decade, that of Laprairie County (particularly the town of Brossard) continued to boom.[31]

6. This great wave of suburban development was matched by much lower population growth and the beginnings of infrastructure disinvestment in the inner cities of many heartland cities.

Consolidation, therefore, brought with it extensive urban growth, the establishment of a twin-hub macro-urban region, and considerable suburban reshaping of urban areas within the heartland.

Downtown Toronto

A Period of Refocussing of the Heartland Population

Population shifts within the heartland in the 1971–1991 period, which reflect the changing economic and demographic situation, have been quite dramatic. Although aggregate population increase within the region has been quite considerable, within the heartland patterns of geographical change have resulted in a core in 1991 that in many ways is quite different from that in 1971.

1. Though there has been considerable population growth within the Montréal region, equivalent to the addition of another metropolis of half a million people in the twenty-year period, the greatest growth by far has been within the large Toronto region (Table 4.10). To the existing population of 3.4 million in the large area around Toronto in 1971 another 1.7 million had been added by 1991 (equivalent to the size of Vancouver). This crystallizes the effect of the relative shift of restructured manufacturing employment toward high value-added industries within the area, the accelerated shift of producer service employment toward Toronto, and the impact of heavy immigration and the baby-boom echo, which have stimulated the growth of consumer and government/public related service employment.

2. There has been a considerable increase in population in the regions involving Ottawa–Hull, and to a lesser degree the city of Québec. These increases are related directly to the growth of government and public service employment within these areas; and secondarily to a small increase, chiefly through back-

ward and forward linkages, in producer service and information processing high-tech industrial activity (Table 4.11).

3. During the 1971–1991 period heartland metropolises became suburban dominated. A further wave of suburbanization, especially during the 1980s, compounded and extended greatly the outward spread of urban areas. It involved generally larger per square-metre homes than were built during the 1960s, and has resulted in even greater metropolitan sprawl. This extensive suburbanization is having a number of repercussions:

    a. Suburban lifestyles and suburban needs are dominating issues relating to local government — as is witnessed by the various moves to reform the structure of local government within the Greater Toronto Area to achieve larger organizational units covering extensive suburbanized spaces.[32]

    b. The trend for some central-city, and many new, manufacturing plants to locate in suburban and "green field" sites during the 1951–71 period accelerated with the restructuring of manufacturing during the post-1971 period. This suburbanization of employment opportunities was reinforced by the continuing decentralization of consumer service and government and public service employment in accordance with the dispersion of population. Furthermore, there has also been significant dispersion of producer service employment, which had hitherto maintained a strong presence in downtown areas, to suburban office and shopping mall complexes sometimes generating "edge cities" of clustered employment opportunities. Heartland metropolises are now multicentred places.

TABLE 4.11   Concentration of the Experienced Labour Force in Government and Public Services in Ottawa–Hull and Québec, 1971 and 1991

CMA	1971 Population	% Share of National Population/Labour Force		
		Govt. & Public Services	Producer Services	Manufacturing
Ottawa–Hull	2.79	6.54	2.83	1.21
Québec City	2.23	3.65	2.01	1.33

CMA	1991 Population	% Share of National Population/Labour Force		
		Govt. & Public Services	Producer Services	Manufacturing
Ottawa–Hull	3.37	6.25	3.88	1.51
Québec City	2.37	3.58	2.47	1.51

c. This increased suburbanization, and more complex dispersed and multi-centred location of economic activities, has changed commuting patterns significantly — the volume of crosstown commuting in most heartland metropolises now exceeds suburban/central city commuting. This has greatly increased automobile use, exacerbated congestion, and impacted negatively on the use of public transit.

d. Given the spread of suburban forms within the heartland area, there is now much less distinction between rural and urban areas within the core. Urban realms reach much farther than they did, and exurbanisation,[33] hobby farms, second homes, and recreational arrangements serving essentially urban–suburban based populations now prevail through much of the less densely populated parts of the region. This trend has been assisted greatly by the decline of family farming, and its replacement with large corporate-managed agricultural units.

4. The older central cities in the heartland are experiencing difficulty in maintaining population and employment opportunities. Given that the population remaining tends to consist of a higher proportion of individuals and households in need, there are strong imperatives to reinforce existing procedures, and institute new financial mechanisms, that reduce tendencies for central city/suburban disparities with respect to access to opportunities and costs arising from the social safety net.

CONCLUSION

Canada has, therefore, moved rapidly from a simple heartland–hinterland structure that existed in 1971, in which southern Ontario and southern Québec formed the core and the rest of the country the periphery, to a more complex situation. The core is contracting to south and southwestern Ontario, with Toronto serving as the focus of financial, service, and industrial development. Although it is important to note that this contraction represents a continuation of trends in the heartland space-economy that commenced prior to 1971, perhaps as far back as the 1930s (see Chapter 3), there is little doubt that the series of post-1971 constitutional crises has reinforced the trend, particularly with respect to the location of producer services. It is unfortunate for Québec, and Montréal in particular, that domestic linguistic and constitutional upheavals occurred at the same time as massive global restructuring of finance and manufacturing.

The heartland–hinterland situation has also become more complex as part of the western periphery appears to be developing as an incipient core based on Vancouver, Calgary, and, perhaps Edmonton. The interaction data in Figure 4.1, and population and economic growth data, all suggest the emergence of an arrangement that in some senses is "hinterland," but in other senses is not. On the one hand these urban areas remain focussed on the organization of primary activities — there has been little manufacturing growth, particularly in fabrication and end products. On the other hand they are centres that are linked directly with the rapidly expanding aerospace and new information-technology producing centres of the Pacific Northwest (centred on Seattle-Tacoma). This link is, perhaps, becoming analogous to the long-standing integration of the economies of southern Ontario and southern Québec with adjacent areas in the United States. Also, as part of this extended Pacific Northwest urban agglomeration, growth and change in this incipient core is becoming more entwined with the burgeoning economies of many countries around the Pacific Rim.

NOTES

[1] The economic role of the heartland by 1971 is defined and discussed in M. Yeates, *Main Street: Windsor to Québec City* (Toronto: Macmillan of Canada, 1975).

[2] Cole, S. "Indicators of regional integration and the Canada–U.S. Free Trade Agreement," *Canadian Journal of Regional Science* 13, 1991, 171–178.

[3] Foot, D.K., and D. Stoffman, *Boom, Bust, and Echo* (Toronto: Macfarlane, Walter, and Ross, 1996).

[4] It should be noted that, unless otherwise indicated, all figures and tables in this chapter are taken from M. Yeates, *Land in Canada's Urban Heartland*, Ottawa: The Lands Directorate, Environment Canada, 1985; idem, "The Windsor-Québec City: Basic Characteristics," *Journal of Geography* 83, 1984, pp. 240–249; and idem, "Urban Canada: Changing Core/Periphery Relations," in L. Bruti Liberati and M. Rubboli (eds.) *Canada e Italia verso il duemilia: metropoli a confronto* Fasano: Schena, 1994, pp. 49–75.

[5] For more detailed discussions of the definition of the heartland see: op. cit., 1975, pp. 4–25; and M. Yeates, "A Note on the Delimitation of the Extent of Urban Development in the Windsor-Quebec City Axis," *The Canadian Geographer* 31, 1, 1987, 64–69.

[6] This is partially due to the declining price of raw materials relative to that of manufactured products during the 1982–93 period.

[7] 34 per cent of Canada's manufactured end products imported in 1993, and 47 per cent of manufactured end products exported.

[8] The political and economic origins of the Auto Pact, the continental integration of the automobile industry, and its part in the formation of the Free Trade Agreement, are discussed in J. Holmes, "The Continental Integration of the North American Automobile Industry: From the Auto Pact to the FTA and Beyond," *Environment and Planning* A 24, 1, 1992, 33–48.

[9] A United States perspective on the geographical impact of this agreement is contained in D. J. Hayward and R. A. Erickson, "The North American Trade of U.S. States: A Comparative Analysis of Industrial Shipments, 1983-91," *International Regional Science Review* 18, 1, 1995, 1–32.

[10] Population projections for metropolitan areas and provinces are summarized in Statistics Canada, "Population Projections for Census Metropolitan Areas, 1995 to 2000," *Canadian Social Trends* Winter, 1996, p. 7; and Statistics Canada, *Population Projections for Canada, Provinces and Territories, 1993–2016*, Cat. No. 91-520-XPB.

[11] Kingston is the only urban area in this list (and in Table 4.3) not defined for statistical purposes by Statistics Canada as a CMA — it is an Urban Agglomeration (UA).

[12] Without wishing to preach, and therefore as a footnote, this contrast between the two highly heterogeneous major metropolises and the rest of the heartland is extremely important. Maintenance of harmony in highly heterogeneous communities requires special tolerance and understanding from all peoples and levels of government, with public policies that recognize and respect difference as well as promote cohesion.

[13] Estimates obtained from: Statistics Canada, *Industry Trends, 1951–1986*, Cat. 93-152.

[14] This, and other traditional manufacturing regions of Québec, are defined in J. C. Thibodeau and T-M Holz, "Etude spatiale de la structure l'industrie manfacturiere au Québec, 1961–1971," *Annuaire du Québec 1977/78*, 988–98.

[15] For a discussion of restructuring see, M. Best, *The New Competition: Institutions of Industrial Restructuring* (Cambridge MA: Harvard University Press, 1990).

[16] The role of the thinking of W. Edwards Deming in the development of the Toyota production system is described in D. Halberstam, *The Reckoning* (New York: William Morrow and Co. Ltd., 1986).

[17] For a discussion of the impacts of restructuring on collective bargaining in the automobile industry see J. Holmes and A. Rusonik, "The Break-up of the International Labour Union: Uneven Development in the North American Auto Industry and The Schism in the UAW, *Environment and Planning, A* 23, 1, 1991, 9–35.

[18] Because of the modification described in note 2 to Table 4.1, these figures should be regarded as estimates. Furthermore, the annual totals (in thousands) in Figure 4.3 also exclude a category listed by Statistics Canada as "other."

[19] The *primary* sector includes agriculture, forestry, fishing, trapping, mining; *manufacturing* includes processing, fabrication, and assembly; *services* include retail and wholesale trade, FIRE (finance, insurance, and real estate), business and personal services, health and educational services, public administration, and defense; *infrastructure* and *interaction* includes construction, transportation, and communication.

[20] Although this point will be developed further in the chapter, greater detail can be found in chapters 6 and 10 of M. Yeates, *The North American City* (New York: Addison Wesley Longman, Fifth Edition, 1997).

[21] See W. J. Coffey, *The Evolution of Canada's Metropolitan Economies* (Montréal: Institute for Research on Public Policy, 1994), p. 13.

[22] Business services are highly diverse. The category includes advertising agencies, providers of information technology functions and systems, personnel supply services, security services, and a variety of consulting firms in such areas as business management, strategic planning, employment equity, market research, and so forth.

[23] Wholesale trade, usually involving the distribution of products from warehousing facilities to end users, is sometimes placed in the consumer service grouping as this is the category with which it is traditionally linked.

[24] W. J. Coffey, "Forward and Backward Linkages of Producer-Services Establishments: Evidence from the Montréal Area," *Urban Geography* 17, 7, 604-632.

25 Though, as is indicated later in the chapter, the Toronto CMA has considerably less employment in this category than might be expected based on its population size — suggesting that considerable dispersion of government and other types of public service employment has occurred to other parts of Ontario.

26 A great deal of work on changing headquarter locations in the Canadian urban system, particularly with respect to the financial industry, has been done by R.K. Semple. For a useful summary of this work see: R.K. Semple, "Quaternary Places in Canada" in J. N. H. Britton (ed.) *The Canadian Space Economy* (Montréal: McGill-Queen's Press).

27 Summary core non-financial concentration percentage estimated from data in M.D. Rise, "Functional Dynamics and a Peripheral Quaternary Place: The Case of Calgary," *Canadian Journal of Regional Science* 14, 1, 1996, 65–82, Table 1.

28 They also correspond to the two periods of growth identified, rather dramatically, by Hobsbawm with respect to the North American and European economies as the "golden age" (1945–1975), and the "landslide" (1975–c. 1994): E. Hobsbawm, *Age of Extremes: The Short Twentieth Century, 1914–1991* (London: Michael Joseph, 1994).

29 For documentation of this 1960s drift of office and financial activities from Montréal to Toronto, see M. Yeates, op. cit., 1975, pp. 320–334.

30 For a discussion of suburbanization, inner city decay, and housing policies, especially in cities in Ontario, see J. Sewell, *Houses and Homes* (Toronto: Lorimer, 1994).

31 The impact of this type of rapid suburbanization on the consumption of rural land, and an analysis of a response from the Québec government to this issue, is discussed in: E. P. Reid and M. Yeates, "Bill 90 — An Act to Protect Agricultural Land: An Assessment of its Success in Laprairie County, Québec," *Urban Geography* 12, 4, 1991, 295-309.

32 In part in order to overcome some of the more egregious aspects of local government, and central city/suburban, competition for acceptable property tax generating economic activities and households. See *Report of the Greater Toronto Area Task Force* (Toronto: Queen's Printer for Ontario 1996).

33 Exurbanisation is extending into even the most agricultural of counties in southern Ontario: S. Davies and M. Yeates, "Exurbanisation as a Component of Migration: A Case Study in Oxford County, Ontario," *The Canadian Geographer* 35, 2, 177-186.

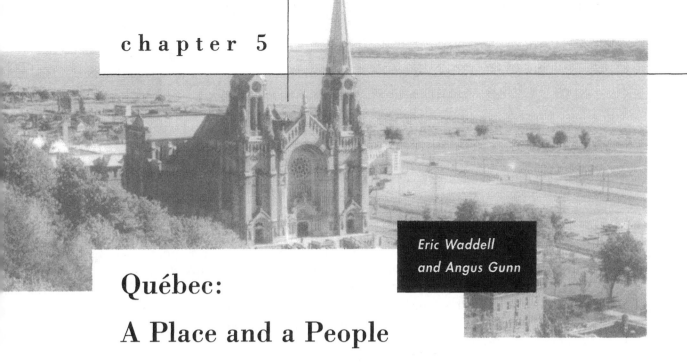

chapter 5

Québec:

A Place and a People

Eric Waddell
and Angus Gunn

At the present time, almost everywhere it seems, there is a recurring clamour for independence within groups of people that feel restricted, even marginalized in some cases, by the larger society of which they have long been a part. The cry for political autonomy by the Scots, the civil war in Sri Lanka as the Tamils fight for an independent state, and the frequent upheavals in African states, all are examples of this phenomenon on the world scale. This is not the place to deal with the causes of these widespread developments. Their consequences, however, are relevant to any study of Canada's regional geography.

World capital is now mobile to a degree that has never been seen before, and the realities of the global economy create demands for capital if enterprises are to remain up-to-date and competitive. But, and this is an important factor, the people who move capital from place to place have a distaste for political instability. They like social tranquillity. Is that why, recently, Québec was less successful than Ontario in attracting capital? (See Table 5.1)

Québec presents a rare example of group dissatisfaction with the larger society, rare because it is not just a simple case of one people seeking freedom; instead, it is a double cry for independence — the Crees of Québec from their province, and the province from Canada. The arguments are strong and persuasive from both parties. It must be said at once that there is nothing inherently wrong with a culturally defined group seeking a greater measure of freedom. After all, many of the world's greatest achievements began with a minority revolting against

TABLE 5.1   Regional Share of Canadian Capital, 1981 and 1991

	1981 (%)	1991 (%)
Atlantic Region	7.8	7.6
Québec	22.2	22.0
Ontario	32.0	34.4
Western Interior	24.5	22.7
British Columbia	13.5	13.3
Total	100	100

an oppressive structure within which it felt trapped. Pejorative language against such group demands is entirely inappropriate.

Aboriginal societies all over the world have a sad history of brutality at the hands of the more developed, better-armed invaders. Their stories are well known in Latin America, in Africa, and in Canada. It is only in very recent times that Canada came to terms with its past mistreatment of Native peoples and sought to make amends. Now, with the prospect of further injustices looming on their horizon, the Crees of Québec are speaking out for a recovery of the freedom they were promised when Europeans first came to this continent, a freedom and an autonomy that still awaits them. In 1995, speaking to the Canadian Club in Toronto, on the eve of the Québec referendum, Matthew Coon, Grand Chief of the Crees, said this:

> *We Crees have lived in the James Bay area since the last ice age. We have always identified ourselves as one people with our own language, laws, beliefs, and system of land tenure and governance. We gave our lands a name long before there was a Canada or a United States. Again and again, ever since 1670, new governments have made decisions about our land without ever consulting us or informing us of their decisions. Now, in the context of yet another plan, that of independence for Québec, I am speaking to defend Cree rights. There will be no annexation of ourselves or our territory to an independent Québec without our consent. We cannot simply be traded from country to country as though we were livestock in a field.*

The province of Québec has a similar argument, based on history like that of the Crees, to support its demand for independence from Canada. The narrow failure of the 1995 referendum only spurs the province into more determined efforts with a view to succeeding next time. This chapter approaches the history and aspirations of the people of Québec by first reflecting on the main events of the past one or two hundred years, and then tackling some of the issues that the province must face if it attempts to change into an independent sovereign state.

Even a cursory review of the major geographical statements on Québec gives the impression of a province distinct from the rest of North America, of a place and a people all too easy to describe and characterize.[1] It was a mixture of Frenchness, and hence sentimental considerations, and a readily manageable settlement history that attracted the spiritual father of Québec geographers, Raoul Blanchard, to the St. Lawrence Valley.[2]

For Blanchard, Québec was synonymous with French Canada; a province in which a distinctive people had settled, survived, and prospered in spite of immeasurable odds. More recently, Paul Claval has provided much the same characterization: Frenchness, Catholicism, and a unique historical experience — hence a distinctive culture — demarcate, if not isolate, the province from the rest of North America.[3] Pierre Biays, too, highlights Frenchness, plus language, and the creation of a distinctive rural society dominated by the seigneurie and the rang.[4]

Time and again scholars invoke the cultural specificity of Québec — the interweaving of a European past with a North American present — to create a region characterized by a distinctive landscape and society in which the long lot and rurality still figure prominently. The land has shaped the actions and decisions of generations of French Canadians. The availability or lack of land has determined whether they have remained in the rural areas, migrated to the cities, or left for other parts of North America. The fact that for generations rural fathers tried to give many of their sons at least some agricultural land has reinforced the importance of a rural tradition. The subdividing of land into smaller and smaller strips has also preserved the traditional seigneurial landscape of long lots.[5]

SETTLEMENT IN FRENCH CANADA

French Canadian settlement began as a society with a marked degree of internal cohesion and solidarity based on such considerations as language, religion, a common historical experience, a referential universe set in the St. Lawrence Valley, and a plethora of "national" institutions. Early in the 1600s, Catholic missionaries and others began to occupy land along the shores of the St. Lawrence River, west of what is now the city of Québec, the same area that now houses most of the provincial population (see Map 5.1).

Most of the best agricultural land in the province is found in this triangular region between Montréal and Québec. It is a relatively small area, about 100 km wide at Montréal, narrowing to less than 30 km at Québec. To the north is the mountainous border of the Canadian Shield and to the south the high ground of the Appalachians. The influx of people from France was initially slow. Then in the last quarter of the seventeenth century came a major push for settlement by King Louis XIV. Under the direction of a governor who represented the king, large tracts of land were granted in perpetuity to seigneurs, who in turn subdivided their territory into smaller parcels for occupation by tenant farmers.

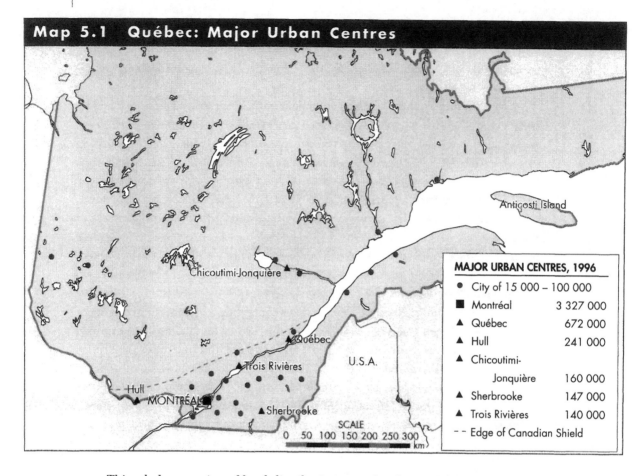

Map 5.1   Québec: Major Urban Centres

MAJOR URBAN CENTRES, 1996

●	City of 15 000 – 100 000	
■	Montréal	3 327 000
▲	Québec	672 000
▲	Hull	241 000
▲	Chicoutimi-Jonquière	160 000
▲	Sherbrooke	147 000
▲	Trois Rivières	140 000
- -	Edge of Canadian Shield	

This whole operation of land distribution was closely supervised by agents of the Crown. The initial allocation to a seigneur was marked by a series of ceremonial procedures at the Castle in Québec: homage was formally paid to the king, promises made to provide regular inventory of people and livestock on the seigneur's lands, assurances given that he would settle as many immigrants as possible, and readiness declared to defend against all intruders. The Crown reserved the right to build forts and roads where needed. The smaller rectangular-shaped parcels of land for tenants had their shortest dimensions on the river. The rest stretched inland from there so that everyone received some good land — on the river front — and some less fertile farther back. The now familiar rang pattern of land use emerged from these beginnings. The strictness of the whole enterprise was like that of the old feudal systems of Europe, especially those of France.

Farming was the economic basis of the new society all through the pre-Conquest period. As late as 1867, at the time of Confederation, when the province's total population was just a little over one million, life was overwhelmingly rural in character. Most of the population — 85 per cent — derived their livelihood from agriculture. Montréal and Québec were the only places worthy of the name urban. Farming continued to be a vital part of the economy right up to the time of World

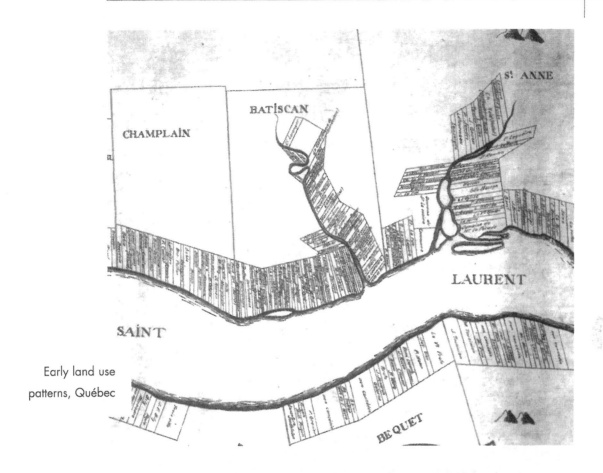

Early land use patterns, Québec

Seigneurial landscape in Québec

War II. Thereafter the impact of industrial development began to interfere with traditional rural living styles (see Table 5.2).

TABLE 5.2   Farm Population and Total Farming Area in Québec, 1951–1991

Year	Farm Population (000)	Total Farming Area (000 km²)
1951	770	42
1971	300	37
1991	132	34

The classic view of Québec's ethnic identity arises from these early formative influences. They were articulated first by the ideologues of the Roman Catholic Church, and subsequently by nationalist historians expressing notions of race and racial purity, of French people separated from their metropolitan hearth, and of the integrating forces of language and religion.[6] Grafted on to these notions are others, of Latin-ness, and of the idea of French Canadians being bearers of a rich cultural tradition, or civilization. In the eyes of the church these qualities rendered the French Canadians a "chosen people" endowed with a "providential mission" in North America, first to civilize and to christianize the Native peoples and later to win Protestants to the Catholic faith. Loyalty was linked with rurality and attachment to the land, evoking the image of a peasantry bound to the soil. For the peasantry, the church and the parish were enduring institutions.

It was inevitable that the regional context of many subsequent analyses of French Canadian society would be limited to a clearly defined geopolitical realm conceived as a stronghold into which a defeated people withdrew to reconstruct its future, inspired by institutions and a rural experience derived from the earlier regime. Conceived in this manner, Québec and French Canada are necessarily synonymous — Québec alone is French America. But a number of social scientists, including geographers, are now expressing dissatisfaction with a vision of a people that is so truncated in time and space. With regard to race, some insist that as much Amerindian–French intermingling occurred in the St. Lawrence Valley during the French regime as took place later in the Western Interior.[7] The only difference was that the situation in Québec did not give birth to a "New (or Métis) Nation."[8]

On the surface, at least, there is a profound contradiction in these two visions of Québec society. One presents French Canadian society as rooted firmly in the past, tightly circumscribed, almost hermetic, and hence profoundly Québécois. The other views the people of Québec as fluid, nomadic, responding to opportunities across the continent, and hence resolutely franco-américain. Yet, on closer reflection, these visions can be seen as two powerful and distinctive forces generating a basic dialectic that has pervaded French Canadian society since its inception. The result is the simultaneous existence of the habitant (the sedentary peasant) and the coureur de bois (the nomad) in the history of a people. This vision also leads to the duality, ambivalence, and conflict that is symbolized in such literary classics as *Maria Chapdelaine* and *Le Survenant*.

In such a universe, where mobility is practised and valued by at least a part of the population, the St. Lawrence Valley emerges less as an ethnic bastion, that is, as a finite or limiting space on an alien continent, and more as a referential hearth — a place to which the migrants may return, or where they have historic roots and distant family connections. Viewed from this perspective, Québec emerges as a cultural hearth having intimate and enduring links with the rest of Canada.

The framework that assures this French unity is typically one of kin and friendship, a solidarity born of a shared experience that, because of the weakness of institutional arrangements beyond the frontiers of Québec, has an almost clandestine quality about it. Ethnic identity and solidarity, instead of being articulated around institutions whose legitimacy is constantly questioned and whose ability to survive is weak, is based on language, food, and music, all of which gravitate around family and kin.[9]

The "family," of course, is infinitely variable in size. At one extreme it is simply a set of indentifiable kin. At another it is the massive *réunions de familles* that the Church or the province have organized from time to time in the city of Québec: the Congres de la langue française of 1912, 1937, and 1952; and the annual Fètes du Retour aux Sources, first organized in 1978. There are also associations of people bearing the same family name that have a continental membership, and are united by a common interest in family history and recognition of Québec, and more particularly of the Île d'Orléans and the Côte de Beaupré, as being their cultural hearth.

To map such a space is almost impossible, for it is a territory without frontiers occupied by a people constantly on the move. At best it can be conceived as a hearth in Québec and a diaspora across the country, a Québec no longer defined in terms of legal or political history and the affairs of government, but rather in terms of a collective ethnic experience.

EXPANSION AND RETREAT IN NORTH AMERICA

Early French colonists forged a new identity in the St. Lawrence Valley, attempting on the one hand to reconstruct their ancestral society based on a loyal peasantry, and turning on the other to the call of the wilderness that opened up invitingly beyond the seigneurial lands. Inhabited by Native peoples and abundant game, the wilderness offered unlimited opportunities for hunting and trading. It was this mixture of autarchy and mercantilism, and the discovery, in the Native peoples, of a new and egalitarian society, that precipitated a sense of being "of this continent," of being Canadian rather than French. To be Canadian to this group meant to be individualistic, mobile, unattractive, to seek the company of Native peoples, to value

leisure and kin, and to fill, gradually, an economic niche as the intermediary between a Europe living in the age of mercantilism and the largely self-sufficient lifestyle of the Native peoples. Such are the roots of a people whose indifference to, if not outright rejection of, authority has resurfaced time and again in Québec.

For such a people, little was changed by the Conquest other than the language and the country of origin of the authorities and the bourgeoisie. If anything, the more aggressive commercial policies of the British and the much less restrictive licensing systems of the fur trade facilitated the diffusion of French Canadians out of Québec and across half a continent. Those who chose to remain in the Great Lakes area after the departure of the French were soon joined by others in the employ of the two major fur trading enterprises, the Hudson's Bay Company and the North West Company. The latter operated out of Montréal and employed, almost exclusively, French Canadians as voyageurs, guides, interpreters, servants, and, to a lesser extent, as clerks and even managers of its trading posts.

Practising aggressive trading policies, the North West Company urged its agents to live among the Native peoples who supply the furs. In the process, these fur traders forged a separate ethnic identity based initially on a simple wintering in the country and later on a cultural and then a racial merging with the Native peoples. The result was a distinctive people, the Métis, who were centred in the west but nevertheless shared a great deal with the people of the St. Lawrence Valley hearth. Although the Métis language was Cree and several other Indian tongues, it was also French. Their religion was Roman Catholicism. They valued the liberty of mobility, of being coureur de bois. They became, in a sense, the ultimate expression of the mobility so highly valued by the Québécois society of the time.[10]

By the early nineteenth century, a Métis "empire" had been created by French Canadians leaving the St. Lawrence Valley either individually or in small groups for the *pays d'en haut* and the *pays des Illinois*. Dozens of small settlements were established around the Great Lakes, along the Red, Assiniboine, and Saskatchewan rivers, and in the Upper Mississippi and Missouri valleys. Their names include Sault-Ste-Marie, Rivière Raisin, Fort Wayne, Prairie du Rocher, La Baie, St-Paul, St-Cloud, Pembina, St-Norbert, Batoche, Prince-Albert, Saint-Albert, Fort Benton, Le Havre, and Lewiston. (See Map 5.2.)

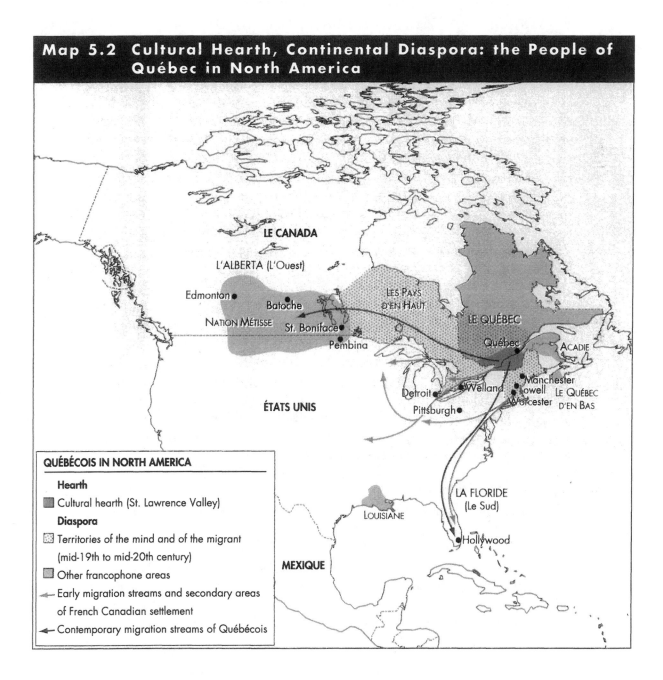

Map 5.2 Cultural Hearth, Continental Diaspora: the People of Québec in North America

QUÉBÉCOIS IN NORTH AMERICA

Hearth
- Cultural hearth (St. Lawrence Valley)

Diaspora
- Territories of the mind and of the migrant (mid-19th to mid-20th century)
- Other francophone areas
- Early migration streams and secondary areas of French Canadian settlement
- Contemporary migration streams of Québécois

The notion of migration was strengthened later in the nineteenth century when successive economic frontiers carried other French Canadians westward, imbued by the continuing tradition of mobility and the possibility of material gain. Lumbering attracted migrants to the Saginaw, Sable, and Muskegon valleys and the towns of Grand Rapids and Bay City in Michigan. Mining in the copper region of the Keweenaw Peninsula and across the iron fields of the Marquette Range in Wisconsin beckoned others. Finally, some were attracted by the farming in such widely scattered places as Georgian Bay (Penetanguishene), Welland, Monroe Country near Detroit, and Bourbonnais in Illinois.[11]

As the agricultural frontier moved westward during the nineteenth century, so did some French Canadian settlers, leaving the Red River Valley to move west across the Western Interior, south into Kansas and Minnesota, and beyond. But these groups never numbered more than a few hundred, and by the second quarter of the nineteenth century the main thrust of migration out of the St. Lawrence Valley had shifted direction, toward the factories and mill towns of New England. Cultural attachment to the hearth in Québec would continue to sustain western settlements, but increasingly the Western Interior became an economic hinterland of Montréal, dependent more on the capital and business organization of the metropolis than on cultural attachment. Hence it slowly became a region distant from and only tributary to Montréal's ethnic elite. The Scots, with the capital and knowledge they had accumulate in the fur trade, became instead the major promoters of western development.

People had been trickling into northern New England from Québec since at least the beginning of the nineteenth century, either to settle as farmers in northern Vermont and the Saint John River Valley of Maine, or to work in the lumbering camps of northern Maine. By mid century, however, this trickle had become a flood, creating by the end of the century a second national foyer, or hearth, in the industrial towns and cities of New England. This *Québec d'en bas* took root above all in the mill towns — Manchester, New Hampshire, Lowell, Fall River, Worcester, Lawrence, and Holyoke, Massachusetts, Lewiston and Biddeford, Maine, and Woodsocket, Rhode Island. By 1930, when Québec had a population of about 3 000 000, New England was home to between 700 000 and 1 000 000 French Canadians who had left Québec to work south of the border. This massive displacement had far-reaching demographic and sociological effects.[12] A largely illiterate rural peasantry, living on the verge of autarchy, riddled by frequent agricultural crises, indebtedness, and overpopulation, viewed with enthusiasm the rapid industrialization of New England and, in particular, the development there of a flourishing textile industry. Significantly, the mill towns were readily accessible from rural Québec. One could even walk there, and certainly travel by cart. With the opening of the Northern Railway Company of New York in 1850, linking Montréal with Boston, and of the St. Lawrence and Atlantic Railway in 1853, connecting Montréal and Portland, Maine, these journeys were transformed into a short and cheap train ride across an open and unpoliced international boundary.

An additional attraction was the availability of work for all members of the family. The manual skills required in the textile mills were more or less familiar, and certainly readily accessible, to a people raised in a social context where carding, spinning, and weaving were common. In the major textile centres of Woonsocket and Lowell, for example, these skills ensured a steady salary and new-found material wealth — the opportunity to dress in a *habit de dimanche* every day of the week. American factory owners were eager to employ what were generally considered to be industrious and docile workers — the "Chinese of the East" — and recruiters regularly travelled to Québec in search of labour. The Depression, the closing of the international boundary, and the erosion of the industrial strength of New England, together with the gradual development of an industrial base in Québec, ended the massive migration.

In the eyes of the Québécois, the mill towns of New England offered prospects entirely different from the resource frontiers of Western Canada and the American Midwest. To go west was to pose an irrevocable gesture, for distance almost eliminated the possibility of return. Moreover, federal government policies did little to ensure the strengthening of French Canadian communities in the Western Interior. Newly arrived immigrants from Europe received free rail passage to the frontier from Québec and Montréal, but no such privileges were offered to those Québécois seeking an alternative to unemployment and rural poverty. Cost and distance forced their attention on New England. Movement between Québec and New England, either to visit or to settle, was entirely feasible and an organic relationship, at the level of family and community, developed between these two foyers.

Such was the scale of the out-migration and the intensity of the relations between francophone communities on either side of the border that by the turn of the century Québec society had been radically changed. Its centre of gravity had shifted away from the St. Lawrence Valley to sit astride the international frontier, thereby providing a link between two distinct, but complementary universes: a rural peasantry in Québec and an urban proletariat in New England. Not surprisingly, political, intellectual, and cultural leadership for this single French Canadian nation came from New England until at least the late 1920s, and lingered on until World War II. At the turn of the century, Woonsocket ranked second only to Québec as the city with the largest proportion of francophones in all of North America. Calixa Lavallee, the author of Canada's national anthem, was a native of Lowell, while Henri Bourassa, founder of *Le Devoir* and an ardent nationalist, frequently crossed the frontier in the course of his career. The internal cohesion of French America was strong.

The integration of the two national foyers into a single ideological realm was facilitated by another critical development — the return of the Roman Catholic Church as a shaping force in society by the mid-nineteenth century. The Catholic Church in Québec suffered as a result of the Conquest, through challenges to its authority, loss of clergy, and damage to property.[13] In 1838, for instance, there were fewer clergy than at the time of the French regime, yet the population had increased from 60 000 to 500 000.

Ironically, this period of isolation encouraged a new, Canadian, identity and hence the emergence of an "Église nationale" that determined the actions of the Church for a hundred years or more. It was to this nascent "national church" that many foreign religious orders gravitated from the late 1830s on, bringing with them human resources and a conservative ultramontane ideology that wedded well with the sentiments of the Catholic Church in Québec. The Frères des Écoles Chrétiennes appeared in 1837, the Oblats in 1841, the Jésuites in 1842, and the Clercs de St-Viateur and the Ordre de Ste-Croix in 1847, initiating a movement away from revolutionary France that was to continue until the end of the century. The number of priests in Québec multiplied accordingly, rising from 225 in 1830 to 2465 in 1910.

The impact of the Church on the population both inside and outside the boundaries of Québec was far-reaching. By their very presence the various religious communities formed an articulate, indigenous elite. Of more importance, the Church provided the structure for a French presence in North America that this elite could serve and promote. On the one hand, the Church provided an explicit ideology that ascribed a vocation to the French Canadians; on the other, it furnished the institutional framework for the establishment, survival, and development of French Canadian communities across the continent. For the architects of Catholicism there quite clearly existed a French race in North America whose origins and spiritual hearth lay in the St. Lawrence Valley. The French were, moreover, a chosen people, having a providential (religious and civilizing) mission first among the Amerindians, but subsequently among all the peoples that inhabited the continent.[14] Evidence of this vocation was furnished by the numerical strength of the people and by their continental mobility. They shared a loyalty to religion and language, and a marked sense of ethnic solidarity.

Inevitably, then the Church sought to cater to the needs of French Canadians wherever they might be, and even to channel and direct their migrations across the continent. What had been a simple parish within Québec became, beyond its frontiers, a *paroisse nationale* that provided the population directly or indirectly with a broad set of social and economic institutions — church, parochial school, convent, orphanage, mutual aid society, credit union, cooperative. Classical colleges were established and national organizations such as the Union Saint-Jean-Baptiste d'Amérique emerged.[15] An individualistic people were thereby provided with an institutional framework that generated not only a collective identity, but also the notions of a foyer in Québec and of a fluid diaspora scattered across a continent.

By the last quarter of the nineteenth century, French Canadians had established themselves across a continent. There were migrations of people out of the St. Lawrence Valley to New England, the American Midwest, and the Western Interior, reinforced by the movement of priests, students, and politicians through the network of institutions, and symbolized in the frequent patriotic rallies in Montréal and Québec.

Yet, at this very time when the idea and expression of a French America reached its apogee, it was also being challenged, and the fact that the francophones outside Québec were generally few in number, scattered, and organized around the church facilitated the challenge. North America was a secular and not a sacred space by the end of the nineteenth century, and power in the Roman Catholic Church was increasingly being shared with another powerful ethnic group, the Irish. The diaspora was progressively weakened and largely assimilated by defeat after defeat: the hanging of Riel in 1885; the elimination of French from the Legislature of Manitoba in 1890; the elimination of French education in Ontario in 1912, Manitoba in 1916, and Saskatchewan in 1929; and the Sentinellist crisis in New England in 1928–29, which led to the elimination of French-language parochial schools.

THE CANADIAN CULTURAL HEARTH

Even as a French America had been taking shape in the late nineteenth century through massive out-migration from Québec and through the creation of an institutional structure comprised of a thousand national parishes, numerous political and religious leaders were voicing their concern.[16] Within Québec voices spoke out increasingly against out-migration for fear that it would irrevocably weaken the demographic, economic, and political viability of the hearth. For almost a century, from 1870 to 1950, the intellectual and clerical elite in Québec promoted a myth of the north, a vision of a promised land where French Canadians were destined to settle, and where spiritual and material regeneration were to be assured through a tactical retreat from Anglo-America. Their project was the agricultural colonization of the Laurentian Shield, a vir-

Lake St. John
farm, Québec

gin territory which for some was limited to Québec, but which for others extended westward to join with the isolated Métis and French Canadian settlements of the Western Interior.[17] In seeking to focus the energies of a nation within a more clearly circumscribed territory, the leaders of this movement aimed to redefine the collective identity. Nomads had to be sedentarized, or at least carefully oriented in their movements. Hence, an old social type re-emerged, the habitant, or peasant, to be glorified by attachment to the land, a land now considerably enlarged (see Map 5.3).

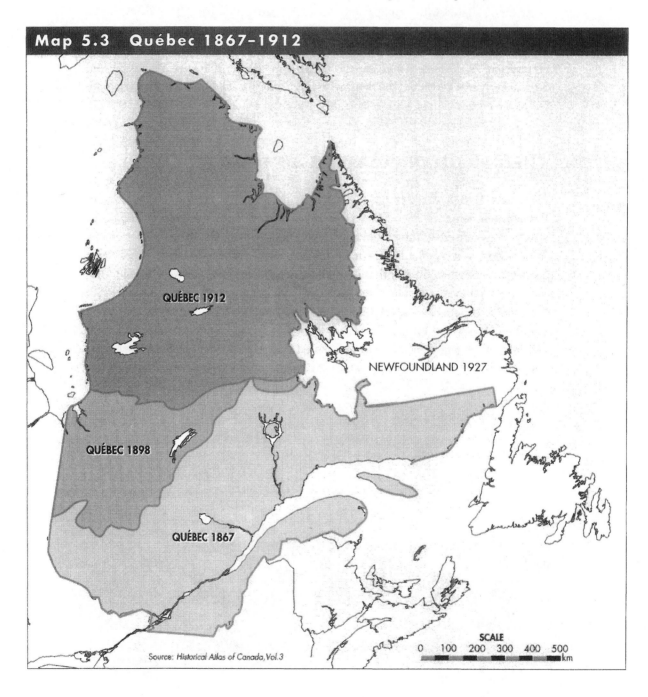

Map 5.3   Québec 1867–1912

QUÉBEC 1912

QUÉBEC 1898

QUÉBEC 1867

NEWFOUNDLAND 1927

SCALE

0   100   200   300   400   500
km

Source: *Historical Atlas of Canada*, Vol. 3

This dream of salvation and regeneration in the north failed to materialize, but there took root the idea of a people established for generations in the St. Lawrence Valley, impermeable both to history and to the dictates of a new continent — the sentiment of a fortress that was Québec itself. By the middle of the twentieth century the diaspora, and therefore the vision of Québec on the continent, had virtually disappeared. Changes in migration patterns did much to account for this. The Depression had defined the boundary between the United States and Canada more sharply, and a rapidly industrializing and urbanizing Québec generated new and abundant employment opportunities in the mining towns of the Canadian Shield and in the Montréal metropolitan area.

World War II reinforced these trends. There was no longer any need for the people of Québec to migrate beyond the boundaries of the province in search of work. With the virtual collapse of the Roman Catholic Church, the institutional arrangements linking Québec with the diaspora were also eroded. Some two thousand priests were ordained in Québec in 1947, but by the early 1970s the number had fallen to less than one hundred. Consequently, the church was less able to play an active role in the education, health, and social welfare of Québec society within the hearth, or to provide tangible aid and direction to the diaspora.

French Canadians were also in the process of redefining their collective identity — as Québécois — to bear witness to the new secular authority, the one North American place in which they constituted the majority of the population. It had become urgent to exercise political power in an aggressive manner "at home" because doubts were being raised about the long-term survival of the French fact even in the St. Lawrence Valley. The farm, the network of kin, and the parish were no longer the main "action space" of the French majority; now it was the factory, the office, and the suburb where the French came into direct confrontation with the English minority. English was commonly the language of work and economic power and was welcomed by most immigrants and even by some French Canadians. Demographers warned that the Montréal metropolitan area, containing almost half of Québec's population, could soon become predominantly English-speaking.

The relationship with a continental diaspora played a significant role in shaping the character of Québec. Its government is not only a provincial government. It also represents a linguistic community whose geographical limits are not clearly defined.[18]

During the late 1970s and early 1980s, a new wave of migrants left the hearth, in part to search for new economic opportunities and in part as a reaction to a Québec that was too tightly circumscribed. Anti-authoritarianism and concern for mobility were reaffirmed as the marks of a collective identity.

Migrants to Florida stem largely from the affluent francophone middle class that established itself during the Quiet Revolution. Those migrants represent an older generation in search of leisure and others seeking new economic opportunity. They resemble those of earlier departures from the St. Lawrence Valley. Starting as tourists spending a week or two in winter in a hotel catering to Québécois, these

migrants become *hivernants*, residing from two to six months in a rented or purchased condominium or mobile home. Finally, they take up permanent residence
and shift all their assets. The bases of group solidarity are much the same as
before: ties of language, kin, and friendship. These ties are further expressed in
Catholic masses, mutual aid, and a host of other social activities.

This new generation of Québécois migrants is concentrated in distinctive
neighbourhoods or mobile home parks in places such as Hollywood, Pompano
Beach, Fort Lauderdale, and Miami Beach. The *Soleil de la Florides*, a weekly
newspaper published by Québécois interests, who claim to have the largest North
American circulation of a French-language newspaper of its kind outside Québec,
estimates that there are at least 250 000 Québécois permanently residing in
Florida. In this new environment, they intermingle with Franco-Americans, many
of whom arrived as early as in 1920s from New England, attracted by economic
opportunity.

QUÉBEC: HOMOGENEOUS OR PLURAL?

If Québec is at once the only province or state in which power is effectively exercised by francophones, it is nevertheless not entirely French. According to the
1991 Census, approximately 80 per cent of the province's population is French and
20 per cent is made up of anglophones and allophones combined. From a political
perspective, as we shall see shortly, this 20 per cent poses a major challenge to
Québec. But first, there are some demographic trends that need to be addressed
because they have long-term significance. In the early years of French Canada,
birth rates were high so that natural increases alone, without immigration, ensured
both a large and a young population. This is no longer the case.

The declining influence of the Roman Catholic Church, which encouraged
large families, the trends of rural depopulation and increasing urbanization,
together with the social changes often described as the Quiet Revolution, have all
tended to reduce the sizes of families. These developments made the government
of Québec accelerate the rate of immigration, but in spite of this move Québec's
place in Canada slipped while Ontario's rose. Between 1971 and 1995 Québec's
proportion of the Canadian population dropped from 28 per cent to 24 per cent
while Ontario rose from 36 per cent to 37 per cent.

Alongside of the drop in population must be placed the shift in age structure.
An aging population is costly to any society because it raises medical expenditure
and lowers the economic contribution of those over 65 years of age. As will be evident from Table 5.3, the smaller natural growth rate has already placed in Québec
in the lowest category of all provinces regarding numbers in the 0 to 14 age category. Projections for the twenty-first century indicate a continuation of this trend.

**TABLE 5.3   Canada and Québec: Percentage of Population
in Different Age Groups, 1971 and 1995**

Year	Area	Population (millions)	Ages 0–14	Ages 15–64	Ages 65+
1971	Canada	22	29.3	62.7	8.0
	Québec	6.2	29.3	63.9	6.8
1995	Canada	30	20.2	67.8	12.0
	Québec	7.3	19.1	69.0	11.9

Following World War II and until the enactment of the *Chartre de la langue francaise* in 1978, the majority of allophones gravitated toward the anglophone population in their choice of a second language and of schooling for their children. The effective leverage of the minorities is enhanced by the fact that they are heavily concentrated in the Montréal metropolitan area, making this city a plural urban society in the heart of a massively francophone province.

The roots of the English and the Scots in Québec can be traced to the immediate post-Conquest period, and those of the Irish to the early nineteenth century. Their pasts are closely interwoven with Québec's past. As seigneurs or settlers, entrepreneurs or labourers, their presence has been felt throughout the province: on the Lower North Shore and The Gaspé, in Charlevoix County and the Beauce, throughout the Eastern Townships, along the Laurentian front and the Ottawa River Valley, and of course in Montréal. The legal system of Québec is based, in part, on English civil law and the parliamentary system is modelled on its British counterpart. At another level, the folk traditions of the Québécois have been inspired by the music of Ireland and Scotland, while the cultural values of the people are rooted as much in the Anglo-Saxon as in the Latin tradition. Moreover, the intellectual elite of the province has been attracted to the universities of Oxford and London as well as to their French counterparts. In other words, the British, by virtue of their historical role, constant intermingling, and intermarriage, have contributed significantly to defining the personality of the province.[19] Despite their numerical strength, economic power, and status in the towns and cities until well in to the twentieth century, the British presence did not significantly alter the structure of French Canadian society. True, in the eyes of the Roman Catholic Church, Protestantism was viewed as an evil force and the cities as places of moral disintegration, but authority was clearly vested in the church. The parish furnished the basic institutional and social fabric of French Canadian society.

The urbanization of French Canadian society within Québec from the late nineteenth century onward, however, meant direct confrontation with the English worlds of Montréal, Three Rivers, Québec City, Drummondville, Sherbrooke, Valleyfield, and Seven Islands. Secularization followed apace. The transfer of power and leadership from the church to the provincial government after World War II transformed the social climate within the province. From a situation of tolerance, expressed by indifference, marked social distance, and occupational divergence, Québec became an arena of ethnic conflict.

The main thrust of the Quiet Revolution, culminating in the election of the Parti Québécois in 1976, was the creation of a powerful national state. Thus, within French Canadian society there was a concern to strengthen and widen the scope of the public domain and to integrate non-francophones into the larger society. The most dramatic expression of this movement was the series of language legislation (Bills 63, 22, and 101) establishing French as the province's sole official language and progressively strengthening its presence in the work place and the educational system. The *Chartre de la langue française* passed in 1978 was the final manifestation of this transition. The process of appropriating anglophone "spaces" began, in fact, in the mid-nineteenth century with the expansion of francophones out of the St. Lawrence Valley onto agricultural land vacated by English, Scottish, and Irish settlers in the Eastern Townships and along the Laurentian front. This process also took place in the neighbourhoods of the major cities. Increased educational opportunities and two world wars accelerated this process. The gradual appropriation of the secondary urban centres (for example, Drummondville and Sherbrooke) followed during the Depression years, and again stemmed from demographic change, increased francophone economic initiative, and a rapidly growing provincial bureaucracy. So, Three Rivers became Trois-Rivières, Seven Islands became Sept-Îles, and Arvida was absorbed into the expanded city of Jonquière.

While recognizing the diversity of cultural minorities throughout the province, the Québec government has sought to promote among these minorities — through language and education practices — a tendency that some sociologists describe as "franco-conformity." In other words, the model or norm to which all members of the population tend to gravitate, irrespective of language or culture of origin, is a French Canadian one. According to this model, Québec is the *patrie*, the territorial expression of collective identity, French is the national language, and it is the history of the Québécois people that determines the course of present and future actions. Québec exists as the hearth of the French Canadian people and as the centre for the continental diaspora.

The English minority is linked to that of the province in a way similar to that of the francophone diaspora scattered through English Canada. While some Anglo-Québecers left the province, including many from the western, English-speaking neighbourhoods of Montréal, others have undergone a shift in territorial affiliation and collective identity. At the same time, the continuing influence of the 20 per cent minority poses a major political barrier to a more homogeneous Québec.

The powerful desire to enhance English usage among anglophones and allophones alike makes it difficult for any political party to obtain a strong majority for policies that erode the status of English. Five provincial ridings have anglophone majorities and ten others have allophone majorities. Of the 125 counties, fifty have linguistic minorities of 10 per cent or more. In one recent election, about half of the counties in Québec gave victory margins of less that 10 per cent. The emotional power of this 20 per cent minority in closely fought campaigns was sharply displayed in the mid 1990s at the time of the referendum on Québec independence.

CONCLUSION

In *The Liberal Idea of Canada*, James and Robert Laxer point to the paradoxical position of Québec within Canada.[20] For them, the province has features of both metropolis and hinterland. Montréal is a Canadian metropolis which, in spite of a major loss of economic power to Toronto, continues to assume essential metropolitan functions. It is a leader in transportation and manufacturing, and is assuming stronger ties with the international business community, particularly with multinationals from France and other European countries. In addition, its resource corporations are linked with markets in the United States, where demand for hydroelectricity and pulp and paper products remains strong.

The heartland–hinterland paradox within Québec, historically, was closely related to the ethnic divisions in the province. Montréal's metropolitan power was led by the anglophone commercial elite. In the course of Québec's development and the building of the metropolis, the anglophone society of Montréal espoused a completely different ideology from that of the province's francophones. The leaders of anglophone society were concerned with entrepreneurship, the development of a transcontinental transportation system, and the accumulation of capital. Their institutions were chartered banks and commercial, industrial, shipping, and railway enterprises, as well as the private clubs where one socialized and created business alliances. Montréal, rather than the larger St. Lawrence Valley or Québec itself, was their home. Rather than being primarily a cultural hearth, it was a centre — an economic gateway to a continent. Because of the sophistication of the English educational system, its Protestant values, and the evident material successes of the anglophone bourgeoisie, this ideology was readily espoused by the anglophone population in general.

Anglophones knew little of the cleavage that marked the francophone community — divisions between urban and rural francophones, between the Church and businessmen, and between businessmen and professionals. These divisions inevitably found expression in different patterns of wealth.[21] However, the relationship between ethnicity, income, and position in the Canadian space economy explains only part of the Québec paradox. If for Anglo-Québecers cultural identity is related to a particular economic role, expressed in a dual attachment to Canada and to Montréal, then for francophones the St. Lawrence Valley and Québec in general constitute a cultural hearth whose sphere of influence has continental dimensions. By virtue of demographic pressure, economic marginality, and the attraction of other regions outside Québec's borders, the people of Québec have often left this hearth in search of work. They, like many of the immigrants to the New World, assumed the historic role of abundant, cheap, and malleable labour in the development of the continental economy, and particularly in the industrialization of New England.

Geographical mobility has thus become an essential feature of the popular collective identity, that is, of Québec's regional consciousness. The francophone elite, while refusing such mobility, has sought periodically to accommodate it in a larger national design. Meanwhile, the gradual peripheralization of Montréal and Québec encourages the departure of a significant component of the anglophone community and a partial substitution by francophones. It also introduces the recurring spectre of a massive departure of francophones in search of employment elsewhere and, once again, the weakening of the cultural hearth. The destiny of Québec remains inextricably linked with that of Canada. At the same time, the character of the cultural hearth and its role in shaping a regional consciousness separates Québec, perhaps more than any other factor, from the rest of the country.

NOTES

[1] The ideas that permeate this chapter owe a considerable debt to the innovative research of Pierre Anctil and Christian Morissonneau. I am also grateful to Dean Louder, with whom I have conducted research on French America.

[2] Raoul Blanchard, *Le Canada français: province de Québec* (Montréal Librairie Arthème Fayard, 1960).

[3] Paul Claval, "Architecture sociale, culture et géographie au Québec: un essai d'interprêtation historique," *Annales de Géographie*, 83 (1974), pp. 394–419.

[4] Pierre Biays, "Southern Québec," in *Canada: A Geographical Interpretation*, ed. John Warkentin (Toronto: Methuen, 1968), pp. 281–333.

[5] See, for example, P. Deffontaines, "Le rang, type de peuplement rural au Canada français," *Cahiers de Géographie* (Université Laval) 5 (1953), pp. 3–32; and R. Cole Harris, *The Seigneurial System in Early Canada* (Québec: Les Presses de l'Université Laval, 1966).

[6] French Canadians are "… un rameau déraciné du grand arbre français. Elle est née de la France. Pendant plus d'un siècle et demi ses sujets sont tous issus du meilleur sang de la France et depuis deux cents ans leurs descendants se sont, pour la plupart, gardes purs de tout alliage. C'est pour eux un premier titre de noblesse." Anon., La vocation de la race française en Amérique du nord Québec: Le comité permanent de la survivance française en Amérique, (1945), p. 25.

[7] See, for example O. P. Dickason, "From 'One Nation' in the Northeast to 'New Nation' in the Northwest: A Look at the Emergence of the Métis," in *The New Peoples: Being and Becoming Métis in North America*, eds. Jacqueline Peterson and Jennifer S. H. Brown (Winnipeg: The University of Manitoba Press, 1985) pp. 19–36. This article was published previously in slightly

altered form, in *American Indian Culture and Research Journal* 6 (1982). The author would like to acknowledge the helpful comments on this article by Dr. Lewis H. Thomas, University of Alberta.

[8] C. Morissonneau, "Mobilité et identité québécoise," *Cahiers de Géographie du Québec*, 23 (1979), pp. 29–38.

[9] Yolande Lavoie, *L'émigration des Québécois aux Etat-Unis de 1840 à 1930* (Québec: Editeur officiel, 1979).

[10] Consider this Great Lakes Métis description of his self-identity in the 1850s: "Où je reste? Je ne peux pas te le dire. Je suis Voyageur — je suis Chicot, Monsieur. Je reste partout. Mon grand-père était Voyageur; il est mort en voyage. Mon père était Voyageur, il mort en voyage. Je mourrai aussi en voyage, et un autre Chicot prendra ma place. Such is our course of life." J. G. Kohl, *Kitchi-Gami, Wanderings round Lake Superior* (London: Chapman and Hall, 1860), p. 259.

[11] Aidan McQuillan, "French-Canadian Communities in the American Upper Midwest during the Nineteenth Century," *Cahiers de Géographie du Québec*, 23 (1979), pp. 53–72.

[12] See, for example, P. Anctil, "La franco-américaine ou le Québec d'en bas," *Cahiers de Géographie du Québec*, 23 (1979), pp. 39–52; and Y. Lavoie, *L'émigration des Québécois*.

[13] These observations on the role of the Roman Catholic Church are derived from N. Voisine, *Histoire de l'Église catholique au Québec, 1608–1970* (Montréal: Fides, 1971).

[14] "Votre mission nationale est la conversion des pauvres sauvages et l'extension du royaume de Jésus-Christ; votre destinée national, c'est de devenir un grand peuple catholique." Cited in Anon., *La vocation de la race française*, p. 37.

[15] For example, Collège Assomption in Worcester, Massauchusetts and Collège St-Viateur in Bourbonnais, Michigan.

[16] Major Edmond Mallett, United States Indian Inspector in the late nineteenth century and hence an influential Franco-American as well as a fervent nationalist and, above all, confidant of Louis Riel and Gabriel Dumont, stressed on the occasion of a St-Jean Baptiste rally in Montréal in 1880 that:

> Notre éparpillement nous est très désavantageux, et nous devrions faire un effort pour concentrer sur quelque point donné: comme à partir du Détroit, où il y a déjà une population canadienne considérable, jusqu'à la montagne de la Tortue, dans le Dakota. Cette position offrirait à nos nationaux (les Franco-Américains) de grands avantages naturels et pourrait devenir d'une suprême importance, à nous, ainsi qu'à nos frères de Manitoba, si certains événements venaient à se produire.

> Taken from Pierre Anctil, "L'exil américain de Louis Riel 1874–1884," *Recherches Amerindiennes au Québec*, 11 (1981), p. 243.

[17] The architects of this geopolitical vision were Curé Labelle, Arthur Buies, and Testard de Montigny. C. Morissonneau, *Le terre promise: Le mythe du Nord Québécois* (Montréal: Hurtubise HMH, 1978).

[18] This fact is not lost on certain political scientists. Consider the following observation regarding the vocation of the government of Québec that appeared in *Le Devoir*, 10 November 1981:

> Ce n'est pas seulement une province que l'Assemblée nationale et le gouvernement du Québec ont pour mission imprescriptible de défendre et d'illustrer. C'est également et surtout une communauté linguistique originale qui a son assise vitale dans cette province et qui, seulement dans cette province, possède des cadres institutionnels suffisants pour lui permettre d'exister comme société particulière...

[19] D. Clift and S. MacLeod Arnopoulos, *The English Fact in Québec* (Montréal: McGill-Queen's University Press, 1979).

[20] James and Robert Laxer, *The Liberal Idea of Canada* (Toronto: James Lorimer, 1977).

[21] See Report of the Commissioners, Volume 3. *The World of Work, The Royal Commission on Bilingualism and Biculturalism* (Ottawa: 1969); and P. Bernard, *L'évolution de la situation socio-économique des francophones et des non-francophones au Québec, 1971–1978* (Montréal: Office de la langue française, 1979).

chapter 6

Graeme Wynn

Places at the Margin: The Atlantic Provinces

This region of Canada, as the title of the chapter makes clear, is a true hinterland. It is useful to begin with the observation, now most generally attributed to the French geographer Henri Lefebvre, that spaces and places are *produced*. They are — in other words and in all sorts of ways — human creations as well as physical entities. To acknowledge as much is to recognize that any account of this, or any other region, must necessarily be concerned with both the form of space and its various representations, with what we might call the topography or more broadly the physical geography of the area, as well as with the ways in which that territory is and has been organized and conceptualized by those who live within and beyond it. This is no easy task. It entails investigation at various scales, from the local to the continental, requires sensitivity to time and circumstance, and forces discussion of often contested and invariably elusive questions of symbolic meaning.[1]

Start with the idea of Atlantic Canada itself. On the Pacific Coast of Canada — where I write — it seems unproblematic enough to most of those I know. It encompasses the country's easternmost provinces: Newfoundland (and Labrador), Nova Scotia, New Brunswick, and Prince Edward Island. But for the fact that Newfoundland keeps a time of its own, these political jurisdictions seem to form a self-evident "region." (Map 6.1) Information about them comes neatly packaged from Statistics Canada and other government agencies. All have shores washed by the Atlantic. They are long-settled and economically troubled. Unemployment rates are high, traditional industries in decline. Bypassed by recent development,

Map 6.1 Atlantic Region

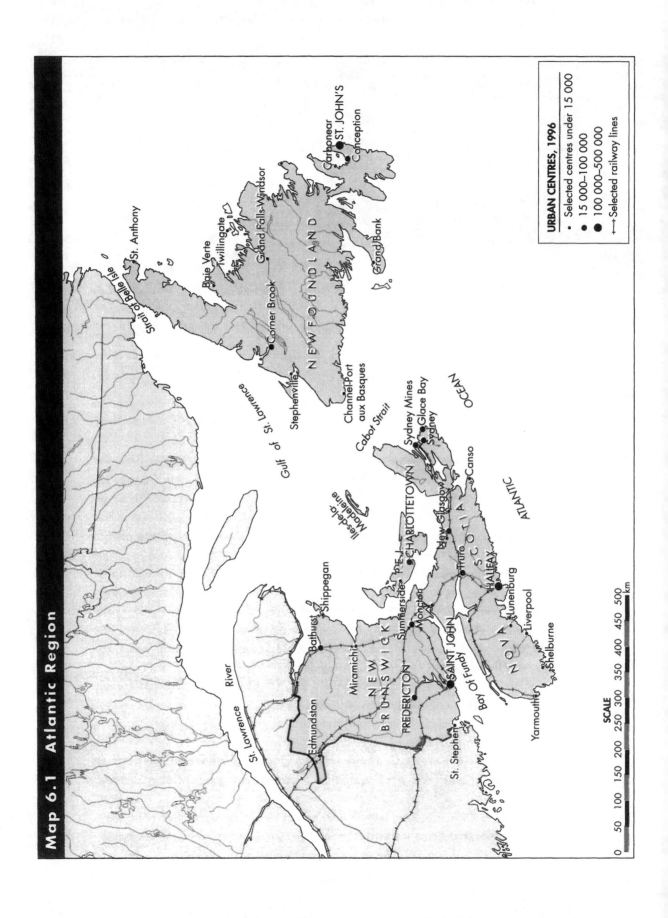

URBAN CENTRES, 1996

- Selected centres under 15 000
- 15 000–100 000
- 100 000–500 000
- Selected railway lines

St. Anthony

Strait of Belle Isle

Baie Verte

Twillingate

Grand Falls-Windsor

Corner Brook

NEWFOUNDLAND

Stephenville

Channel-Port aux Basques

Cabot Strait

Gulf of St. Lawrence

St. Lawrence River

Edmundston

NEW BRUNSWICK

FREDERICTON

St. Stephen

SAINT JOHN

Bay Of Fundy

Miramichi

Bathurst

Shippegan

Moncton

Summerside

P.E.I.

CHARLOTTETOWN

Îles-de-la-Madeleine

Carbonear

ST. JOHN'S

Conception

Grand Bank

Sydney Mines

Glace Bay

Sydney

Canso

New Glasgow

Truro

HALIFAX

Lunenburg

Liverpool

Shelburne

Yarmouth

NOVA SCOTIA

ATLANTIC OCEAN

SCALE

0 50 100 150 200 250 300 350 400 450 500 km

they are rich in "heritage," good places to visit on vacation. These are provinces whose people have retained qualities and values slowly but surely eroded from the everyday lives of those caught up in the "rat race" of the contemporary world.

Find confirmation of this in the songs of Stan Rogers, the stories of Alistair McLeod, the images of the regional tourism industry. Rogers' songs are powerful reminders of earlier times. They explore the consequences of changes that ensued as foreign trawlers swept the oceans, young people moved to the city, and those left behind found it impossible to continue traditional ways.[2]

There is even greater poignancy in McLeod's powerful story "The Boat," the richly textured and deeply moving reminiscence of a professor in "a great Midwestern university" who is haunted, a decade after the drowning of his father, by memories. Memories of his childhood in a village of half-a-hundred houses. Memories of a place created by successive generations of exiled Celts around the horseshoe of an unnamed Nova Scotia harbour. Memories of the tramp of shore-bound boots on gravel at 4:00 in the morning; of the boat — named for his mother, manned by his father and uncle — around which life and conversation turned; of the salt and wind-burned physicality of the fishermen who put to sea each day; of the time his father sang for hours, in Gaelic, for uncomprehending but grateful tourists; of sisters leaving the settlement as brides of young men from Boston, Montréal, or New York; of the realization that his gnarled, white-haired, book-loving father had spent an uncomplaining lifetime in an occupation for which he had no affection; of the day that proud, impassive man fell from the boat into the cold, grey waves of the November Atlantic; of the quiet conversation a few months before, suggesting that he, the only son, pursue his love of books and learning when his father died; of the body hurled and slammed on the rock-strewn shore.[3]

None of the complexity and sadness, the sense of human struggle, disappointment, hope, and tragedy that so enrich McLeod's story is evident in the rhetoric of the modern tourist industry. Although they often trade in similar images — of lobster traps and fishing boats, of houses strung along the shore, of weather-beaten men and women skilled in traditional crafts — these accounts seek to persuade prospective vacationers that the region is a simpler place than the rest of North America. Thus Prince Edward Island becomes "a kind of dreamy never-never land that most of us haven't known since childhood — unsophisticated, slow-paced and satisfying . . . ," and Newfoundland is a place where "there is no artifice, just an air of remoteness and peace," a place in which "the 20th century has not quite arrived." No matter that streets and villages are reconstructed, gentrified, "heritaged" to mirror these verbal pictures, or that poverty and despair lie hidden behind images of happy contentment. Produced with all the skill of the publicists' art, the images are powerful and entrenched. Their messages, simplified in the extreme, make a pastiche of the past and offer little room for nuance. "Just like the Maritime Archaic Indians 7500 years ago. Like the Vikings in 1000 A.D. And John Cabot 500 years later. The world keeps discovering Newfoundland and Labrador.

And, in 1997, the whole world is dropping by." More often than not — and for all their differences in intent and fidelity — the cumulative impact of such portrayals seems to bolster rather than undermine the widespread conviction, identified by Ralph Matthews in 1983, that "the Atlantic provinces are poor; they always were and always will be — that's just the way it is."[4]

Under closer scrutiny, the matter is much less straightforward. The very notion of an Atlantic region, so naturalized and accepted in contemporary discourse, is a relatively recent and still rather questionable conception. Newfoundland's claim to Labrador, disputed since 1902 by Québec, but granted by a judgment of the British Privy Council in 1927 and confirmed by Canada in 1949, continues to be contested by the latter province. Residents of the Island of Newfoundland, which remained a British colony until dire economic circumstances led to amalgamation with Canada in 1949, still assert their singularity within Confederation. Decades after the colony became part of Canada, scholars remarked on the existence of a deep-seated sense of Newfoundland nationalism that drew its strength from the island's geographic, cultural, and economic distinctiveness. The three neighbouring provinces, which joined Canada between 1867 and 1873, are more readily, and more traditionally, thought of as a unit than the Atlantic provinces, but inhabitants of Nova Scotia, New Brunswick, and Prince Edward Island hold strong provincial (as well as local) loyalties. In historical perspective the idea of the "Maritimes" as a distinctive entity is also a relatively recent phenomenon. In broad terms it is associated with both the effects of modern communication technology on local life and sentiment, and the Maritime Rights movement, a sectional protest against difficult economic circumstances and declining political influence on the national stage in the first quarter of the twentieth century.[5]

If residents of all four provinces have been touched by the sea — "These are the fellows who smell of salt to the prairie" reflected Charles Bruce, one of the region's most perceptive writers and poets — the connection has, generally, been more spiritual than economic. Born in Saskatchewan, and raised in Alberta before his parents (descended from generations of Nova Scotian Gaels) moved back to Cape Breton when he was ten, Alistair McLeod has lived most of his life in Windsor, Ontario. "Maritime Stan," the hearty balladeer full of evocative songs of the sea, was in some sense a token; although Stan Rogers' parents came from Nova Scotia, he was born, raised, and made his home in Hamilton, Ontario. Substantial proportions of the population in each of the four provinces live in urban areas, forest-dependent employment is high in two jurisdictions, growing numbers of workers are engaged in tertiary activities, and, beyond Newfoundland, agriculture and secondary manufacturing are of considerable economic significance (see Map 6.2). Neither a homogeneous nor a functional region, the "Atlantic region" is perhaps best regarded as a construct of convenience, less a coherent entity than a means of ordering space and place for the purposes of classification, description, and memory. Or to put it even more starkly, as does Ann-Marie MacDonald in her sprawling saga of a Cape Breton family, *Fall on Your Knees*: "There is no such place as 'down home' unless you are 'away.'"[6]

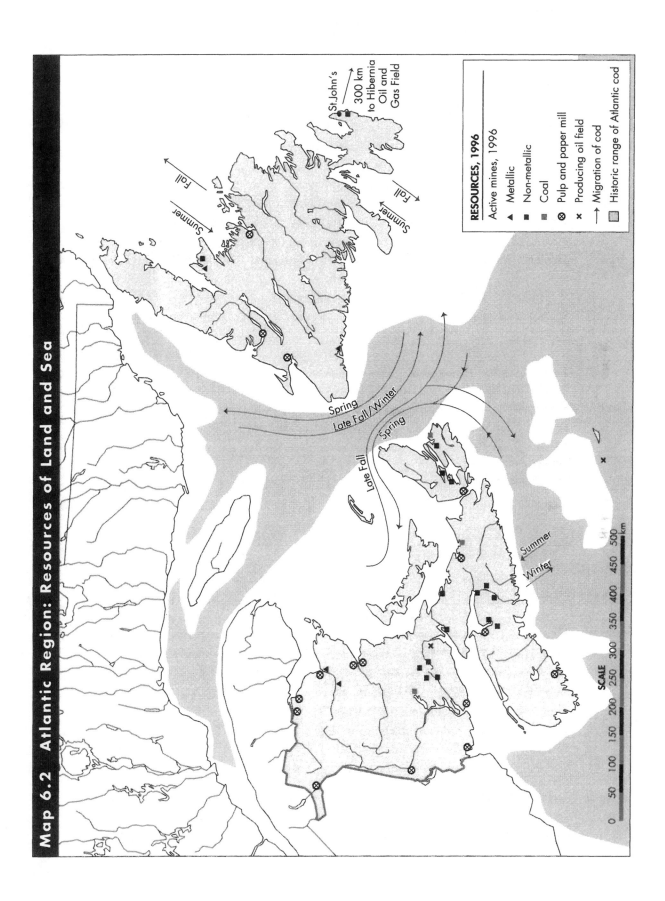

Map 6.2 Atlantic Region: Resources of Land and Sea

St. John's

300 km
to Hibernia
Oil and
Gas Field

Fall
Summer

Summer
Fall

Spring
Late Fall/Winter

Late Fall
Spring

Summer
Winter

RESOURCES, 1996

Active mines, 1996

▲ Metallic
■ Non-metallic
■ Coal
⊗ Pulp and paper mill
✕ Producing oil field
→ Migration of cod
▭ Historic range of Atlantic cod

SCALE

0   50   100   150   200   250   300   350   400   450   500
km

HISTORICAL DEVELOPMENT

Understanding this rather indeterminate "region" requires a strong sense of time and change, and of the ways in which concepts such as space and place are shaped and reshaped by circumstances. It demands careful grounded analysis, a willingness to narrow the scale of inquiry from the "regional" to the local, and an acceptance of the possibility that coherent general interpretations may prove elusive.

From Cape Spear to Madawaska, from Cape Sable to the Torngat Mountains, this area forms a mosaic of striking physical, economic, and social variety. Narrow valleys dissect generally forested uplands; steep slopes divide settled lowlands from sparsely occupied plateaus. Heavily indented coastlines and centrifugal drainage patterns enhance isolation and distinctiveness by separating province from province and inlet from estuary: northeastern New Brunswick faces the Gulf of St. Lawrence; southwestern Nova Scotia has had strong ties with New England; Labrador has been and remains the disadvantaged hinterland of a peripheral region; insular Cape Breton, Prince Edward Island, and Newfoundland remain distinct in outlook and identity. Variations in climate, soils, and vegetation reflect differences in location, terrain, surface material, and precipitation, and further tessellate the natural scene. Nothing less than thorough engagement with the details of lives lived and places made in a generally niggardly environment can serve to illuminate the human geographies created by the struggles and sacrifices of unsung generations of settlers, by the decisions of politicians, by the often dimly understood consequences of technological change and time-space compression, by changing individual and societal aspirations, by market forces, and by seemingly ineluctable ecological processes.[7]

The Roots of Distinctiveness

A century and more ago, this area was markedly different from the place connoted by the phrase "Atlantic Canada." Colonies of a distant British metropolis in the mid-nineteenth century, New Brunswick, Newfoundland, Nova Scotia, and Prince Edward Island were shaped by the pre-industrial circumstances of the great age of wood, wind, and water. Their economies turned upon their farms, fisheries, forests, and commercial shipping fleets; and their societies mirrored the isolation and diversity of this fragmented territory. Commercial ties linked the region with Britain, the United States, and the West Indies. Lumbering and fishing were basic enterprises. Forest products made up well over half of New Brunswick's exports; wooden sailing vessels accounted for 50 to 75 per cent, by value, of all other exports from the province. Nova Scotia possessed a more diverse export base, but fish often exceeded one-third of total exports. Farm produce, sawn lumber, and local manufactures such as furniture and shoes diversified the cargoes sold from provincially built schooners in scattered Caribbean harbours. In the Gulf of St. Lawrence, Prince Edward Island concentrated on farming, sending its surplus to the lumberers of New Brunswick and the schooner masters and merchants of Nova Scotia. In Newfoundland the transition from a migratory to a resident fishery had begun in the last years of the eighteenth century. It had produced a wide scatter of small coastal settlements — outports — from which small boats fished the inshore for the cod (marketed through all-powerful merchants in St. John's) upon which the economy of the island depended.

In the countryside and along the shore a mixed economy prevailed, although dependence upon the fishery, and the cod in particular, was especially marked in parts of Newfoundland. Generally, however, fishermen "farmed," or at least kept a patch of potato ground, a small garden, a pig or two, and perhaps a cow. Many farmers or their sons had worked a season or two in the woods, and subsistence was generally the farm family's prime concern, although a barrel or two of salted meat, a bushel of apples, a firkin of butter, or a load of hay might be delivered to a local storekeeper in payment for sundries purchased on credit during the year. Domestic industry provided many necessities. Tallow candles, homespun cloth, boots, shoes, and harnesses were made by wives, children, and local artisans. At the census of 1861, there were 13 230 handlooms in Nova Scotia; the value of cloth manufactured in New Brunswick exceeded $700 000. Beyond Newfoundland, in particular, village artisans and small-town manufacturers met the demands of a nascent market economy. Blacksmiths, skilled makers of tools and implements, thrived. Tanneries and carriage works dotted the provinces. Saw and grist mills at many river falls provided for local needs.[8]

The colonies formed a patchwork quilt of cultural variety. Overland movement was slow and expensive; life was fundamentally local. Religious, economic, ethnic, and physiographic diversity were reflected in varying agricultural practices, in the multitude of churches, and in the proliferation of educational institutions with different denominational affiliations. At a more general level, the basic physical expression of this diversity was in a mosaic of distinctive subregions. Recognizably Irish traits marked areas in which Irish settlers predominated; elsewhere, Scottish Highland origins were revealed in place names, widespread use of the Gaelic language, and a concentration upon livestock rather than crop farming. There were even regional variations in the generic names assigned to physical features. Minor rivers, "brooks" in most of Maritime Canada, were commonly known as "creeks" in areas of Loyalist settlement; low-lying, seasonally flooded riverine tracts were known in different parts of the colonies as "flats," "interval," or "bottom land"; and the "marshes" of some locales might be "lowlands" or "meadows" in others.

People were both divided and united by isolation, ethnicity, and religion. In the small scattered settlements of Newfoundland, people spoke English variously, with the rich accents of the eighteenth-century English West Country, the brogue of southern Ireland, or the inflections of those whose family origins lay in the coastal fringe of southern England. Catholics and Protestants formed separate settlements, or created a distinct, divided social geography in a few well-favoured coves, by congregating around their respective churches. On the island's French Shore, first defined by the Treaty of Utrecht in 1713, long-standing rivalries between England and France continued to shape animosities and settlement patterns. Elsewhere, Gaelic-speaking Highland Scots, both Catholic and Presbyterian, predominated in Prince Edward Island, Cape Breton, and the Northumberland shore of Nova Scotia. Irish clustered in the major ports, in parts of the St. John Valley, and in the lumbering areas of northern New Brunswick. Settlers of New England descent — including both Anglicans and Baptists — established their influence in Halifax, on both Fundy coasts, and along the St. John River. Acadians settled in discrete clusters in Nova Scotia, New Brunswick, and Prince Edward Island, the nuclei of more extensive territories occupied today. These ethnic hearths were shaped by the timing of migration flows, the routes of available transport, and variations in economic opportunity across the region. Once established, they were sustained by chain-migration, by the distinctiveness of language and religion, and by the spatial separation attributable to the physical fragmentation of the provinces.

Despite their trans-Atlantic outreach, these were profoundly local societies. In 1857, eighty per cent of all settlements in Newfoundland had fewer than 200 residents, although well over half of all Newfoundlanders lived in places of more than 500 people, and approximately one-third in places of more than 1000 (among which St. John's, with twenty per cent of the population – 25 000 people – was pre-eminent). Four-fifths of those who lived in Nova Scotia, New Brunswick, and Prince Edward Island were rural dwellers. Although some 40 000 lived in Saint John, there were fewer than 30 000 in Halifax, and Fredericton and Charlottetown had populations of only 6000 and 8000 people, respectively. Locally important, these scattered centres were also provincial outposts of British economic and political influence. They served their immediate hinterlands, but none controlled the entire region.

In the 1850s and early 1860s, international circumstances were broadly favourable to the colonial economies of British North America. Inflated British lumber demands during the Crimean War doubled spruce deal exports from New Brunswick between 1850 and 1860. Free trade with the United States, established by the Reciprocity Treaty of 1854, opened new markets for fish, minerals, and forest products. In the early 1860s, the American Civil War stimulated exports while providing new opportunities for the carrying trade that engaged the sizable fleets of New Brunswick, Nova Scotia, and Prince Edward Island. But the end of the Civil War, and the abrogation of reciprocity, both in 1866, undercut the expansion and prosperity of the preceding years. In 1873, a major international depression constricted remaining American markets for Maritime lumber, and seriously reduced the region's sales of wood and ships to Britain. At the same time, the Maritimes' West Indian trade, which carried fish and lumber to the Caribbean and returned with cargoes of molasses and rum, was jeopardized by the vast surplus of refined sugar dumped on a glutted world market by European and American refineries. The Caribbean economy took a downturn and trade was reduced. This situation in turn threatened the viability of Nova Scotia's fishery as it lost a major market.

Beyond these developments, less clearly perceived changes with telling long-term effects began to undermine the old economic order. Fast and reliable steamships began to force sailing vessels from prime trans-Atlantic routes. Iron was winning favour in shipbuilding and replacing wood in many other types of construction. Territorial expansion carried the interests of the new Dominion of Canada westward. New technologies of communication and production were harbingers of a more integrated and extensive space economy. Political decisions began to reshape Canada's economic environment and the terms of its trade. Initially, these developments created opportunities for many in the country's eastern provinces; within a few years prospects of industrial ascendancy in Canada danced in the minds of regional leaders; but time and circumstances ultimately exposed the chimerical quality of these dreams. The region was reasonably endowed with the resources essential to the age of wood, wind, and water; it possessed vital advantages in the early stages of the epoch of coal, iron, and steam; but an eccentric location within a country increasingly orientated toward the west, declining traditional resource endowments, and a modest potential for expansion based on the twentieth century's turn to electricity and the internal combustion engine, left the Maritime provinces on the economic margins of Canada by the 1950s. At the time when Newfoundland joined the Confederation in 1949, the decline of its traditional cod-based economy, and the limited benefits of efforts to diversify through import-substitution and the development of new resources, had reduced its economy to crisis. All four provinces benefitted from national fiscal policies that ameliorated the worst consequences of economic decline into the 1970s, but in the assessment of some observers they retained little control over their own destinies. They had become, in effect, "client states of the federal government."[9]

Looking Inward

Confronted with the effects of world recession in the 1870s, the government of Sir John A. Macdonald sought to create a strong, relatively self-sufficient Canadian economy by promoting interregional commerce and fostering domestic manufacturing. Under the avowedly protectionist National Policy, new duties were imposed on imported manufactures after 1879. Common consumer goods incurred the heaviest tariffs: in 1879, for example, furniture, woollen clothing, and cotton piece goods were charged approximately 35 per cent. Fully manufactured farm implements, such as castings, were levied at about 25 per cent. Duties on semifinished goods — pig iron, rolled steel — were from 10 to 20 per cent. Protection was extended to primary iron and steel in 1883, and increased when duties were raised again in 1887. Thereafter, a $3.50 per ton advantage to Canadian-made iron further sheltered producers of iron and steel goods. Finally, a 50-cent per ton levy on imported coal (raised to 60 cents in 1880) protected Canadian producers of this prime fuel of the new industrial age.

These altered circumstances reoriented the economies of Nova Scotia and New Brunswick. With the Dominion's only known and viable coal and iron deposits, with maritime access to raw materials, and with the Intercolonial Railway (completed in 1876) offering year-round connection from Halifax to the expanding markets of central Canada, provincial optimists envisaged a significant place for the region in Canada's emerging industrial order. The general economic upsurge of the early 1880s saw the rapid expansion of secondary industry in the two provinces. Cotton mills, sugar refineries, rope works, steel and rolling mills, and iron and steel manufacturing plants were established or expanded to serve national markets.

Tariff protection, and rate schedules that markedly favoured the long-distance movement of goods on the Intercolonial Railway, allowed Maritime manufactures and primary products into the St. Lawrence region. But under the umbrella of the tariffs, rail-borne central Canadian flour and manufactures replaced American and British goods in the Maritime market. Before long, the sugar, coal, cotton cloth, iron, and fish sent westward from the Maritime provinces comprised only one-third of the traffic on the railroad link to the St. Lawrence. Yet so long as the sea borne trades in wood and fish prospered, and back-haul cargoes from the sugar islands and southern United States provided raw materials for Maritime manufacturing, these trading patterns rested in precarious balance.[10]

This was upset in the 1880s by a steep decline in the English lumber market and the re-imposition of American duties on fish. At the same time, central Canadian markets for cotton cloth and sugar from the Maritimes were threatened by declining demand and falling prices. These circumstances were compounded by American dumping of devalued cloth on a Canadian market barely able to absorb the output of its own mills. The resulting commercial uncertainty led to the development of trade associations that attempted to regulate the cotton industry by controlling production; their failure, precipitated by the refusal of some Maritime producers to limit their output according to association quotas, had far-reaching consequences. Well-financed Ontario and Québec cotton manufacturers decided to stabilize production and secure their interests by industrial consolidation. Their aggressive strategies and locational advantages made them formidable competitors with the generally undercapitalized Maritime enterprises. Central Canadian interests rapidly gained control of most Maritime mills.

Similar developments occurred in other manufacturing industries. Some Maritime companies disappeared; others worked within larger corporate structures. By 1895, only confectionery production and manufacturing tied to the iron and steel and staple industries remained under local control. With this decline of local autonomy, there was a growing tendency for management decisions to be based on profit and loss calculations rather than concern for the fortunes of

Maritime communities. From the 1890s on, the Maritimes held a shrinking proportion of a Canadian consumer goods industry that was increasingly concentrated in the Lower Great Lakes–St. Lawrence region, where ready access to both capital and markets enabled producers to realize economies of integration and scale. There was a parallel decline in the Maritimes' tertiary sector as upper echelon service and financial functions concentrated in Montréal and Toronto. Even the region's workshops and artisans producing for local markets gradually succumbed to the competition of mass-produced articles from central Canada, distributed across the region by branches of the Intercolonial and Canadian Pacific trunk lines. Together, these changes undermined the economic vitality of the province's larger urban centres, and left the Maritimes without a regional metropolis.[11]

In the iron and steel industry, large local coal reserves, small quantities of local iron ore, and ready access to the rich Bell Island ore deposits in Newfoundland, opened in the mid 1890s, gave Maritime producers an initial advantage over their central Canadian counterparts. The early twentieth century was a period of industrial expansion. Primary production was based on the coal fields of Pictou and Cape Breton; iron-using industries — making bridges, railway cars, wire, nails, and stoves — clustered in towns along the railroad linking them to the emerging Canadian heartland. With the expansion of western railways and Prairie settlement after 1896, demand for such products soared. Nova Scotia's pig iron output increased from 32 000 tons in 1896 to 425 000 tons in 1912.

Once again, however, competition among eastern and central Canadian firms resulted in consolidations and the eventual transfer of control out of the Maritimes, thus reaffirming the region's hinterland status. Beginning with the amalgamation of secondary iron manufacturing industries and culminating in the consolidation of Nova Scotia steel producers to form the British Empire Steel Corporation in the early 1920s, ownership of the region's iron and steel industries slipped into international hands. With this loss of control, elimination of the government bounty on iron and steel production in 1912, changing product demands (which required different plant capability), and distance from the growing concentration of industrial output in Ontario, the position of the Maritimes' iron and steel industry declined. Whereas Nova Scotia produced 43 per cent of Canadian pig iron in 1913, it accounted for less than 30 per cent in 1929, even though output increased.[12]

An extractive industry, coal mining was less footloose than manufacturing, but it, too, was subject to mergers and consolidations engineered for the benefit of expanding corporate capitalism. By 1920, virtually the entire Cape Breton coal field was controlled by the British Empire Steel Corporation. Inevitably, the fortunes of the coal mining industry reflected developments in iron and steel production. But the collieries were never entirely dependent on local markets. Tariff protection made Cape Breton coal competitive in Montréal, and expanding shipments to central Canada through the 1890s employed many Nova Scotian vessels previously displaced by the disruption of Atlantic trade. In 1914, almost 2.4 million tons of Nova Scotian coal, thrice the quantity of 1896, entered the Québec market. And for a brief period during the 1890s, before a smoke abatement law and revisions to the American tariff schedule excluded it from the Boston market, Cape Breton coal was also sold in New England.

In turn-of-the-century Newfoundland, concerted efforts were made to bring the fishery-dependent economy on to the trajectory defined a century before in Britain and widely emulated in the nineteenth century — seeking modernization through industrialization. Here, as elsewhere, the development of railroads was seen as a *sine qua non* of development. Early in the 1880s, construction of a narrow-gauge track began. Sixteen years of escalating costs and political debate ensued before the line was completed between St. John's and Port aux Basques. Profits were elusive. Twenty years after completion of the Newfoundland Railway, freight rates were ten times higher than tariffs for equivalent movement on the mainland Canadian Pacific Railway. Accumulating deficits helped push the colony toward bankruptcy in the 1930s. Concurrently, but independently, the Nova Scotia Steel and Coal Company began exploitation of iron-ore deposits on Bell Island in Conception Bay, utilizing Newfoundland labour in two six-month shifts to capitalize on prevailing patterns of occupational pluralism in the colony. One group of men mined in the winter and fished in the summer; the other mined in the summer and worked in the woods through the winter. Wages were low, and until 1910 no royalties were levied for utilization of the resource. Returns to Newfoundland were correspondingly modest. Early in the twentieth century, the Newfoundland government encouraged the Anglo-Newfoundland Development Company to build a pulp and paper mill at Grand Falls. As the geographer Michael Staveley observed: "Generous incentives were offered to the company: complete water and timber rights over an area of 2000 square miles at a rental of $2.00 per square mile…; a renewable 99 year lease; full mining rights for a scant 5 percent of net profits from mineral operations; no municipal taxation; and no duty for 20 years on materials imported for use in construction and development." In return the company was charged a small royalty on wood cut. The pattern was repeated again a few years later, when pulp and paper operations were established in Corner Brook and Deer Lake, and mining began at Buchans. Resources were exploited, some employment was created, and the colony's gross domestic product was increased, but little attention was given to resource conservation, and neither municipalities nor the Newfoundland government gained significant revenues from these developments. To a substantial degree such profits as were made accrued to individuals and companies beyond Newfoundland.[13]

Meanwhile, the fishery declined disastrously. Although it accounted for 86 per cent of all Newfoundland exports by value in 1900–1901 (and cod sales earned almost 80 per cent of this total), the engine of the colonial economy was sputtering badly. Population growth had generally outstripped improvements in productivity and prices since the mid-nineteenth century. While there were some relatively good years before 1911, overall returns from the fishery declined by more than 2 per cent per annum through the next quarter century or so as the industry clung to traditional methods and lost ground to competitors such as Norway and Iceland. Moreover, the sorry social and economic consequences of this decline were highly uneven. Through the four decades spanning the turn of the century, the problems of the fishery were particularly acute on the northeast coast, the relatively densely settled area between Conception Bay and Twillingate. In community after community, if not everywhere, in this region (regarded by some as "the economic core and cultural hearth" of traditional Newfoundland), where participation in the migratory Labrador fishery and the spring seal hunt had been grafted on to the inshore cod-fishery in the archetypal adaptation to the challenges of making a living in this difficult environment, the per capita productivity of men engaged in catching and curing fish fell dramatically. According to an analysis by Michael Staveley that uses an index number of 100 to characterize the per capita value of fisheries production for the colony as a whole at five dates between 1884 and 1920, productivity in Carbonear fluctuated widely between twice the colonial average in 1884 (index # 201) to approximately half the average in 1910 (index # 54). In Port de Grave, index numbers for 1884 and 1910 were 114 and 38; in Harbour Main, 61 (1884) and 23 (1920). If, in contrast, per capita output in Fogo and Twillingate increased during this period, it was only in relative terms; there, index numbers of 83 and 52 in 1884 stood at 92 and 69 in 1920. All of these numbers betoken low returns to capital and labour; most of them indicate declining returns that would not be broken without the introduction of new technologies, massive increases in the price of fish, or the withdrawal of labour from the fishery.[14]

Nfld seal hunt, 1910

Conception Bay, 1890

St. John's, 1900

At the same time, evolving technologies and changing product demands substantially altered the traditional fishing, forest, and farming industries of the Maritime provinces. Cod fisheries declined. Long-established European markets were lost as improved shipping and advances in the techniques of packing, canning, and refrigeration brought meat from Argentina and Australia into competition with salt fish from Nova Scotia. Caribbean markets for cod also declined as cane sugar producers suffered stiff competition from an expanding sugar beet industry in Europe. Steamships took much of the remaining export trade of the West Indian Islands from provincial schooners.

In home waters, large expensive Banks trawlers, operating from fewer ports and requiring fewer workers, heightened competition in the cod fishery. Rail links from the Gulf shore brought the fishermen's catches to the holds of steamers in Halifax and bypassed local intermediaries. Yet there were some improvements. Lobster, in demand for live shipment and for sale to small canneries, offered inshore fishermen another catch. Half-decked, gasoline-powered "Cape Island" boats (named for the area of southwestern Nova Scotia in which they were developed) became more common after 1900, making access to the fishing grounds easier, and increasing the range of those fishermen who could afford them. Moreover, special trains and new tariffs led to the development of cold-storage warehouses and opened central Canadian markets after 1908. Nonetheless, the real value of Nova Scotia fish production declined between 1896 and 1913.

Lumbering also reached a plateau during these years. Railroads opened new timber stands in northern New Brunswick. Employment and production increased in provincial sawmills, but with most of the valuable pine gone from the forest, much of the cut was small spruce. Low-valued box shooks and spool wood made up a significant proportion of the industry's output. Heavier capitalization and mechanization set the forest industries of New Brunswick apart from those in Nova Scotia and Prince Edward Island; in the former province the impetus was toward pulp production. Here, the heavy investment necessitated by new technologies is apparent. Three small plants producing $108 000 worth of pulp in 1890 were valued at $298 000; in 1911, a dozen plants manufactured pulp worth $1.4 million.

On farms, the area planted in grain decreased; wheat from the Western Interior produced better and cheaper flour than that grown in the eastern provinces. Subsistence potato growing also declined with the increasing availability of imported foods in the region. Wool and cheese exports to Britain, the latter fostered by the development of cooperative cheese factories after 1870, were challenged before 1900 by large, high-quality shipments from Australia and New Zealand. Retrenchment and specialization resulted. Increasingly, in many older settled areas, marginal upland farms once dependent on sheep raising grew back to forest.

But there were improvements. Apple orchards, concentrated in the Annapolis and St. John valleys, yielded a profitable crop for the British market, and annual exports tended upward. Prince Edward Island farmers formed cooperatives to market butter, eggs, and bacon in the expanding urban centres of New England; a few made fortunes in a fox-ranching industry that reached its most prosperous level on the eve of World War I. More generally, Maritime farmers concentrated on those crops least susceptible to outside competition — fodder and roots for the lumber camps, dairy products for local consumption — or ran their holdings to provide the bulk of their own needs. The combination of farming with other employment remained common. Still, the Maritime provinces lost 20 000 of 134 000 farmers between 1891 and 1911.

Together, the tariffs and the railroads — the building blocks of the National Policy — transformed the Canadian economy between 1879 and 1914. Yet the Maritime provinces gained a relatively small share of the benefits accruing from a quarter-century of national expansion and prosperity generated by industrial development, an upswing in the world economy, and the rapid settlement of the Prairies. They became increasingly peripheral to the developing St. Lawrence core of the Canadian economy after 1889. Growing external control of primary and secondary manufacturing in the region took important decision making out of local communities. Parallel developments in the financial sector saw the region's mercantile banks amalgamated into national corporations between 1900 and 1920. By 1914, the provinces were clearly losing ground in the expanding continental economy. Their traditional export industries languished and they slipped far off the pace of manufacturing expansion in Canada.[15]

Interwar Difficulties

Wartime circumstances reversed many of these trends and restored a degree of prosperity to the economy of the Maritimes. Coal and steel industries expanded as the demand for munitions and the growth of railroad and steamship transportation generated markets for their products. Lumber prices soared. With the disruption of world trade, Maritime fishermen increased their sales of dried codfish in the western hemisphere. Fresh fish shipments to central Canada grew as meat prices rose and the government allowed rebates on transportation costs. Distant producers offered less intense competition to local producers, and sheep farming and cotton cloth production increased to meet military needs.

Precipitous economic decline followed the war-induced boom, however. The entire region was affected by the shift of industrial dependence and economic development away from coal, iron, and the railroad to electricity, oil, and the automobile. The coal industry faced a static market in 1918. Cut off from their largest customer, the St. Lawrence market, by the disruption of shipping between 1916 and 1918, Cape Breton's mines now had to compete with the better coal of more efficient American producers. These difficulties increased as inflationary price rises reduced the effective protection of the duty (53 cents per ton) on imported coal. Moreover, local coal consumption was down. As railway expansion declined, Nova Scotian steel mills (geared to the production of rails and ingot steel) and their dependent rolling stock manufacturers worked far below capacity.

Meanwhile, Ontario mills, nearer the market, with larger supplies of scrap metal, and favoured by tariff adjustments, adapted to new demands for structural steel and lighter grades of plate. As these mills expanded, the likelihood of takeovers and the eventual phasing out of Maritime production increased. Virtually all established manufacturing plants in the region declined in the 1920s. In 1925, the net value of regional manufacturing output was less than half the 1919 level. Expansion in the newer industrial spheres also passed the region by. Robbed of their energy-cost advantage by the turn to new resources, with few good hydro-electric generating sites, and without known reserves of minerals such as those newly discovered in the Canadian Shield, the Maritimes could not match the growth of Canada's Industrial Heartland.

The business recession that followed spiralling postwar inflation in 1920 also had a profound impact upon the regional economy. Falling prices and protection-ist policies affected many Maritime industries. Of the traditional activities, the fishery was perhaps hardest hit. Technologically superior European vessels, sell-ing their catch under the shelter of national tariffs, squeezed Nova Scotian and Newfoundland producers out of many markets. World demand for dried cod fell; even West Indian purchases declined as sugar prices dropped again. By 1921 prices for salt cod were one-half those of two years earlier. In 1919, government subventions for the shipment of fresh fish to Ontario were withdrawn, and two years later the United States re-established tariffs on the import of fresh and frozen fish. With the general decline of the economy, and the lack of alternative opportunities within the region, employment in the Nova Scotian fishing industry fell by almost 25 per cent between 1920 and 1929.

Lumbering and farming also felt the impact of recession and external compe-tition. After 1918, the Panama Canal allowed Pacific Coast lumber into the mar-kets of eastern Canada and the seaboard states. In 1920, the downturn of the British business cycle limited trans-Atlantic sales; and the decline of local con-struction reduced another outlet. Sawmill employment fell dramatically in 1920–21, and lumber production in Nova Scotia and New Brunswick remained far below prewar levels during the ensuing decade. Some related industries did pros-per, however. Improvements in hydroelectricity generation and transmission allowed development of larger pulp and newsprint mills, which absorbed labour and logs from the failing sawmilling industry, although the American capital and corporations (such as the International Paper Company) that lay behind this expansion gradually undermined the leadership of local entrepreneurs. Pulp out-put doubled in the 1920s. By 1930 this capital-intensive industry employed 3000 men; many others cut pulpwood for sale to the mills.

Renewed competition from western producers and processed foodstuffs led to a steady decrease in field crop acreage in New Brunswick and Nova Scotia during the 1920s. Dairying remained important for local consumption. Apple exports to Britain continued, although profits fell in the face of competition from British Columbia and the United States, and as a result of disease in Nova Scotian trees. Potato exports rose in the wartime decade, only to encounter higher American tariffs in the 1920s. Further specialization in seed potatoes maintained the viability of New Brunswick and Prince Edward Island growers. But on the whole, Maritime agriculture was in difficulty. The general recession, and the hardship common in the declining industrial towns of the region, reduced local markets for farm produce, pared away many a farmer's slender source of operating cash, and by reducing the diversity of their enterprises, left even those selling export produce more susceptible to external market fluctuations.

Adjustments to the rate structure of the Intercolonial Railway compounded the consequences of economic and technological change for the region. In the late nineteenth century, the rate structure was broadly beneficial to the Maritimes. In an effort to overcome chronic financial difficulties, the railroad implemented rates that encouraged long-haul traffic. From a basic rate some 20 per cent lower than that on central Canadian lines for hauls of up to 100 miles (160 kilometres), mileage levies decreased proportionately with distance, until shipments of 700 miles or more (1 126 kilometres) were charged only half the prevailing central Canadian rate. In addition, traffic bound beyond Montréal was charged an arbitrary rate for the stretch from the Maritimes to Montréal. Westbound rates were approximately 12 per cent below those on eastbound goods. Clearly these arrangements extended the westward range of Maritime commodities while providing them with some local protection. They were instrumental in allowing the region's manufacturers access to large central and western Canadian markets before World War I.

Long a focus of contention, the Intercolonial's rate structure came under increasing criticism early in the century. The east–west differential was eliminated in 1912. Management of the railroad was transferred from Moncton to Toronto in 1919, with the creation of the Canadian National Railway. Soon after, the arbitrary and special commodity rates (applied to sugar and coal) were eliminated. Maritime rates were set at par with those in central Canada, and in 1920 they were included in a general increase necessitated by the rapid rise in prices after World War I. The net effect of these adjustments was to raise rates on the Intercolonial between 140 and 216 per cent. With deflation, and the fall in prices that followed, this was a devastating blow. In effect, it crippled the manufacturing sector of a regional economy already lame from the impact of economic change and technological adjustment.[16]

Freight rate adjustments and modifications to trade and tariff policies lessened the Maritimes' difficulties somewhat in the late 1920s. Following the recommendations of a royal commission inquiry, railway rates within the Maritimes and on westbound traffic moving beyond them were reduced by 20 per cent in 1927. Federal support was offered for the construction of coking plants. Government subventions and bonuses allowed Maritime coal back into Québec and Ontario markets. To encourage Canadian port and railroad traffic, federal harbour commissions were established in Halifax and Saint John; freight-handling facilities were further upgraded; and imports from specified origins were allowed tariff reductions if they entered the continent through selected Canadian ports. Dominion subsidies also fostered a regular steamship service to the Caribbean although a reciprocal trade agreement with the West Indies did not significantly increase Maritime exports. With this government support — a portent of future reliance upon federal investment in the region — and expansion in tourism and primary production, the regional economy turned upward after 1926. Construction boomed. By 1929 manufacturing employment in the region was approximately 45 per cent above 1921 levels. Coal sales to central Canada exceeded prewar totals.

With the Great Depression, however, manufacturing employment fell from 40 000 to 24 000 in four years, and salaries and wages declined by 40 per cent. By 1936 less capital was invested in Maritime manufacturing than in 1917. Coal production dropped by more than 40 per cent in the three years after 1929. By 1933 lumber production in the region was 75 per cent below the output of 1920. The difficulties of the fishery also continued. In 1933, when Halifax dry codfish prices were barely one-third those of 1920, the total value of fisheries production was little more than one-half the 1929 level and only 43 per cent of that in 1918.

Farming was just as hard hit. Total farm income in 1932 was barely one-half that of 1929. Yet farming and fishing sustained a growing proportion of the Maritime provinces' population. Individual small-scale exploitation of land and sea augmented or replaced failing incomes in other sectors. New land was broken in northern New Brunswick; by 1933 there were 10 per cent more fishermen than in 1929, and three years later numbers were 20 per cent above the pre-Depression level. Fishing and farming were combined to provide a meagre subsistence. Farmers — 25 per cent of whom reported a principal occupation other than farming at the 1931 Census — eked out livings by cutting pulpwood or taking other work as it was offered.[17]

After fifteen years of almost unbroken recession, many Maritimers were in distressed circumstances. Overall, the economies of the three provinces were severely afflicted. Largely dependent upon the returns of a few vulnerable export staples, their manufacturing industries decimated by problems of cost, scale, and competition, and their best resources depleted or devalued, the Maritime provinces could provide the majority of their people with no more than a meagre, hard-won, and unreliable subsistence.

Economic conditions began to improve in the mid 1930s. Lumbering recovered slowly with the introduction of Imperial preferences and trade agreements with the United States, and more quickly when World War II eliminated competition and spurred domestic construction. Military demand expanded the region's pulp and paper markets. Iron and steel production doubled in value between 1939 and 1942; the gross value of regional manufactures increased 140 per cent between 1939 and 1944. Cash incomes from the sale of farm produce more than doubled in the same period. Returns from fish packing and curing virtually tripled. Even in constant dollars, gross values of production rose from 10 to 50 or 60 per cent in almost all sectors of the regional economy between 1940 and 1945.[18]

But again, these were temporary benefits of abnormal circumstances. Chronic economic problems reappeared. Although the maturing Canadian economy expanded rapidly after 1945, the Maritimes shared little of the national prosperity. Unemployment rates ran considerably above national levels. Earned incomes in the region, 75 per cent of those in the rest of the country in 1945, slipped to 66 per cent by 1955. In Prince Edward Island, incomes at the beginning of the decade were a mere 58 per cent of the national average, and they fell to 50 per cent by 1955. Employment in the region's primary industries declined, and there were no compensating increases in other spheres. As prospects faded, migration from the region increased.

In Newfoundland, the situation was equally dire. The spread of protectionist economic policies during the Depression years of the 1930s, and the comparative rise in value of the American dollar (in which Newfoundland purchased most imports) drove individual and aggregate returns from the fishery downward, precipitated financial crisis, and led to the establishment of a caretaker Commission of Government. When the population voted, narrowly, to join the Confederation of Canada, federal largesse flowed into the new province. Family allowances, which averaged approximately $16 a month, more than any other jurisdiction in the country because of the preponderance of large families, and old-age pensions seemed to confirm the wisdom of union. But the widespread enthusiasm for these payments simply underlined the difficult circumstances many had faced before 1949. Before Confederation, approximately 3000 Newfoundlanders received "pension" payments of $72 per year; immediately afterward, the Canadian government paid $30 a month (and a little later $40 a month) to some 10 000 old-age pensioners.

Despite subsidies, mechanization, and changes in control, Nova Scotia's coal and steel industries declined. When the railroads switched from steam to diesel locomotives in the 1950s, coal production fell by one-third. Mining ceased on the Cumberland coal field. Mines were closed in Pictou and Cape Breton. Mechanization increased productivity in the industry by almost 50 per cent between 1945 and 1955, but extraction costs made it impossible to compete with the selling price of American coal in Ontario. Without market expansion, jobs disappeared. Employment in Nova Scotia's mines declined from 13 200 in 1946 to 9200 in 1957. Sydney steel production suffered high costs — the burden of outmoded plants — and the loss of established markets. By 1957 industrial Cape

Breton was in serious economic difficulty, and the future was dark. Nine thousand people left this area between 1951 and 1956.

Large investments in the New Brunswick forest industry by local-born industrialist K. C. Irving paralleled growing American, European, and British Columbian involvement in pulp and paper production. Within a decade of World War II, the value of pulp and paper production in Nova Scotia and New Brunswick was up approximately 85 per cent. In lumber- and wood-using industries, employment fell in both provinces, almost completely offsetting the gains in pulp and paper operations. Later, in the 1980s, it was a similar story across Canada (Table 6.1). In other manufacturing, there was little expansion in the postwar decade, especially by comparison with manufacturing growth in central Canada. After 1945, Maritime manufacturers competed less and less effectively with their St. Lawrence rivals. Increases in railway freight rates — applied equally across the country but offset by concessions to combat the competition of truckers in Ontario and Québec — enhanced the burden of distance for Maritime shippers through the 1950s. Low traffic intensity and poor road networks (unimproved because provincial revenues were already sorely stretched) restricted the growth of a regional trucking industry whose competition with the railroads might have lowered transport costs. While the number of manufacturing jobs in Canada increased by some 310 000 between 1949 and 1957, the manufacturing work force of the Maritime provinces declined.

Fishing and farming were also affected. Large increases in the output of the fishery (stimulated by high prices and a strong domestic market) were achieved by increased capitalization and improved productivity. Employment in the fishery remained more or less constant between 1945 and 1955. In farming, too, mechanization maintained output, but limited markets, high transport costs — on fertilizers and equipment brought into the region as well as on produce shipped out — and the lure of urban amenities, often beyond the Maritimes, led to widespread abandonment of small holdings. In New Brunswick, agricultural employment fell by 45 per cent in the postwar decade. In Nova Scotia, the decline was one-third; in Prince Edward Island it was one-fifth. In all, 36 000 farm jobs were lost in ten years.

Table 6.1   Numbers Employed in Selected Canadian Forest Industries, 1982 and 1992

	1982	1992
Logging	40 000	41 000
Sawmills	54 000	52 000
Plywood	10 000	8 000
Pulp / Paper	83 000	70 000

Source: Statistics Canada, 1992

Federal transfer payments to individuals — including unemployment insurance, old age pensions, and child allowances — alleviated the worst consequences of this postwar decline. Between 1949 and 1956 these payments accounted for some 10.5 to 13 per cent of personal incomes in the Atlantic region. Augmented by new federal–provincial fiscal arrangements to provide more uniform health and education services across the country, they prevented a catastrophic decline in the quality of regional life. Paradoxically, they also permitted the survival of marginal lifestyles as part-time fishermen substituted unemployment insurance payments for subsistence farming or seasonal work in the woods. The federal payments were no more than a papering over of the weaknesses of a hinterland economy. With the downturn of the national economy in 1957, the region's problems were revealed once more. Despite increases of 3 to 4 per cent per capita in transfer payments to the region in 1957 and 1958, the personal incomes of residents of the Atlantic region remained substantially below the national average.

Toward Equalization

Persistent regional disparities within the country prompted new federal initiatives to foster structural changes and economic expansion in the lagging regions. In 1957, equalization payments were introduced. In the 1960s, a series of federal–provincial programs addressed the problems of unequal development. Among these, the Agricultural Rehabilitation and Development Act (1961) and the Agricultural and Rural Development Act (1965) were intended to alleviate rural farm and non-farm poverty across the country. They offered low-interest loans to farmers, and established make-work programs for community pasture development. A Fund for Rural Economic Development (1966) offered assistance to primary industry, tourism, and manufacturing in Prince Edward Island and parts of New Brunswick, as well as in Québec and Manitoba. Development incentives were offered to firms locating in high-unemployment areas. In 1962, the Atlantic Development Board was established to advise on the economic problems of the region and to finance infrastructure development within it. In 1967 the Cape Breton Development Corporation was charged with rationalizing the coal industry and stimulating economic adjustment in that economically depressed island.

Regional development efforts were rationalized by the creation of the federal Department of Regional Economic Expansion in 1969. DREE and its successors, the Department of Regional Industrial Expansion (DRIE) and the Atlantic Canada Opportunities Agency, have provided incentives to private industry, and have contributed to the development of transport networks, utilities, energy supplies, research agencies, and the infrastructure needed for economic expansion and social adjustment. Between 1969 and 1975, DREE spent $520.5 million in the Maritime provinces. Prince Edward Island received approximately $800 per capita, New Brunswick $390, and Nova Scotia $230. By the department's estimates, these efforts created or maintained 20 000 jobs. Yet critics condemned DREE's "passive" approach to industrial development, dismissing it as a policy of subsidized laissez faire lacking the radical impact of more aggressive strategies. As DREE evolved into DRIE and was then transmuted into ACOA, there was little indication of greater risk-taking, or a significant commitment to financing unproven but potentially rewarding industries such as new forms of pulp and paper production or marine technologies.

Provincial Crown corporations also provided concessions, capital, and services for industrial development. Naive, poorly coordinated strategies and unscrupulous grant-seeking entrepreneurs often vitiated their efforts, and they were, at best, only partially successful in transforming the spatial organization of manufacturing. But, together with federal departments, they stimulated industrial development. Their combined interest in the creation of growth poles — concentrated development centres — and the encouragement of sophisticated, footloose, high-technology manufacturing initiated a reorientation of regional industry.[19]

General-purpose transfers (equalization grants) and specific-purpose payments for the provision of health, welfare, and education also flowed into the region from Ottawa. In the early 1970s, the federal government directed well over one billion dollars a year into the Atlantic region. By 1982 the figure had effectively doubled. In sum, transfer payments exceeded provincial revenues in Prince Edward Island, and amounted to approximately 80 per cent of those in New Brunswick and Nova Scotia. Had federal transfers been spread equally among Canadians in the early 1980s, they would have amounted to some $595 per head; on a per capita basis the three Maritime provinces received $1540, $1134, and $1157, respectively. In addition, federal government transfers to individuals — old age security and guaranteed income supplements, unemployment insurance, family and youth allowances, veterans' pensions, and employment mobility and training allowances — were above average in the region. They averaged almost $540 per capita in Atlantic Canada in the mid 1970s, when the national average was $410. In the early 1960s, Nova Scotians derived perhaps 8 per cent of their personal incomes from federal transfer payments. In the 1980s the proportion was closer to 10 per cent.[20]

Demographic patterns reflected fluctuating economic circumstances. The region held a declining share of Canada's population as late nineteenth- and twentieth-century tides of immigration ran more strongly to Ontario and the west than to Atlantic Canada. In the three Maritime provinces, numbers increased by 25 per cent and 16 per cent in the two decades after 1851. In the next ten years, the increase was only 13.5 per cent. Between 1881 and 1901 numbers were almost static; not until 1921 did the population of the Maritimes exceed one million. Between 1871 and 1941, growth in the three provinces was less than 45 per cent; in the same period, Canada's population rose by 212 per cent. Increases of 11 and 15 per cent per decade in the 1940s and 1950s were well short of national levels, and growth rates in the Maritimes slipped back in the 1960s. For relatively remote

Newfoundland, percentage figures convey a more robust picture. The island's population grew by approximately 80 per cent between 1857 and the end of the century, and by a further 45 per cent by 1945. But numbers were relatively small — just over 120 000 in 1857, approximately 200 000 in 1890, slightly fewer than 322 000 in 1945.

Massive out-migration was the major cause of slow growth in the Maritime provinces. At least 100 000 people left the Maritimes each decade between 1881 and 1931. Between 1871 and 1901, departures of former immigrants exceeded immigrant arrivals, and the exodus of Canadian-born people almost offset the natural increase of the resident population. According to one intriguing case study of the Canning district of Kings County, Nova Scotia, this movement ran slowly in the decade or so after Confederation but increased dramatically in the mid 1880s. It was dominated by the young and the single, and comprised slightly more men (55 per cent) than women. It was also structured by family ties, and took the majority of migrants to Massachusetts. Thus, argues Alan Brookes, the author of this analysis, there was "established a tradition of leaving school, leaving home, and ...leav[ing] the region in search of employment and autonomy" that remains an integral part of the experience of many of its young men and women.[21]

Migration continued through the early twentieth century. In the difficult decade of 1921–31, more than 147 000 people left the Maritime provinces. The massive exodus began in 1920 and most out-migration occurred in the worst economic times, before 1926. That net out-migration fell to almost one-third of 1920s levels during the next decade was due more to the general depression of economic conditions elsewhere than to any significant improvement in the Maritimes' economy (at least until the end of the decade). Since World War II the loss of population has continued; although provincial population totals have risen, the increment has been well below the rate of natural increase. During a postwar peak of out-migration between 1961 and 1966, some 80 000 Maritimers left home. They were joined, in the booming cities of Ontario and on the oil fields of Alberta, by a swelling tide of mostly young Newfoundlanders, who shared with their Maritime cousins the hope of gaining experience and prosperity by "going down the road" — although many soon found the only work they could get unfulfilling, questioned the wisdom of going "away," and dreamed of the day when they might return to the familiarities of home.[22]

Among the Maritime provinces, Prince Edward Island was most affected by emigration. Its population declined at every census between 1891 and 1931. Despite subsequent increases, it did not reach the 1881 total until 1966. In contrast, New Brunswick increased its population each decade between Confederation and 1971. In Nova Scotia, the 1920s were the only decade of actual decline. Yet out-migration from the mainland provinces began earlier than that from Prince Edward Island, involved more people, and was compounded by the consequences of widespread and significant internal migration. Almost half of all counties in the Maritime provinces reached their maximum population in 1901 or before. In some of these, out-migration slowed growth rates even in the 1860s. By 1881, population growth in the Annapolis Valley, rural Cape Breton, and neighbouring Antigonish and Guysborough counties was less than the natural increase. People were leaving the lower St. John Valley and Charlotte and Northumberland counties in appreciable numbers, and growth was slowing in Prince Edward Island. Of the provinces' thirty-six counties, only nine returned a larger population at every census between 1871 and 1941, and only five of these sustained a numerical increase through each census to 1971. Of the nine, three included urban concentrations (Halifax, Fredericton, and the Cape Breton coal field). The remainder comprised six of seven counties in an arc through northern and eastern New Brunswick from Victoria and Madawaska through Gloucester to Westmorland. The seventh, Kent County, diverged slightly from its neighbours because its population declined in the 1920s and 1930s. Including the last agricultural frontiers of the region in Victoria, Madawaska, and Restigouche, these counties also contain the majority of the Acadians whose sustained growth is largely attributable to their high fertility and their tendency to remain and make do in L'Acadie. Natural increase, however, has been complemented by migration from Québec to Madawaska and Restigouche since 1871. Today, Acadians comprise almost 40 per cent of New Brunswick's population.[23]

The patterns of population change are revealed even more clearly by analysing urban trends. Before 1880 the Maritime provinces were essentially rural; small centres provided services for local settlers, but apart from Halifax and Saint John there were few large urban places. Indeed, the absence of a dominant regional metropolis is often cited as both a cause and consequence of Maritime underdevelopment. Yet as overland communication improved, and as mass production and distribution encouraged centralization, the region's urban population did increase. From less than 12 per cent in 1871, it reached 38 per cent in 1941 and 53 per cent in 1971. Urbanization was most evident in Nova Scotia, where urban dwellers increased from 8 to 43 per cent of the population between 1871 and 1921. Until 1921, urban expansion reflected the growth of tariff-sheltered industry and focussed on the region's two major ports (with Halifax exceeding Saint John in its rate of growth) and along the line of the Intercolonial Railway. Moncton, a key node in the region's rail network, grew from 1650 to 19 500 in fifty years after 1871. On the Pictou coal field, where iron and steel and secondary manufacturing grew rapidly after the National Policy, Pictou and Trenton with approximately 3000 people, Stellarton and Westville with some 5000, and New Glasgow with 9000, formed a dispersed urban cluster of almost 25 000 people by 1921. In Cumberland County the manufactures of "Busy Amherst" supported 10 000 people; nearby Springhill on the coal field had 5600. And in Cape Breton County, 63 000 urban dwellers lived in Sydney (22 500), Glace Bay (17 000) and the satellite towns of Dominion, New Waterford, North Sydney, and Sydney Mines.[24]

Urban patterns since 1921 may be summarized briefly. Concentration in Halifax, Saint John, and Moncton continued. Slow growth, stagnation, or decline marked the other important centres of 1921. New nodes of growth emerged on the periphery of the region. The populations of Dalhousie, Edmundston, and Fredericton tripled in half a century; that of Campbelltown doubled. Bathurst's population grew slowly until the late 1950s, then tripled in the 1960s. Late in the same decade, Port Hawkesbury more than doubled in size. These developments are attributable to the expansion of pulp and paper production, the expansion of the civil service, the growth of metal mining, and the development of high-technology industries. In general, they reflect the resource- and government-dependent character of regional growth since the 1950s.

In sum, the relatively stability of provincial population totals since 1871 concealed a great deal of local movement. Virtually everywhere, the fluctuations of county and parish populations reflected the limitations of Maritime resources. By the end of the nineteenth century, there was insufficient land in the older settled agricultural regions to provide for the third and fourth generations of early nineteenth-century immigrant families. As economic conditions made farm subdivision less viable, settlers' sons and daughters moved on. Change and hardship in the fishery forced people from that traditional occupation. In southwestern and northwestern New Brunswick, the declining lumber industry further limited the capacity of local economies to employ the offspring of those who had developed the area. Again, migration was forced on many of the young. Of those who moved, many gravitated to the region's growing urban areas. There, in time, manufacturing decline or mine closures often necessitated another relocation. Ultimately, the majority of migrants departed for central and western Canada, the United States, and overseas destinations. Propelled by the relative decline of the regional economy, this exodus has robbed the region of many of its most productive people, for a disproportionate share of migrants were young adults, many of them skilled and ambitious. Moreover, migration, whether within or beyond the region, fractured communities and demoralized families. Although out-migration from the region slowed in the 1970s, and was reversed in the early 1980s by the collapse of the Alberta oil boom and yet barely realised the hopes, in Newfoundland and Nova Scotia, for the development of offshore oil and gas reserves, the turnaround was hardly decisive. In 1983–84, the Maritime provinces counted slightly more than 6000 in-migrants, but Newfoundland lost 2500 residents.

THE CONTEMPORARY SCENE

Since the 1960s, income transfers, development strategies, modern industrial growth, and the improvement of communications within and beyond the Atlantic region have transformed economy and society within it. But these changes have been grafted upon geographies shaped by earlier decades of development and decline. They have moved, at varying speed, through countless channels, and their influences have been felt in a thousand ways. They have not been uniformly deleterious, although they have sometimes been seen that way. Among the more poignant expressions of their pervasive impact is the following description of an everyday scene in Neil's Harbour, a Cape Breton fishing village, in the 1950s:

> *[the] . . . mail order catalogue . . . arrived and every family got a copy of the Wish Book or Winter Bible . . . . "I believe it's the catalogues that's causing all the trouble in the world," [Matt] said . . . . "They sees all the foine things hinto it and wants to have 'em . . . ."*[25]

From another perspective, the availability of catalogue shopping in such remote communities was a significant boon. It made new goods available, reduced the price of many necessities, and broke the monopolies (and all they entailed) of local merchants. But in its several manifestations, such encroachment of modern materials, attitudes, and ideas upon older, more local ways of life has been a basic process in the shaping of the intricate and diverse landscapes of the Atlantic region at the end of the twentieth century.

Today, new and old exist side by side and are often in conflict. Economic dualism is a marked feature of the region; it is most clearly reflected in the juxtaposition of capital-intensive, technologically complex industries tied directly to national or international markets with small-scale family enterprises with few, and generally indirect, links beyond their immediate area. In the regional capitals and growth poles of the Maritimes, more than "Upper Canadian concrete and glass" have made their mark. The attitudes and practices of modern industrial society have replaced traditional local ways. National bank executives and corporate lawyers in Halifax and St. John's are provincial outriders of Toronto's Bay Street. Government bureaucracies apply heartland regulations to the hinterland circumstances of the eastern provinces. School curricula reflect pressures to conform to national aims and ideals. Mass-produced goods from distant factories line the shelves and crowd the racks of the region's stores which bear, to substantial extent, the names of national and international companies. Corporate advertisers increase hinterland awareness — and sometimes envy — of heartland standards of living. And rather paradoxically, tourists in the Atlantic region demand standards of accommodation and service akin to those generally found in such metropolitan centres as Boston and Toronto, from which they seek escape. Here modernity challenges tradition; old ways retreat, but often begrudgingly; images of revitalization shimmer, but often as mirages.

There is much of interest to the geographer in this, for the region is a theatre of contrasts, of revealing juxtapositions and interesting tensions between lingering past and uncertain future. Halifax, the regional centre of Nova Scotia, backs into scrub forest. High-rise office towers, multifunctional redevelopment complexes, shopping centres, urban freeways, and sprawling suburbs — the characteristic features of modern cities — give way, abruptly, to a rough, rocky, and sparsely settled upland. Along the spine of the province, infertile soils on unconsolidated glacial drift and geologically ancient granites, quartzites, and slates carry a low forest, broken only occasionally by roads, marginal farms, and abandoned clearings. To the west, a bleak interior passes, equally abruptly, into patches of productive farmland. Orchards, pastures, cropland, poultry-sheds, dairy- and tobacco-barns, neat farmsteads, and sedate, tree-shaded villages crowd the narrow Annapolis Valley, flanked by upland brow to the south, and sheltered from Fundy winds and fogs by North Mountain, a 213 m escarpment of Triassic lava.

South and west, beyond Digby, the coastal settlements fringing St. Mary's Bay form another strand of complexity — Acadia. Here life seems less prosperous than in the Annapolis Valley. There are frequent signs of occupational pluralism: school buses parked alongside farmsteads; lobster traps "out back" on a 10 ha holding; appliance servicemen who also farm. Families are also generally larger. The bald landscape and close, continuous line of settlement is marked by large, distinctive Roman Catholic churches and variegated by brightly painted dwellings. Distinctive place names — Belliveau Cove, Comeauville, Surette's Island — define this territory, and blood ties bind its communities. Most inhabitants of this area can trace their origins to the seventeenth-century settlement of Port Royal, to resettlement after the infamous deportations of the 1750s, and to on-going migrations within the Acadian realm. The essential elements of their landscape, their sense of community, and their attachment to place are replicated wherever Acadians predominate in the three provinces, from Caraquet to Cheticamp and from the Memramcook to Madawaska.

Cape Breton Island offers sharp contrasts within its 10 360 square kilometres. Acadian and Scottish settlers occupy markedly different landscapes north and south of Margaree Harbour. On the island's southern shore, fishermen — hardy, versatile, and intimately aware of their local environment — set their simple, homemade gear from small boats pitching in the wake of supertankers bound for Port Tupper's modern oil refinery, where acres of plant and complex chemical processes are controlled from a single room by a dozen workers. Between the pristine grandeur of the Cabot Trail and Bras d'Or Lake, the dishevelled, polluted townscape of Sydney intrudes. A narrow, slow road winds east along the edge of Sydney Harbour past now-closed coke ovens, a steel mill that recycles scrap-metal, and the toxic tar-ponds (or catch basin) into which the chemical wastes — benzopyrenes, polynuclear aromatic hydrocarbons, perhaps even polychlorinated biphenyls, of the plant were poured for decades. It passes unkempt fields and seemingly endless strings of tiny, narrow-windowed houses lining damp, ill-paved streets. There is little pretension in these old coal mining towns. Despite their age, *no past hangs in the air as in Halifax, and there is little romance in the streets, as in the seafaring towns such as Lunenburg and Liverpool and Bridgewater. All of those towns, or at least some of their people, made money and built homes with rolling lawns and columned steps. But in [New Waterford, Reserve, Dominion, and] Glace Bay the money went mostly out of town to the men who owned the mines.*[26]

In recent years, with Cape Breton mines closing, people retiring, youth migrating, and the tax base shrinking, local governments have been hard pressed to sustain even minimal levels of service. The restored eighteenth-century fortress of Louisbourg, an industrial park near Sydney, and an ill-fated, enormously expensive, heavy water plant on the outskirts of Glace Bay reflect efforts to provide work for unemployed miners, to attract tourists to Cape Breton, and to modernize the manufacturing economy of Nova Scotia. Yet these projects are incongruous and the island still lags far behind Nova Scotia and national levels in the provision of many basic amenities. On average, Cape Breton's dentists serve many more patients than their counterparts across the country. A decade or two ago, diabetes and influenza were 40 per cent more common on the island than in mainland Nova Scotia; the incidence of pneumonia was 175 per cent greater. Until recently, "backhouses" (unplumbed, outdoor toilets) remained as common as flush toilets in some fishing villages. The hard lives of those who mine and fish are exceedingly remote from those of the business-suited, fashionably clad men and women employed in the banks and offices of Halifax. Yet each day television brings the lifestyles and expectations of affluent North Americans into the living rooms of Glace Bay, Ingonish, and Arichat.

In New Brunswick, farmland is rare. More than 80 per cent of the province is productive forest. Settled land — a mere 8 per cent of the total — skirts the coasts, flanks the St. John River, and forms a distinctively open landscape only between Sussex and Moncton. Yet there is diversity even within the forest. In the northeast, bogs and barrens are common; "swamps" of black spruce and tamarack alternate with birch and jack pine stands; a red spruce–yellow birch–beech–sugar maple forest occurs on better sites. In the cool north-central uplands, where the frost-free period is but half of Prince Edward Island's 140 days and the mean annual snowfall exceeds 254 cm, a spruce–fir forest predominates. Here, as in other mature stands of timber, budworm infestation causes high rates of tree mortality, despite repeated aerial spraying with insecticide. Between the cluttered, moss-strewn spruce–fir forests of the Fundy coast and the left-bank watershed of the St. John River, a well-developed — though now severely culled — mixed forest is characteristic. Sugar maple, hemlock, pine, and ash are its dominant species. Basswood, beech, butternut, red oak, and red spruce also occur.

On the fringes of New Brunswick's forest, abandoned farms are common. Each encapsulates a common regional scene. Old fields succumb to the march of aspen, birch, spruce, and fir; barren orchards and exotic ornamental trees straggle unpruned; shabby buildings decay. On average, surviving New Brunswick farms are small and little more than a third of their land has been improved. Mixed crop-livestock farming that reduces risk from the consequences of poor seasons and uncertain markets remains common, but farm consolidation and specialization proceed apace — as in other agriculture districts of the region — propelled by new technologies and economies of scale. Overall, there is considerable regional variation in the major commercial orientation of New Brunswick's farms. Dairying is the main source of income in the Sussex and Fredericton areas; poultry enterprises cluster near Fredericton and Moncton; most blueberries come to market from northeastern and southwestern corners of the province; potato and, less markedly, feed grain production is concentrated in the Upper St. John Valley. Other features compound the fragmentation of this compact 75 520 square kilometre territory and its 700 000 people: the peripheral foci of pulp, paper, and metallic mineral production; the lobster fishing of the Gulf shore; the ethnic differentiation of the province with Acadian hearths in Gloucester, Madawaska, and Westmorland counties; the economic disparities between south and north; and the poor articulation of the urban network. All enhance diversity, yet each reflects broad patterns characteristic of the region.

Measured by the imagery of the tourist industry recently encouraged in all of the Atlantic provinces, Prince Edward Island is a land apart, an island retreat from the pressures and demands of modern existence. But this is misleading. Old-settled landscapes, embellished, white-painted, wooden houses, pastoral vistas, and picturesque fishing villages are certainly common. Yet signs of the tension between traditional ways and the pressures of modernity are everywhere. On the island, as elsewhere in the region, tractors have replaced horses, and though har-

PEI farming

vest efficiency has increased, farmers are less independent and more susceptible to economic and political fluctuations beyond their control. Shopping centres and chain stores proliferate, bringing the goods and advertising practices of North American mass-production and franchise-marketing into competition with the local retailer. Tourism spawns motels. Golf and recreation complexes are created from island farms, and public beaches have been developed for the 600 000 or 700 000 visitors who come to the province, whose population is only 120 000, each year. Plans call for the consolidation of fish-processing plants to increase efficiency, but there is often little concern for the fishermen's preferences or requirements. Similar trends have transformed potato farming in the province. Mechanization has increased productivity, but the industry that employed 15 000 in the late 1940s now occupies perhaps one-fifth of that number. Many growers are now under contract — a Malpeque Bay processing plant has expanded its payroll and its "stable" of farmers by becoming the supplier of frozen french fries to the northeastern division of North America's largest hamburger chain. Island society is at a divide.[27]

In Twillingate, in northern Newfoundland, the fish plant that once employed 500 workers stood idle through 1996. Families whose roots in the community ran back generations worried about maintaining the population of 3000, about preventing cuts to the hospital, about whether the town's stores would remain open. In April 1997, one-quarter, perhaps one-third, of the men in the town were about to embark on the seal hunt. With the cod-fishery closed indefinitely because of the catastrophic decline in fish stocks, the seal hunt and summer tourism are, suddenly, the lifeblood of the community. Reopened by the federal government in 1996 after years of closure in the face of pressure from animal-rights groups, the hunt is a shadow of what it was at its prosperous peak in the mid-nineteenth century. With the quota set at fewer than 300 000 animals, the most optimistic sealers hoped to earn $8000. In the first five years of the 1990s, Newfoundland was the only Canadian province to decline in population.

Effects of Modernization

Behind these scenes there is a pattern. Established forms of rural and small-town life have been challenged, eroded, even eliminated by the march of modern techniques, goods, and ideas into the region from continental and trans-Atlantic hearths. Redefining the nature and purpose of work, modernization undermined the traditional bases of rural settlement in the provinces. And as the scale and cost implications of these changes were felt, family-centred independence (so long secured by the possession of one's own land, the provision of the bulk of one's needs, and freedom from onerous financial obligation) became increasingly difficult to realize in its traditional form.

The Forest Industry

In the forest industry, mechanization and specialization, coupled with a relative decline in the importance of lumbering and sawmilling, reduced the forest's role as a source of part-time employment and additional capital for settlers (see Table 6.1). The industry's work force became increasingly, if not exclusively, full-time; working mainly for corporations and with corporate equipment, it became, in effect, an industrial working class, or proletariat. Cordwood production from private woodlots remains an integral element of the region's pulp and paper industry; it sustains a number of self-designated "farmers" for whom cordwood sales are the major source of "farm" income, but price and demand fluctuate widely, and the risks of dependence on cordwood sales are high. There are significant capital costs, such as trucks and chainsaws, entailed in all but the smallest woodlot operations. Wood prices, held down by the mechanized efficiency of large-scale production, are generally low. Individual returns are often small. Today, the part-time cordwood operation is as often a component of economic marginality as adequate subsistence.

The ravages of the spruce budworm also threaten the incomes of small cordwood producers and the profitability of the New Brunswick forest industry in general. For sixteen years after 1952 New Brunswick sprayed its forests with DDT in an effort to control the budworm; when this insecticide was banned, chemical larvicides were substituted and remain in use, despite rising concern about their effects on the health of children. Small gains were made by a vigorous lobby of concerned parents in the 1970s: the times and places of aerial spraying are indicated in advance, and spray-planes are supposed to stay away from settlements. Still the infestation remains. Some contend that chemical intervention has only prolonged a problem whose solution lies in less clearcut logging, which encourages the growth of the balsam fir favoured by the budworm, and better silvicultural practice. In 1982, members of the New Brunswick Federation of Woodlot Owners, who supply approximately one-quarter of the province's pulpwood, adopted an anti-spray resolution and recommended the development of a more diverse mixed-species forest.

Spraying is, at best, a short-term palliative; forest management may be a long-term remedy, but the former has been favoured by the industry and the government because it reduces logging costs and saves trees for immediate use. Allied problems surround the use of herbicides to destroy young hardwoods and encourage the softwoods required by the mills. The defoliants 2, 4, 5-T and 2, 4-D, otherwise known as Agent Orange, were used on the forests of Nova Scotia and New Brunswick in recent decades. Individuals and community groups have objected, but control remains with the large integrated producers and government bureaucrats who are battling with issues of short-term gain versus ecological balance on the one hand, and increasing capitalization with growing marginality on the other.

Agriculture

Similar concerns have confronted agriculturalists in the Maritime provinces, especially since 1945. Changing values and falling returns have undermined the viability of most small family farms in the Maritimes. Improved communication — exemplified by the mail-order catalogue — brought mass-produced goods into largely self-sufficient communities at prices that made home production unnecessary. Western meat and flour replaced home-barrelled pork and family-harvested grains on farm tables; rising costs of production and competition in local markets kept farm incomes low; long hours of work and lives with few modern amenities appeared less attractive; sons and daughters, if not their parents, left rural life for the cash incomes and apparent attractions of city life.

Specialization and commercialization drastically altered the nature of agriculture for those who remained on the land. Well into the twentieth century, farm ownership was an attainable goal; modest, landed independence was within reach of the common family when farming was unmechanized and its main purpose was subsistence. But in recent decades, competition and the commercialization of agriculture have brought soaring costs. Land prices have risen and product specialization has usually necessitated heavy investment in buildings and machinery. With steady improvements in farm technology, farmers have had to purchase new equipment and/or enlarge their operations with increasing frequency to remain competitive. Operating costs have escalated with the shift from animal to mechanical power and increased off-farm purchases of feed, fertilizers, fuel, and pesticides.

Together, these changes have multiplied the difficulties of farm financing. Mortgage debt has increased, and short-term credit is a common necessity. An analysis of potato farming — among the most profitable of Maritime agricultural enterprises — undertaken in the 1960s reveals the general pattern. Late in that decade, 160 acres [65 hectares] of cropland was optimal for potato farming in New Brunswick. The capital cost of such a farm, allowing $150 an acre for land, $42 000 for farm buildings, and $40 000 for equipment, including two tractors, a harvester, and planters, sprayers and cultivators, was $106 000. Annual expenditures amounted to some $33 000. Cropping inputs accounted for the largest part of this total. Seed potatoes cost the farmer $5000; fertilizer, applied at a ton per acre to the thin, long-cropped soils in the sloping fields of Carleton and Victoria counties, cost over $6000; spraying to reduce weeds and against blight required a further $1800. The necessary hired hand earned $3000. Seasonal workers, employed to cut seed, to harvest, and to grade and load the crop, raised labour inputs to $7800. Machinery and equipment costs, including depreciation, exceeded $6000. Miscellaneous expenditures such as building depreciation, repairs and insurance, taxes, utility costs, and interest charged on loans to cover cropping inputs amounted to an approximately equivalent sum. At year's end, the net income of an enterprise of this size was $12 000 or $13 000. Allowing a 6 per cent return on capital, the farmer's labour income was no more than $8 500. Costs and returns have climbed over the years, but in real terms, farm incomes are no larger.[28]

The experience of many a small farmer in the region during the 1980s was typified by that of Brewster and Cathie Kneen, of Pictou County, Nova Scotia. Their farm, they said, was

> *as large as can be managed as a family unit, and we're not doing the kind of job that we'd like to be doing with it. We're using 12-year-old machinery which has to be kept going because there's simply not enough income to replace it. This means you take care of it — which is good — but you can't use it as you might, and you haven't got the resources to match things up better.[29]*

Confronted by rising interest rates and escalating costs, many farmers were forced to adjust. In areas with special locational advantages, such as the Annapolis Valley with good access to the Halifax market, some of the more prosperous and entrepreneurial farmers specialized in market gardening or the production of broiler hens. In other areas many simply abandoned agriculture. Smaller, less efficient producers moved into nearby towns, leaving the land to deteriorate and buildings to decay. Others remained in the countryside but produced little and faced poverty. In such agriculturally marginal areas as northeastern New Brunswick — which was said in the 1960s to have only enough good land for one hundred viable commercial farms — small, crowded, poorly serviced houses, dilapidated barns, broken fences, and overgrown pastures are graphic testimony to the poverty that resulted.

In better areas, small farmers were able to sell their properties to large, integrated agribusinesses. Yet others tried to alleviate financial problems by sowing specified crops under contract to corporate buyers. McCain Foods of Florenceville, New Brunswick, is pre-eminent among those operating in the region. Founded by local entrepreneurs, it has become a major multinational corporation, marketing frozen french fries, vegetables, pies, pizzas, and other food products around the world, and controlling fertilizer, cold storage, transport, and other industries, as well as thousands of hectares of farmland in New Brunswick. The smaller C. M. McLean Company on Prince Edward Island also increased its landholdings by purchasing family farms, and placing hundreds of potato growers under annual contract. Reflecting the essentially industrial cast of its operations, the firm held an annual banquet for its growers and offered premiums for productivity.

The Fishing Industry

The fishery went through similar changes. In the first half of the twentieth century, the traditional whalers, dories, and schooners of the Maritimes fishery were challenged inshore by motor-driven Cape Island boats and offshore by steam trawlers. Fishermen's protests, and a government ban on further development of the trawler fleet in 1930, allowed diesel-powered schooners to continue hauling longline trawls on offshore banks into the 1960s; but otter-trawlers, working from side or stern and equipped with electronic navigation and fish-finding gear, have swept them from the high seas in the last twenty years. In parallel fashion, vessels of intermediate size — smaller and less technologically sophisticated than Banks trawlers but larger and appreciably more expensive than half-decked boats — began to open up exploitation of the "mid-shore" in the late 1950s. These 60 foot [18 metre] longliners, wooden-hulled draggers, and seiners work longer seasons and in wider areas than those formerly fished by inshoremen. Independently owned by their skippers, they employ crews of three to five, and work the Nova Scotia Banks and Gulf of St. Lawrence waters in particular. Costing several hundred thousand dollars, these vessels are beyond the means of traditional inshoremen. Few of them became owners of such boats. But many found work on them; those who spent sixty to one hundred days a year at sea on a longliner might earn two-thirds or more of the income obtainable from a year of work hauling lobster pots and handlines.[30]

Still, the majority of Atlantic fishermen in the early 1990s used small (9 to 15 metre), versatile diesel-driven Cape Island or Northumberland Strait boats for lobstering, longlining, seining, handlining, and gill netting. Crews of one or two put out with the dawn and returned before nightfall, to sell their catch to buyers at the wharves or to the local fish plants. But returns were low. Data collected for the Kirby Task Force on Atlantic Fisheries (1982) show that the average full-time fisherman earned $11 907 annually. Highliners — the top 10 per cent of income earn-

Atlantic fishing
boats

ers, almost all of whom worked the mid- or offshore — made net incomes of
$23 350 or more. In a few rich fishing areas such as Cape Sable Island, incomes
of over $40 000 were not uncommon. Still, one-third of the households of full-time
fishermen (about 8000) had total incomes below Canada's official rural poverty
line. A low Canadian dollar, higher than usual landings, and good prices for lobster in the middle of the decade improved the circumstances of many. But in the
1990s declining fish stocks, market difficulties, and new regulations making it
more difficult for seasonal workers to draw unemployment insurance only wors-
ened the lot of most inshore fishermen. Ironically, though, a few among them prospered, their good fortune a product of location and regulation as much as any
special skill. In one community in northern Cape Breton, for example, seventeen
men licenced to catch crab each grossed approximately $200 000 in ten days of
fishing in 1994; for the other thirty or so families in the village without crab-
licences, there was no such bonanza. Why, they asked, could access to the crab-
fishery not be shared equally among all in the community. Each family might then
have realized $70 000 before expenses, dependence would have been reduced, and
envy and anger might have given way to self-respect.[31]

For Newfoundland fishermen, the post–World War II decades were particularly
tumultuous. In its enthusiasm for modernization and efficiency, the Smallwood
government initiated a "resettlement program" under which hundreds of families
from small, isolated communities were relocated to "growth centres." Here planners projected the establishment of industries and the provision of centralized services (that were arguably too costly to offer residents of scattered outports).
Detached from their familiar places, resettled individuals found their local knowledge of weather and tide and land-based resources devalued, and the scale and
temper of their lives changed irrevocably.[32]

At the same time, efforts to regulate and organize the fishery implemented
under the Commission of Government were set aside as European salt fish markets
declined, and the province became increasingly dependent upon North American
consumers who preferred fresh-frozen fish. Finally, the marketing organization
established in 1947 as Newfoundland Associated Fish Exporters Limited was
deemed in violation of the British North America Act, and did not long survive
Confederation. Without the capital or the expertise to adjust to these new circumstances, the dispersed, labour-intensive, small-boat inshore fishery upon which so
many Newfoundlanders had depended for so long foundered. Meanwhile, the number of large foreign trawlers operating offshore — especially on the Grand Banks,
then as for centuries before one of the richest fishing grounds in the world — began
to increase markedly. Cod stocks were subject to increasingly aggressive exploitation. By the 1960s there was growing concern that they were being seriously diminished. Canadian vessels working offshore took barely 40 per cent of the catch in
the early 1970s and annual yields declined sharply. In 1969, Canada's Atlantic
waters — the Labrador Sea, the Grand Banks, the Nova Scotia Banks and the

Gulf of St. Lawrence — yielded 2.43 million metric tons of groundfish. In 1978 the catch was 1.41 million metric tons. Where foreign trawlers fished most vigorously, on the Grand Banks and especially in the Labrador zone, the decline was most severe. Massive exploitation of the offshore severely reduced shoreward migrations of feeding fish in the spring and early summer, and pushed the inshore fishery (especially on Newfoundland's northeast coast) deeper into depression. With rising protest against the seal fishery by environmental and animal rights groups in the 1970s, and the closure of this traditional supplement to fishing families' incomes in the 1980s, conditions were difficult indeed.[33]

The establishment by the Canadian government in 1977 of a 200-mile zone for control and management of the fishery, and the exclusion of foreign trawlers from these expanded "Canadian" waters, was a belated, and contested, response to these circumstances. Initially, it seemed to work. With a decrease in the number of vessels on the Banks, fish prices rose. Encouraged by federal policy, Canadian companies invested heavily in the expansion of deep-water trawler fleets. With the advantages of earlier establishment and refrigerated truck access to New England and New York markets, Nova Scotian entrepreneurs were at the forefront of these developments. By the 1980s, capital had substantially replaced labour and modern technologies had substantially discounted the value of traditional environmental knowledge in the off-shore fishery. Steel stern-trawlers, operating from a few ports in which processing facilities were concentrated, brought in hundreds of tons of iced groundfish after twelve or fourteen days at sea. Both vessels and freezing plants were owned by large organizations that controlled all facets of the trade from catching to marketing. Among

Outport of
Nfld

them, National Sea Products of Halifax and H. B. Nickerson and Sons Ltd. of North Sydney (merged into a single corporate entity) were regional leaders in the 1980s. The former had some fifty vessels serving its fourteen processing plants in Atlantic Canada (it sold its products internationally under the brand name "High Liner"). It was the country's largest-volume fish processing firm. Despite the increasing cost of operating large modern fishing vessels, representatives of the integrated firms insisted that sophisticated new freezer trawlers were essential to realize the opportunities presented by the 200-mile limit. Whatever its accuracy, this was also an argument for further concentration of ownership and control of the fishing industry. It was resisted by inshore fishermen, who feared elimination of the remaining small, seasonal processing plants and depletion of inshore fish stocks by increased offshore catches, and by the Newfoundland government, which saw the benefits of the new fishery policies accruing disproportionately to Nova Scotia and recognized that they did little to address the plight of their province's large number of increasingly impoverished outport families. Through the 1980s, the Newfoundland government sought increased provincial jurisdiction over offshore resources, but their efforts realized few gains.

The balance between inshore and offshore fisheries is still debated. Corporate representatives argue the importance of economic efficiencies that come with increases in the scale of operations and the need for investment and integrated management. Proponents of the inshore point to the importance of sustaining families, and the capacity of the labour-intensive small boat fishery to provide work for far more people than its offshore counterpart. They have also, on occasion, claimed that the lower investment levels required in the inshore fishery mean that returns to capital are greater there than in the offshore industry. Whatever their merits, the arguments of both sides are essentially moot in the late 1990s. A disastrous decline in cod stocks — seemingly unanticipated despite significant investments in fisheries research and monitoring — forced a moratorium on the fishery. In settlement after settlement across the region, and especially in Newfoundland, boats were pulled out of the water, families faced the loss of their livelihood, and governments scrambled to implement job-training and employment schemes for people who, in many cases, had neither known nor desired any other way of life than they had.

Through the last several decades then, the impact of modern technologies and corporate structures has been to undermine the traditional ways of the fishery, foster frustration and restlessness in the region's small fishing communities, and provided ready material for nostalgia. Today, dwellings still cling to the shore in fishing settlements such as those in the Ragged Islands area of southern Nova Scotia and along much of the Newfoundland coast. But kitchen gardens and small fields beyond are no longer productive of basic necessities. In Nova Scotia in particular, houses that sheltered generations of fishermen and their families are now summer places for distant urbanites. Beyond a few centres in which specialized operations persist there are generally few obvious signs of the fishery; perhaps a small wharf or two; a few boats, moored or drawn up on the beach; and lobster pots stacked on the strand.

In sum, industrialization has transformed the traditional industries of Atlantic Canada. In lumbering, farming, and fishing, expensive machinery has replaced manual labour; sophisticated technology has superseded local knowledge. The capital costs of large-scale production have drastically reduced prospects of independent ownership. Wages and contracts have imposed a set structure where before there was at least the illusion of freedom to work when and as intensively as one wished. Fishermen, who once shared their profits with crews of kith and kin, now resemble wage-workers in other industries. Divorcing home and work to the extreme, full-time trawler workers spend twelve of every fourteen days away from their families. Lumberers, who once set out with sons and brothers to fell tall pines in the vicinity of their homes, now denude swaths of company-controlled forest during a 40-hour work week spent in the cab of a wheeled skidder or Beloit harvester. Farms on which parents and children followed the ox plough and adapted to the rhythm of the seasons are now mechanically sown with crops specified by marketing agencies or multinational corporations.

All of these developments have had a profound effect on the attitudes of people in traditional sectors of the regional economy. Many of them feel a sense of isolation and intrusion — an unease that machines have come between people and their environment and that the decisions of remote and anonymous authorities have impinged upon local life. The changes have reached deep into the texture of rural life. What price the dory fisherman's slowly acquired and intimate knowledge of local seas and skies in all their moods when radar, sonar, and decca navigators find fish and port in the foulest weather? How common now the farmer's satisfaction that his land "fits . . . [him] loose and easy, like . . . [his] old clothes"? How rare the comforting reflection: "That rock there is one my father rolled out, and my son's sons will look at these rocks I am rolling out today"? For those who have lived in small communities steeped in shared experience, where memories and names are kept alive by their connection to everyday features, there is both sadness and resistance at their passing. For them,

> the 'Bart Ramsey place'...[will] always be the 'Bart Ramsey place', however often it change[s] hands....The fire that plundered the forests for miles around when it escaped from George Rawding's pipebowl the day the falling hemlock knocked the pipe from his mouth in the tinder August...[will] always be known as 'the George Rawding fire'. The brook through Peter Herald's meadow...[will] be 'Pete's Brook' as long as water [runs].[34]

Against such a sense of attachment to place, it is hardly surprising, for example, that official insistence on common access to marine resources is disregarded in some fishing communities because "the lobster bottoms were distributed...before any of us can remember and the grounds my father fished were those his father fished before him and there were others before and before and before."[35]

Urban Centres and Industrialization

In the urban centres of the region, on the other hand, the drive, since the 1960s, to create a modern, more viable economic base has largely banished such sentiments. Faced with the decline of older established manufacturing industries and the communities that had grown up around them, politicians and planners attempted to sustain what they could of existing infrastructures, and to attract new industries to the region. Industrial parks were developed and sought prestigious "high-tech" tenants, primary industries won tax concessions, and tourism was promoted as a potentially lucrative tertiary activity. Meanwhile, the number of bureaucrats administering the new programs soared. Where these strategies met with success — as they generally did in the larger cities of the region — townscapes were transformed, new jobs were created, and the range of services and facilities available to residents and visitors alike expanded. Where they did not, places continued to decline, unemployment remained high, and disillusionment grew.

Recent decades have been difficult in industrial Cape Breton. Facing the continuing unprofitability of both steel and coal industries in the 1960s, the Dominion Steel and Coal Company (DOSCO, which succeeded BESCO in 1930 and subsequently became a subsidiary of Hawker Siddeley Canada Limited) decided to end its involvement with both sectors. Ten thousand workers faced redundancy. To stay this setback, Crown corporations took over DOSCO's operations. The federal Cape Breton Development Corporation (DEVCO) assumed control of the mines; the provincial Sydney Steel Corporation (SYSCO) took over the mill. Both sectors continued to decline. In the mid 1970s SYSCO's losses exceeded $40 million a year and debts continued to mount even though a thousand jobs were trimmed in five years after 1973. Employment in the coal fields fell by one-half in the decade after 1967. In an effort to arrest the decline, SYSCO fired a new blast furnace and negotiated new long-term sales agreements beyond the region; mining operations were modernized and returned a profit for the first time in decades. By the mid 1980s, however, federal and provincial governments agreed that the coke ovens in which iron-ore, limestone and coke had been smelted since the end of the nineteenth century would be closed; an electric arc furnace would allow steel production to continue, using recycled scrap metal and raw steel purchased elsewhere to supply the rail mill. Several hundred — in some estimates as many as one thousand — jobs were lost. In twenty years the number of steelworkers in Cape Breton fell from four thousand to fewer than five hundred.[36]

Taking the larger view, and recognizing its responsibility as a Crown corporation for the economic fortunes of Cape Breton as a whole, DEVCO attempted to provide work for redundant miners by offering grants and subsidies to attract new industry to the island. Other programs encouraged local entrepreneurship and small-scale manufacturing. Plants were built to produce products ranging from ornamental welding to machinery formerly manufactured beyond Cape Breton. Fishery- and marine-related activities were fostered. There were also efforts to stimulate the rural economy by offering incentives for sheep farming in the hope that the wool might be used in the revival of "traditional" local crafts; and for the development of bed-and-breakfast accommodation in island homes as a means for residents to supplement their incomes without large capital costs and the problems of underutilized investment that beset more conventional provisions for the highly seasonal tourist trade.

These were significant efforts to turn an economy long dominated by a single industrial complex and vulnerable to the fluctuations of external markets into one using diverse local resources to yield considerably more general prosperity and self-reliance. But success was limited. Such initiatives as the construction of a heavy water plant to serve the needs of the nuclear power industry never returned the jobs or profits anticipated. Companies attracted to this peripheral location by "tax holidays" and other favourable arrangements all too often remained only long enough to capitalize the concessions they had been awarded. The economic downturn of the 1980s, the continuing unprofitability of the mill, a conflagration that closed one of the area's pits, and the destruction of a Glace Bay fish plant by fire pushed unemployment in industrial Cape Breton to excessively high levels. And the situation has hardly improved in the 1990s. Unemployment levels remain unacceptably high; after decades of failure to achieve substantial, sustained improvement morale is low, and economic and social hardship are endemic in this remote location, scarred by the industrial developments of the nineteenth century, and substantially left behind by the course of Canadian growth in the late twentieth century.

In Newfoundland, too, dozens of small places charted similar trajectories of enforced change, adjustment, and decline. The settlement called Riverrun by one who grew up there, stands in a very real sense for them all:

> *Riverrun is now [in the early 1990s] a town almost entirely reliant on its education system, government offices, the service sector and work opportunities in a larger nearby town to sustain its residents. The sounds of chainsaws felling timber and the sight of large woodtrucks transporting the resources disappeared for good in the seventies.*[37]

Most people had come to this settlement from one or another of the small islands in a nearby bay, forsaking the fishery for work in the lumber camps or on the railroad. In doing so they entered a realm "where labour was more individualized than in the collective traditions of shared labour and interests within the extended fisher family." As a result, "family-focussed labour, independence and subsistence" were shunted aside by the new conception of the individual as wage earner. At much the same time, such local events as church- or school-organized plays, concerts, and dances waned as people embraced the popular cultural forms newly available to them through radio, print, and television. Opportunities for inclusive community-building became scarce, and society began to fragment and atomize.

In Saint John, Fredericton, Halifax, and St. John's, meanwhile, there was considerable growth. With rapidly expanding service and construction sectors, these centres attracted migrants from the declining urban and rural areas of the region and other Canadian cities. In Saint John, demolition, new construction, and the development of urban freeways recast the downtown area. Harbour facilities were upgraded and now include a container port, a general-cargo handling area, and an oil-tanker berth serving one of the continent's largest refineries. Nearby thermal and nuclear generating plants offer power to a number of newly established heavy and energy-intensive industries located in specially designed industrial parks. Research-based firms in agriculture and forestry were drawn to Fredericton by the university and the federal agricultural research station. The capital city of New Brunswick prospered, as did other provincial capitals, by the growth of government bureaucracies and the contracts and other opportunities that they generate.

Halifax has become the premier city of the Maritimes, a leader in tertiary activities including research, defence, finance, and transportation. In the 1980s, the servicing of oil and gas drilling ventures on the Scotian Shelf added a further dimension to the Halifax economy. Much has been made of the actual and potential impact of these developments. Claims that almost a quarter of the exploration budget was spent in Nova Scotia likely overestimate the benefits, however. Government figures suggest that some 1400 Nova Scotians worked on the drilling rigs and in support industries (including rig construction) in 1983. Including jobs created by the expenditure of offshore wages in the province, the impact, in terms of employment, probably accounted for less than 1 percent of the provincial labour force.

New office towers built in Halifax by Canada's leading chartered banks reflect the city's importance as a regional financial centre. Government and commercial facilities created by redevelopment (Scotia Square) and restoration (Historic Properties) have added vitality to, and changed, the character of formerly run-down fringes of the urban core. In Halifax as in Saint John and St. John's, late nineteenth-century business blocks of brick and stone have been replaced by glass and concrete high-rise architecture. Suburbs of single-family detached dwellings have spread, often haphazardly, across peripheral subdivisions. In older residential areas on the fringes of the urban cores, row houses and walk-up apartments are still common. Built mostly of wood in the more compact pedestrian cities of the nineteenth century, many of them have declined in quality, and now house the sizable poor and unemployed populations of these major urban centres.

Elsewhere, urban population growth and the expansion of the tertiary sector have had less dramatic impact. Beyond the region's pulp-and-paper and metal mining towns, light industry, warehousing, and distribution have been the most common bases for growth. Investment incentives, lower wage rates, and a largely unorganized work force have attracted foreign firms, such as the Michelin Tire Corporation, to a few of the region's small towns. Moncton, at the crossroads of the three provinces, and the so-called "hub" of the Maritimes, has attracted transportation and warehousing activities to its industrial parks, and has benefitted from an expansion of federal government employment as well as activity associated with the construction of the "fixed link" to Prince Edward Island.

Slight concentrations of secondary manufacturing in Amherst, Truro, and the New Glasgow area have diversified local economies. Industrial parks, a handful of new commercial buildings, a shopping centre or two, and trailer parks marking the outskirts of most urban places in the region mirror the recent surge of urbanization. But many an effort to attract industrial development has followed a familiar pattern. Offered tax breaks, free water and power, nominal rent, and interest free loans, many companies opened plants with much fanfare but ceased operation only a few years afterwards. Thus Clairtone, Bricklin, Gulf Garden Foods, and a General Instruments electronics assembly plant that were offered development incentives in the 1960s and 1970s all closed in short order amidst disappointment and, often, controversy. Far more tragic was the short history of the Westray mine in Pictou County. Opened without government-recommended in-depth evaluations in September 1991, after the last of the old coal mines in the county had closed, Westray offers a harrowing example of what can go wrong when regulations and responsibilities are downplayed in the quest to provide

Halifax
waterfront

jobs in hard times. According to Shaun Comish, who worked there, miners toiled in an atmosphere of "animosity and intimidation." Health and safety concerns were ignored in the push to meet production quotas. Miners were afraid to report violations for fear of reprisal. On 9 May 1992, the mine was closed by a devastating explosion. Twenty-six men were killed. Families and friends were robbed of kin and companions. Hope and trust suffered equally grievous blows. Hundreds had died in Nova Scotia's mines before the Westray explosion. But in 1992 the deaths bore an especially steep price; the blood on the coal might never have been spilt had those in authority paid more attention to workers' concerns and safety regulations.[38]

So, too, in Newfoundland the ill-fated oil refinery at Come by Chance, and the failed linerboard mill at Stephenville, were only the most high-profile examples of the pitfalls awaiting politicians desperate for development, and ready to concede more than they otherwise might, to secure the appearance of success and the promise of improvement. Earlier ventures that had depended upon a familiar readiness to concede resource rights to encourage industrial growth also caused difficulties for the Newfoundland government in the 1980s. Citing aging equipment and a downturn in world markets, for example, Bowater Mersey threatened closure of their pulp and paper mill around which the regional centre of Corner Brook had grown up since the 1920s, and government assistance was necessary for the plant to continue in operation under new ownership.

More complex were the circumstances surrounding the development of hydro-electric power potential on the Upper Churchill River in Labrador in the early 1970s. One of the largest and most economical producers of electricity in Canada at the time of its completion, the Churchill Falls project operated under an agreement (necessitated by Québec's refusal to allow Newfoundland to transmit power across its territory) that required the sale of its electricity to Hydro Québec at the provincial border. The rate, established for 65 years by a contract signed in 1967, was less than three mills per kilowatt hour. Within little more than a decade, the market price was ten to fifteen times this, and it has risen since. Québec continues to capture an enormous windfall from this agreement, which remains a focus of severe contention between the two provinces. Newfoundland's efforts to seek resolution through the federal government have failed, and the province's argument that the interprovincial movement of oil and gas by pipeline establishes a precedent for the unhindered transmission of electricity across Québec territory has fallen on deaf ears.

Against this backdrop, and a report from the Economic Council of Canada that concluded, in 1980, that "huge sums of money in the form of foregone revenue from natural resources are leaving the province," it is hardly surprising that Newfoundland fought so tenaciously for control of the seabed resources of the continental shelf once gas and oil reserves were discovered on the Labrador Margin and in the Hibernia field in the 1970s. By some accounts, however, much of what the province fought for was lost when the Atlantic Accord was signed in 1985. By this agreement, the federal government retained legal title to offshore oil and gas, and Newfoundland was guaranteed some control over exploitation of the resource and the revenues flowing from it. Critics suggested that the provincial government actually gained too little, that the agenda was set by the oil industry, and that it mattered little whether regulations were federal or provincial if both were shaped by corporate concerns rather than the social and economic interests of Newfoundlanders. Certainly the accord deprived Newfoundland of revenues that it hoped to derive from the front-end of the project through the Newfoundland and Labrador Petroleum Corporation. A decade later, the benefits of this arrangement have yet to be felt, at least in part because world prices have generally remained below the estimated break-even point of US $27 barrel for Hibernia oil, and production has not yet begun in earnest.[39]

THE FUTURE: AT THE CROSSROADS

In the 1980s it seemed clear that the Atlantic region and its people were at a crossroads. Simply put, the question before them was whether economic growth should be pursued in an effort to overcome the region's underdeveloped hinterland status, or whether the benefits of underdevelopment should be recognized as the foundation for a distinctly different path into the future. Rephrased, the question turned on whether the region should break with its historical roots (by becoming more and more similar to countless other places in the modern world), or nurture and protect its distinctive legacy. On the one hand there was the promise of more employment, higher wages, a more broadly based economy, and a better standard of material life — and with it, almost certainly, environmental pollution, urban concentration, and a general decline in the quality of Maritime living. On the other hand there was a sense of the rootedness of people in place, of the satisfaction that familiarity with one's setting can bring. As one of Nova Scotia's industrial development agencies posed it, the question was how to equate the crash of industry with the cry of the loon.

Each side of the debate had its forceful proponents. By and large, politicians, business people, developers, and planners embraced economic expansion. Growth had been the underlying assumption of North American life for generations, after all. The Industrial Heartland owed its prosperity to its strong, broadly based, and expanding manufacturing economy; surely it was time for the hinterland to secure similar benefits by promoting similar development? "Catch up" was the game plan. "In Province House and on the hustings," wrote Malcolm Ross (an unreconstructed Maritimer, critical of this position, who taught English at Dalhousie University), "the cry is 'Forward' — on to that brave new world of super-port and oil spill, with offshore the nuclear island, and on-shore the twelve-lane highway obliterating the village and valley it takes the tourist to."[40]

By the 1980s, however, a growing number of North Americans had began to question the wisdom of continued economic growth. Rachel Carson, Paul Ehrlich, E. F.

Schumacher, the scientists of the Club of Rome, and others raised public consciousness about desecration of the environment, the exhaustability of resources, the fragility of our planetary ecology, and the need for conservation. At the same time, many turned away from the technological materialism and pressure-filled existence of twentieth-century North American society to live lives at a less punishing pace in surroundings less austere than those of the large modern metropolis. Some among them (writers, professionals, and members of the counterculture) had settled in the Maritimes, where they added weight to local unease at the passing of traditional livelihoods and the decline of community life. The counterpoint implicit in Ross's sarcastic comment spawned its own strong criticisms of destructive progress, and resistance to the idea that life in the region should be patterned on developments in the rest of North America. After all, observed those on this side of the argument, "Chicago, Detroit, Jersey City, Hoboken and even Hamilton, Ont. look nothing at all like the New Jerusalem."[41]

Central to this position was the recognition that the underdevelopment of the Atlantic region, and the Maritime provinces in particular, had its roots — as this chapter has shown — in those improvements in transport and technology that favoured large producers over small, fostered centralization, and based the organization of land use and productive capacity on economic principles that assumed the benefits of development at the sub-continental scale, and the efficacy of the state in implementing or encouraging this. As Canada developed, the eastern hinterland fell under the sway of the centre; it contributed resources — both people and products — to the expanding national economy and felt the growing impress of Heartland dominance. With transport costs rising steeply in response to the first oil crisis of the 1970s, it seemed, in the early 1980s, that these circumstances might well change. If interregional exchange were to become more costly, the competitive reach of heartland manufacturers would decline, producers would gain a measure of protection in local markets, and a tendency toward industrial decentralization might be encouraged by the capacity for small-scale production conferred by new technologies.

Agriculture and the fishery would also respond to changed circumstances. Regional specialization based on the optimum use of soil and climate, and predicated on the assumption that needs and preferences for locally "marginal" crops could be met by cheap imports, would decline and there might be a resurgence of diversified production for local requirements. Thus might the region's farms be revitalized. With less need to compete with and against world producers, reliance upon expensive technology could give way to less capital-intensive production. With financial requirements reduced, family farms would no longer be threatened by corporate consolidation; abandoned land might be reclaimed; and the decline of rural population might be reversed. In the fishery, smaller processing plants, more responsive to local needs, might emerge, and more fishermen's cooperatives might be established. No longer, in short, would it be necessary for industrial development strategies to depend on attracting large, highly specialized activities, or offer massive incentives to more commonplace industries, to be competitive in the national market. Instead, a variety of smaller plants, dispersed evenly within the region, might provide for local demands.

Decentralized small-scale production would benefit both environment and society. Urban growth might slow, heavy applications of chemical fertilizers could be reduced, exploitation might be tempered by a sense of stewardship. Small places, once considered redundant, could again become the foci of daily lives. Opportunities to participate in government and decision making at the local level would increase the attachment of people to place. Integration of residents into their communities would occur in many ways. Thus, ran this line of thought, the region possessed a peculiar advantage in its relative underdevelopment. Backwardness by the standards of the day embodied prospects of a head start for growth attuned to late twentieth-century circumstances. Relatively untrammelled by the technological infrastructure of big industry, Maritimers were well set to improve their economic position without destroying their countrysides and their beaches, without undermining the distinctive qualities of regional life, without, in short, welcoming "the Trojan Horse of troubles that has already wrecked much of the pleasure of being alive in so many cities of the world."[42]

The passage of time has exposed both the difficulties and dangers of attempting to divine the future. Little of this optimistic vision has been realised. Steeply rising oil prices, seen as a powerful instrument of transformation in the 1980s, soon declined and remain far below once-projected levels. Meanwhile, significant gains in efficiency have helped control costs in most oil-dependent activities. "Small" may still be beautiful in many eyes, but the centralizing tendencies of modern industrial society have hardly faded, and the community-orientated, environmentally-sensitive forms of production envisaged in the aftermath of the oil-shock are by no means common, although new developments are often held up as success stories of one sort and another.

Response to the New Economy of the 1990s

Perhaps the most noticed of the new initiatives has been the development in New Brunswick of what Premier Frank McKenna calls "high quality, highly skilled, high paying, pollution free jobs" in the high-tech communications, information, and business-service sector. In four years after 1992, more than thirty companies created over 4000 jobs in "1-800 Call Centres" in the province. Attracted by NBTel's advanced telecommunications infrastructure, the availability of bilingual workers, and government incentives such as tax-free 1-800 numbers, training subsidies, and low pay-roll taxes — as well, perhaps, as a low-wage, largely non-union work force — such companies as Purolator Courier, UPS, IBM, Xerox, Air Canada, Canadian Pacific, the Royal Bank of Canada, and Northern Telecom established customer-service centres using toll-free telephone numbers in Moncton, Fredericton, Saint

John, St. Stephen, Bathurst and elsewhere. Workers — of whom 60 to 70 per cent are women — earn $15 000 to $25 000 and competition for jobs has been strong.[43]

Elsewhere, innovative responses to new opportunities have also provided jobs and income to the region. Among several possible examples, the manufacture of high-quality stainless steel cookware on Prince Edward Island, the production of organic foodstuffs, and the international trade in blue-fin tuna from the Bay of Fundy suggest the ingenuity of many local efforts to find and develop market niches in a fast-changing world. Started by an Italian company as Paderno (Canada) when the Prince Edward Island government was encouraging manufacturers to locate in a new industrial park on the outskirts of Charlottetown in the late 1970s, Padinox Inc. was created when the parent company encountered financial difficulties and disposed of its island assets to Canadian buyers. Today the company has between fifty and one hundred employees, manufacturing and selling cookware under the Paderno and Chaudier brand names. Stainless steel, sourced from the United States, is purchased from a distributor in Québec. With the introduction of "Factory Sales" in stores across the country (95 per cent of sales are Canadian, the remainder in the eastern United States) and the development of six Paderno factory outlets in the 1990s, production increased. In the late 1990s sales are five or six times greater than they were in 1979. New Brunswick's Speerville Flour Mill is a worker-owned cooperative that opened in the 1980s, and found markets for its organically grown products around the Maritimes and in northern Maine. Catering to the Japanese consumer's desire for top-quality fresh fish, a small number of individuals harpoon individual blue-fin tuna (rather than catch them with nets or hooks) in order to kill them quickly and avoid discoloration of the flesh. The catch is small, sometimes only one or two fish a week, and the investment in boats and gear is substantial. Costs are also high. Captured and cleaned, the fish are packed in ice and carried, quickly, back to shore. Then they are flown in special containers directly to market. Less than forty-eight hours after being hauled from the waters of the Fundy basin, they will lie on the auction floor of the Tsukiji market near Tokyo. At 9000 or 10 000 yen per kilo, even moderately sized fish bring several thousand dollars and one remarkable giant sold for $50 000.[44]

These are the cases that exemplify what is possible. They are by no means the norm. Economic, political, and ecological forces have continued to buffet Atlantic Canada and its peoples. Unemployment remains exceptionally high across the region, and egregiously so in many parts of it. Declining towns and boarded-up stores are everywhere. The fishery, the forest and mining industries, and much of the region's agriculture confront various problems of marginality and declining resources. Yet many see the road to future improvement as an extension of the path through the recent past. "J-O-B-S," the four letters Newfoundland Premier Joey Smallwood once speculated might be inscribed on his heart, are still the priority of regional politicians, and in this capital-poor area, the idea that they will be created through the big investments of foreign firms is deeply entrenched.

Consider, by way of illustration, the case of Long Harbour, on the eastern shore of Placentia Bay, Newfoundland. In the late 1960s, an elemental-phosphate processing plant was built there on the promise of cheap electricity. Four hundred jobs were created. But the environmental consequences were severe. Months after the plant opened, part of Placentia Bay was closed to the fishery, and vegetation for miles around was destroyed or damaged by fluoride emissions. Pollution-control devices were installed but great damage had been done. Then, when rising energy costs led to renegotiation of the original power contract, the $40-million plant ran into difficulties. In 1989 it was closed. A year or so later the community began to consider the possibility of finding economic salvation in the plans of another foreign-owned company. With funding for a feasibility study provided in part by ACOA, Long Harbour councillors sought to attract an American company to build a waste-incinerating, power-generating plant in their community. When protestors from beyond Long Harbour questioned the wisdom of this initiative, and proclaimed the indignity of Newfoundland becoming a disposal site for up to seven million pounds of American garbage each day, the ready response of Long Harbour residents framed the issue in no uncertain terms: "Dignity," they observed, "does not put food on the table."[45]

On the face of it, the McKenna government's attempts to link the future of New Brunswick to the rapidly expanding high-technology industries of the 1990s are a radical departure from the oft-repeated regional pattern. Yet even this initiative has its critics. Although some would suggest that it bears comparison with efforts to alleviate the consequences of the precipitous decline in the Massachusetts textile industry after World War II through the development of high-technology manufacturing along the "Route 128 corridor" around Boston, others are less certain. The "electronics parkway," they point out, profited from its connections to Boston's universities, from the inventive talents of highly educated researchers, and from the long-standing commitment to high-quality education in the state. It created things — electronics, aerospace technologies, instrumentation, computers — and had an inherent capacity to spawn spin-off companies. By contrast New Brunswick's Call Centres provide jobs with limited growth potential. Perhaps, the argument goes, they are more akin to Mexican *maquiladoras* — free trade zones to which transnational companies are attracted by tax incentives, a lack of tariffs, and a relatively low-wage labour force — than to the Boston miracle; here, footloose companies perch so long as it is in their economic interest to do so.

Which Path?

What then of the future? The question is never far below the surface in Atlantic Canada. How might the secession of Québec affect the region? Will deficit-cutting, social-program-trimming federal policies produce ever greater immiseration? What will happen when eastern forests no longer provide a viable basis for the pulp industry? Is there a future for communities once dependent on the cod fishery? Will Nova

Scotia and Prince Edward Island, in particular, be turned into a mix of extensive tree farms and "comfortable rustic" retreats for the wealthy by non-resident landowners and local developers? Will "the last person to leave the island turn out the lights" as requested, some years ago, by a concerned Newfoundlander distressed that the conversion of the province into "a therapeutic space for post-modern tourism" left little for locals to do but reminisce and entertain a fading nostalgia for the past. There are no firm answers. But there are possibilities that promise a different and brighter road ahead than that lately travelled by many residents of the Atlantic region.[46]

At the end of the millenium, as in the 1980s, some still place their faith in the ability of people to improve their lives and the capacity of "alternative economics" to allow them to do so. They point to the long and strong tradition of cooperative enterprise in the Maritime provinces, epitomised by the Antigonish movement, and look to the success of the cooperative movement in Cheticamp as a model of what might be possible.[47] Since the establishment of a fish marketing co-op in 1917, this small Cape Breton community has supported the development of local food co-ops, a credit union and most recently a co-op dinner theatre that caters to the tourist industry. Together these locally owned and operated enterprises provided three hundred jobs and made a payroll, in the mid 1990s, of $2.3 million, almost all of which circulated in the community. Activists and community workers also promote the development of alternative currencies and barter systems to address the cash-flow problems many people confront under conventional economic practices, and envisage the formation of community land trusts whereby residents can contribute funds to the development of affordable housing or community facilities.

In similar vein, the Toronto-based non-profit Calmeadow Foundation established a pilot peer-lending program in Shelburne County, Nova Scotia. Under this plan, four to ten people form a Partnership Assistance for Rural Development group that is responsible for approving members' small-business start-up loans funded by the Foundation. Loans are granted on the individual's "good name" in the community rather than on collateral or credit rating, and default rates in the first two years were zero. Peer lending, says a Calmeadow representative, is "the Trojan Horse method of community development....it isn't going to solve a community's problems, but once people get together in groups they start to organize."[48]

Environmental concerns are another focus of debate about the future in Atlantic Canada. Growing global anxiety about the consequences of increasing environmental damage is shared by many in the region, where the closure of the Atlantic cod fishery provides an arresting local example of what can go wrong when long-term vision and effective resource management are lacking. Paradoxically, however, a dearth of obvious alternatives to current patterns of resource dependency, and the economic marginality that has resulted from inadequate understanding and short-sighted exploitation of the environment, militate against the implementation of significant change; fishers and other primary producers sorely affected by degradation of the resource base on which they depend are generally in no economic position to restrict their demands on the environment. As the leader of the Cape Breton coal miner's union put it after listening to David Suzuki speak about the magnitude of the environmental crisis confronting the world, he had come to understand why coal mining would soon have to cease, but he had no idea what to say "to his 2500 union members who had no other way of making a living." Environmentalists advocate more local production for local consumption as they point out the foolishness inherent in the calculation that in North America on average food travels 3200 km, most by truck, before it is consumed. Specialty products, such as organic vegetables or stone ground flour, appealing to wealthier consumers can find a viable niche in this equation. But so long as imported food is cheaper than local produce on supermarket shelves, families on tight budgets are hard pressed to pay the premium that environmental good sense and local patriotism require. A sounder environmental future will not be gained by exhortation alone. Concern for the planet will have to be spliced with other movements for political and social change if old-established patterns of profligate resource use and unacceptable levels of environmental pollution and degradation are not to run deep into the new millenium.[49]

Political activists have seen opportunity in this. Recognizing the hold that arguments for a cleaner, healthier environment — characterized by historian Samuel P. Hays as a crusade for beauty, health, and permanence — have taken on the public imagination, they call for a new synthesis of ideas. According to Rick Williams, a member of the Nova Scotia New Democratic Party (NDP) and associate editor of *New Maritimes* — a magazine of culture and politics that offered an independent per-

New bridge
to PEI

spective on regional affairs between 1981 and 1997 — the political left has to come to grips with the limits to growth.[50] Democratization of the economy and the redistribution of wealth might remain fundamental goals, but residents of both the Atlantic region and the country at large, he argues, need "to combine the popular politics and radical humanism of the socialist left, the creativity and person-to-person effectiveness of feminism, and the energy and future-mindedness of the greens" to realize a better world than most of those in the region now know. Yet there is little sign that this is likely to occur in Atlantic Canada. Support for the political left has been enigmatic. Until the 1997 federal election (when the leadership of Halifax lawyer Alexa McDonough and visceral rejection of the Liberal party produced an unprecedented gain in NDP support), strong backing for militant action in the workplace and in the interests of community survival has rarely translated into votes and electoral success for NDP or Green parties in the formal political arena.[51]

In this context, there is, perhaps, some point in considering the possibilities of moving forward by turning back to an old idea — Maritime Union. Last considered seriously in the 1960s, when the idea of creating a substantial political entity from the three provinces foundered on the unwillingness of people, politicians, and legislatures to give up power and access to it, the idea of a full-fledged, political and economic amalgamation of the Maritime or Atlantic provinces still lacks powerful wings. On current appearances at least, any proposal to create one big province is bound to run into resistance from scores of small interest groups. Yet there may be room — even the absolute necessity — for changes in the organization of regional political and fiscal powers in the face of circumstances now confronting the country.

As much of the preceding analysis makes clear, the Maritime provinces and Newfoundland have been shaped over the last one hundred and fifty years by technologies, policies, settings, and opportunities that have far less potency and pertinence at the end of the century than was once the case. The National Policy, improvements in communication, economies of scale, the growth of population in central and western Canada, the consolidation of corporate and financial power, all worked to concentrate manufacturing, wealth, and political clout in the centre. The country divided into core and periphery, heartland and hinterlands. Regional disparities increased. After World War II these tendencies were ameliorated to some extent by income transfers and equalization payments. Much as they helped people and communities economically, however, they simply reinforced the power of the centre over the margin.

At the end of the twentieth century, the conditions and assumptions upon which these arrangements were founded have changed dramatically. Modern communications, or more precisely, the capacities they offer transnational corporations and money traders to effectively ignore political boundaries, are widely seen to be undermining the ability of nation states to control the most basic facets of their economies. The North American Free Trade Agreement is refocussing tendencies toward the concentration of wealth and capital at the continental scale. The country-wide social compact upon which social security and equalization programs were built has been significantly weakened if not dismantled by the deficit-reduction campaign of neo-conservative economists and their political followers. The country is being reshaped, and the Atlantic provinces — hardly the foci of prosperity under the old regime — will be forced to readjust.

If chances lie in change, there are opportunities ahead. They exist not in the dead horse of formal political union, but perhaps in the creation of new political structures that capitalize on modern technologies to establish a tiered system of regional administration that is at once effective and sensitive to the need for balance among regional, subregional and local powers and interests. The challenge is to establish an arrangement capable of defining widely acceptable regional goals yet able to respond to changing needs and local circumstances. The careful allocation of responsibilities is essential. Although massive decentralization of administration and services might seem attractive as a means to eliminate faceless bureaucracies and move control of institutions closer to the people, it is not a universal panacea. Some things are simply too costly, too complex, too abstract to be dealt with effectively at the local level. Taxation, the development and maintenance of some infrastructure, the defence of human and cultural rights, some broad policy and economic planning questions — these are all matters possibly better handled at the regional than the local scale. Other concerns — some social/medical services, economic development, cultural questions — may be dealt with most effectively at an intermediate scale, at the level of sub-districts defined to reflect historical, cultural, and economic differences within the region. Finally, the provision of many services and the implementation of work-site practices (both conceivably defined in general terms at the regional or sub-district level) might best be done by communities.

Consider fisheries management for example. Formally trained fisheries scientists might best be employed by the regional authority to consult with and advise fishing communities, and to establish overall parameters for management of fish stocks. But with catch-quotas established for each community or small district, responsibility for managing and controlling the fishery might be very effectively devolved to the local level. Existing highly centralized licensing and policing arrangements (resented by fishermen and perhaps subverted as often as supported by them) could be replaced by mutual surveillance programs. Members of each community would feel a collective interest in effective management of the fishery; they might opt to develop a longliner-based rather than a dragger-based industry for the extra on-board jobs that would result, the prospect of serving more lucrative markets, environmental reasons (longlining does less damage to the seabed and kills fewer undersized fish), or to achieve all three of these ends. In any event, the decisions, under this scheme, would be theirs, although they might receive assistance in making them and have market connections facilitated by organizations structured at the sub-district scale to carry out market and product development work and encourage cooperation among individuals and communities.[52]

The precise details are unimportant here. They would need to be worked out to reflect the goals and circumstances of people across the region. Far more significant is the general direction in which these musings trend. They speak, in the end, for the development of a more democratic and autonomous society; for the validation of local knowledge; for the fuller empowerment of ordinary citizens; and for the conviction that people deserve the opportunity to play meaningful and important parts in shaping their communities. To realize these goals will not be easy. Faith and commitment as well as a substantial investment in people, through the provision of high-quality education and other services that will allow them to reach and utilize their potential, will be essential. Yet if these proposals, or something like them, were to be implemented, they would reverse a century or more of declining regional and communal control over the economic affairs of the region. They might also break the defensive shell that imparts a negative cast to Atlantic (or more precisely Maritime and Newfoundland) regionalism, and contribute toward the emergence of a place whose future holds more promise than that offered its people for many a decade. In this scenario, the Atlantic region would come to be seen as a place in which people had "got it right" rather than as an area in decline because things had "gone wrong." It might serve as a model of what is possible when the energy, wit, resilience, commitment, and competence of a people are liberated and directed to productive ends, rather than be constructed over and over again in the popular imagination as a romantic Neverland of nostalgic escapism. It is profoundly to be wished.

NOTES

[1] Henri Lefebvre, *The Production of Space* (Oxford: Blackwell, 1991); see also E. R. Forbes, *Challenging the Regional Stereotype: Essays on the 20th Century Maritimes* (Fredericton: Acadiensis Press, 1989).

[2] See Chris Gudgeon, *An Unfinished Conversation: The Life and Music of Stan Rogers* (Toronto: Viking Penguin, 1993).

[3] Alistair MacLeod, "The Boat," in *The Lost Salt Gift of Blood* (Toronto: McClelland and Stewart, 1976), reprinted in G. Peabody, ed. *Best Maritime Short Stories* (Halifax: Formac Publishing, 1988), pp. 2–22.

[4] *The Financial Post* (Toronto), 2 May 1970, p. 16; Sandra Gottlieb, "Newfoundland: Canada's Place Apart," *Newfoundland Lifestyle*, 2, 1 (1983), p. 29, quoted in James Overton, *Making a World of Difference: Essays on Tourism, Culture and Development in Newfoundland* (St. John's: ISER, 1996), p. 146. This book includes several thoughtful essays on the general point of this introduction. See also Ian MacKay, *The Quest of the Folk* (Montreal and Kingston: McGill-Queen's University Press, 1994) for more on this theme; Ralph Matthews, *The Creation of Regional Dependency* (Toronto: University of Toronto Press, 1983); Newfoundland and Labrador, *500th Anniversary Festivals & Events Calendar 1997* (np: nd).

[5] The best account of this movement is E. R. Forbes, *Maritime Rights: The Maritime Rights Movement, 1919–1927. A Study in Canadian Regionalism* (Montreal and Kingston: McGill-Queens' University Press, 1979).

[6] Charles Bruce, "Words are Never Enough," in R. Cockburn and R. Gibbs, eds. *Ninety Seasons: Modern Poetry from the Maritimes* (Toronto: McClelland and Stewart, 1974), p. 62. Stan Rogers, "Fisherman's Wharf" from the album *Fogarty's Cove* (Dundas, Ontario: Fogarty's Cove Music, 1977). For a brief discussion of some of these issues, see Graeme Wynn, "The Mark of the Maritimes," in R. Berry and J. Acheson, eds. *Regionalism and National Identity* (Christchurch: ASCANZ, 1985), pp. 555–68; Ann-Marie MacDonald, *Fall on Your Knees* (Toronto: Alfred A. Knopf Canada, 1996), p. 241.

[7] Some of what follows in this section occurs in original and often fuller form in Graeme Wynn, "The Maritimes: The Geography of Fragmentation and Underdevelopment," in L.D. McCann, ed. *Heartland and Hinterland: A Geography of Canada*, 2d ed (Scarborough, Ontario: Prentice-Hall Canada Inc, 1987), pp.186–210. The most recent and useful general accounts of the history of the Atlantic region through the period covered in these paragraphs are to be found in E. R. Forbes and D. A. Muise, eds. *The Atlantic Provinces in Confederation: A History* (Toronto: University of Toronto Press, 1993) and P. A. Buckner and J. G. Reid, eds. *The Atlantic Region to Confederation: A History* (Toronto: University of Toronto Press, 1994), which might also be consulted for bibliographic purposes. For a geographical account of rural change in the region, see Graeme Wynn with Robert MacKinnon, "The Countryside of Atlantic Canada," forthcoming in Brian Osborne,ed. *Canada's Countryside*.

[8] Rusty Bitterman, Robert MacKinnon and Graeme Wynn, "Of Inequality and Interdependence in the Nova Scotia Countryside," *Canadian Historical Review*, 74 (1993), 1–43; T. W. Acheson, "New Brunswick Agriculture at the End of the Colonial Era: A Reassessment," in Kris Inwood, ed. *Farm, Factory and Fortune. New Studies in the Economic History of the Maritime Provinces* (Fredericton: Acadiensis Press, 1993), pp. 37–60.

[9] T. W. Acheson, "The Maritimes and 'Empire Canada'", in David J. Bercuson, ed. *Canada and the Burden of Unity* (Toronto: Macmillan, 1977), p. 103.

[10] The classic article here is T. W. Acheson, "The National Policy and Industrialization of the Maritimes, 1880–1910," *Acadiensis*, 1 (1972), 2–34.

[11] For a useful overview and analysis, see David G. Alexander, "Economic Growth in the Atlantic Region, 1880–1940," *Acadiensis*, 8,1 (1978), 47–76.

[12] David Frank, "The Cape Breton Coal Industry and the Rise and Fall of the British Empire Steel Corporation," *Acadiensis*, 7, 1 (1977), 3–34.

[13] Michael Staveley, "Newfoundland: Economy and Society at the Margin," in McCann, *Heartland and Hinterland*, p. 272. I am indebted to Dr. Staveley for his generous invitation to make such use as I wished of his original chapter in this necessarily much recast treatment of the Atlantic region in a single essay. See also James K. Hiller, "The Origins of the Pulp and Paper Industry in Newfoundland," *Acadiensis*, 11, 2 (1982), 42–68.

[14] Staveley, "Newfoundland," pp. 266–9. See also David G. Alexander, "Newfoundland's Traditional Economy and Development to 1934," in J. K. Hiller and P. Neary, eds. *Newfoundland in the Nineteenth and Twentieth Centuries: Essays in Interpretation* (Toronto: University of Toronto Press, 1980); Shannon Ryan, *Fish Out of Water: The Newfoundland Saltfish Trade, 1814–1914* (St. John's: ISER, 1986), and J. K. Hiller, "The Newfoundland Seal Fishery: An Historical Introduction," *Bulletin of Canadian Studies*, 7, 2 (1983–4), 49–72.

[15] L. D. McCann, "Metropolitanism and Branch Businesses in the Maritimes, 1881–1931," *Acadiensis*, 13, 1 (1983), 111–25, and J. D. Frost, "The 'Nationalization' of the Bank of Nova Scotia, 1880–1910," *Acadiensis*, 12, 1 (1982), 3–38. Neil C. Quigley, Ian M. Drummond and Lewis T. Evans, "Regional Transfers of Funds through the Canadian Banking System and Maritime Economic Development, 1895–1933," in Inwood ed. *Farm, Factory and Fortune*, pp. 219–50, resist the implication that bank operations hindered economic growth by discriminating against the Maritimes.

[16] The strongest statement of this interpretation is E. R. Forbes, "Misguided Symmetry: The Destruction of Regional Transportation Policy for the Maritimes," in D. J. Bercuson, ed. *Canada and the Burden of Unity* (Toronto: Macmillan, 1977), pp. 60–86. Ken Cruikshank has called for a more nuanced assessment of the role of the enigmatic railroad in " The Intercolonial Railway, Freight Rates, and the Maritime Economy," in Inwood, ed. *Farm, Factory and Fortune*, pp. 171–96.

[17] Bill Parenteau, "Pulp, Paper and Poverty," *New Maritimes*, VII, 4 (March/April 1989), 21–26.

[18] Still, these improvements were less significant than those in central Canada; see E. R. Forbes, "Consolidating Disparity: The Maritimes and the Industrialization of Canada During the Second World War," *Acadiensis*, 15, 2 (1986), 3–27.

[19] Donald J. Savoie, *Regional Economic Development: Canada's Search for Solutions.* 2d ed. (Toronto: 1992); Roy E. George, *The Life and Times of Industrial Estates Limited* (Halifax: 1974).

[20] A useful, accessible basic reference for comparison of economic, demographic, and social conditions in the Atlantic region with those in the rest of the country is The Economic Council of Canada, *Living Together: A Study of Regional Disparities* (Ottawa: Ministry of Supply and Services, 1977).

[21] Alan A. Brookes, "The Golden Age and the Exodus: the Case of Canning, King's County," *Acadiensis*, 11(1981), 57–82. See also P. A. Thornton, "The Problem of Outmigration from Atlantic Canada, 1871–1921: A New Look," *Acadiensis*, 15, 1 (1985), 3–34.

[22] As most effectively portrayed in Donna Gallant's short story, "What's a Fella Gonna Do?" in Michael O. Nowlan, ed. *The Maritime Experience* (Toronto: Macmillan, 1975), pp. 89–91.

[23] Jean Daigle, ed. *The Acadians of the Maritimes: Thematic Studies* (Moncton: 1982).

[24] D. A. Muise, "The Great Transformation: Changing the Urban Face of Nova Scotia, 1871–1921," *Nova Scotia Historical Review*, 11, 2 (1991),1–42.

[25] Edna Staebler, *Cape Breton Harbour* (Toronto: McClelland and Stewart, 1972), p. 94.

[26] Kenneth Bagnell, "The Evening Town I Knew as Morning," *The Globe Magazine*, (Toronto) 10 October 1970.

[27] Verner Smitheram, David Milne, and Satadal Dasgupta, eds. *The Garden Transformed: Prince Edward Island, 1948–1980* (Charlottetown, 1982).

[28] Tom Murphy, "The McCain's Revolution: The New Capitalism in New Brunswick Potato Farming," *New Maritimes*, 5, 6 (1987), 7–11.

[29] *New Maritimes* (Enfield, Nova Scotia) 3, 4 (December 1984–January 1985), p. 10.

[30] Peter A. Sinclair, *From Traps to Draggers: Domestic Commodity Production in Northwest Newfoundland, 1850–1982* (St. John's: ISER, 1985); Anthony Davis, *Dire Straits. The Dilemma of a Fishery: The Case of Digby Neck and the Islands* (St. John's: ISER, 1991).

[31] Women's Fishnet/Lori Cox, "Health of Ocean, Community, Society," *New Maritimes*, XIII, 3 (January/February 1995), 13–14.

[32] Parzival Copes, *The Resettlement of Fishing Communities in Newfoundland* (Ottawa, 1972); Farley Mowat, Parzival Copes, Noel Iverson, D. Ralph Matthews, "The Fate of the Outport Newfoundlander: Four Views of Resettlement," in Peter Neary, ed. *The Political Economy of Newfoundland, 1929–1972* (Toronto: 1973); C. Grant Head, "Settlement Migration in Central Bonavista Bay, Newfoundland," in R. L. Gentilcore, ed. *Canada's Changing Geography* (Scarborough, Ontario: Prentice Hall, 1967), pp. 92–110.

[33] Staveley, "Newfoundland," pp. 274–9.

[34] Ernest Buckler, *The Mountain and The Valley* (Toronto: McClelland and Stewart, 1961), p.157.

[35] MacLeod, "The Boat," p. 19.

[36] Rick Williams, "Shifting Gears at SYSCO," *New Maritimes*, VII, 4 (March/April 1989), 17–19; Don MacPherson, "Tar Pond Tango," *New Maritimes*, VIII, 4 (March/April 1990), 16–19.

[37] Ursula Kelly, *Marketing Place: Cultural Politics, Regionalism and Reading* (Halifax: Fernwood Publishing, 1993), pp. 51-5.

[38] Shaun Comish, *The Westray Tragedy: A Miner's Story* (Halifax: Fernwood Publishing, 1993); Harry Glasbeek and Eric Tucker, "Death by Consensus: The Westray Mine Story," in David Frank and Gregory S. Kealey, eds. *Labour and Working-Class History in Atlantic Canada* (St. John's ISER, 1995), pp. 399–439.

[39] Staveley, "Newfoundland," pp. 279–83; Brian O'Neill, "Victims of Hibernia: Truth and Development," *New Maritimes*, IX, 3 (January/February 1991), 13–21.

[40] Malcolm Ross, "Fort, Fog and Fiddlehead: Some New Atlantic Writing," *Acadiensis*, 3 (1974), 10.

[41] Rachel Carson, *Silent Spring* (Boston: Houghton Mifflin, 1962); Paul Ehrlich, *Ecocatastrophe* (San Francisco: City Light Books, 1969); E. F. Schumacher, *Small is Beautiful: Economics as if People Mattered* (New York: Liveright, 1975); Ross, "Fort, Fog and Fiddlehead," p. 121.

[42] Atlantic Provinces Economic Council, *The Atlantic Economy: Sixth Annual Review* (Halifax, 1972), p. 58.

[43] Joan MacFarland, "Many Are Called, But What Are the Choices," *New Maritimes*, XIV, 6 (July/August 1996), 10–19.

[44] George Emerson, "A Yen for Tuna," *Saturday Night*, 112, 3 (April 1997), 31–4.

[45] Andy Pedersen, "Devastated by Development," *New Maritimes*, XI, 6 (July/August 1993), 6–12.

[46] Jim Foulds and Stephen Manley, "'Toothpicks' and the Forests of Tomorrow," *New Maritimes*, VIII, 6 (July/August 1990), 12–3; Kathryn Morse, "Sorting Through the Fisheries Tangle," *New Maritimes*, VIII, 2 (November/December 1989), 16–25; The Land Research Group, "Whither Our Land? Who Owns Nova Scotia and What Are They Doing With It?" *New Maritimes*, VIII, 6 (July/August 1990), 14–25; the "lights" quotation is from the St. John's *Evening Telegram*, 6 May 1977, in Overton, *Making A World of Difference*, p. 41, and the "therapeutic" phrase is Overton's.

[47] The Antigonish Movement began in the early twentieth century, led by M. M. Coady and J. Tompkins, both priests associated with St. Francis Xavier University. Directed from the University's Extension Department, the movement involved large numbers of farmers, fishers, and coalminers in cooperative ventures to alleviate economic distress in eastern Nova Scotia. See R. James Sacouman, "Underdevelopment and the Structural Origins of the Antigonish Movement Cooperatives in Eastern Nova Scotia," *Acadiensis*, 7 (1977), 66–85.

[48] For brief comment on the Cheticamp and Calmeadow cases, see Erin Goodman, "On the Trail of Choice and Change," *New Maritimes*, XII, 5 (June/July, 1994), 18–20.

[49] See "Our Surroundings: The Maritimes and the Environmental Crisis," a Special Issue of *New Maritimes*, VIII, 4 (March/April 1990).

[50] The paragraphs that follow owe a special debt to *New Maritimes*, which was a leading source of thoughtful and thought-provoking writing on the region in the 1980s and 1990s, and to the various contributions of Rick Williams, among which the following warrant special mention: "What's Left? Environmentalism and Radical Politics," VIII, 4 (March/April 1990), 13–15; "From Here to Oblivion: The NDP in the 90s," VIII, 6 (July/August 1990), 8–10; "Maritime Unity. 'The World is in Chaos...'" IX, 5 (May/June 1991), 30–3; "The Politics of the New Age," IX, 6 (July/August 1991), 31–3.

[51] Samuel P. Hays, *Beauty, Health and Permanence. Environmental Politics in the United States, 1955–1985* (Cambridge: Cambridge University Press, 1987).

[52] A useful analysis that bears on these suggestions is provided by D. Ralph Matthews, *Controlling Common Property: Regulating Canada's East Coast Fishery* (Toronto: University of Toronto Press, 1993).

Iain
Wallace

Canadian Shield: Development of a Resource Frontier

The regional identity of the Canadian Shield is rooted in its distinctive physical environment. The rugged terrain of ice-scoured granitic rock, supporting extensive boreal forest but little farmland, contrasts sharply with the agricultural landscapes that abut it to the south and west. The region's northern boundary is less distinct, as climatic influences on vegetation and ground conditions (extent of permafrost) gradually become more constraining polewards, and as the ancient rocks of the Shield dip below the more recent sedimentary deposits exposed by postglacial isostatic uplift around Hudson Bay. The human geography that has been developed through exploitation of the Shield's forest, mineral, and water power resources reinforces its regional distinctiveness. Limited options have resulted in similar patterns of livelihood across a vast expanse — fully 40 per cent — of the Canadian landmass.

Unlike other hinterland regions, however, the Shield has no real voice in national affairs. There is not one provincial government unequivocally committed to advancing the interests of its inhabitants. The Shield extends across five provinces (six, if a small area in northeastern Alberta is included) and underlies much of the Northwest Territories (see Chapter 10). But even in Ontario and Québec, where it constitutes by far the largest physiographic region in each province, it is a sparsely populated hinterland. Its scattered residents are able to exert only limited influence on their respective legislatures; and the political fragmentation of these minority "provincial norths" has hindered the emergence of effective advocacy for the Shield at the federal level to match that of the Maritime and Prairie hinterlands. Influence in the world of business is equally lacking: only four of the country's two hundred largest corporations are based in the region.[1] The frustrations and alienation which can arise in this remote setting have not been, in themselves, sufficient to galvanize constructive region-wide initiatives.[2] Despite some recent growth in decision-making capacity within the region, external demands and external perceptions are profoundly influential in shaping its human geography. Hence, this chapter focusses primarily on the resource development undertaken in the Shield by external heartland interests, and considers the impact of this activity on the character of the region's settlements and the life of its inhabitants.

The fur trade was the first staple industry to bring Europeans into contact with the Shield's natural resources and with the indigenous societies that had adapted to its environmental potentials. British demand for timber in the early nineteenth century laid the basis for the next staple economy, which expanded in scale and geographical extent as the needs of the industrializing United States created a closer market after mid century. By the turn of the twentieth century new industrial staples, in the form of pulp and paper, hydroelectricity, and non-ferrous metals were rapidly attracting international investment in the Shield's resources. These sectors have remained the backbone of the regional economy to the present day. Together with the transportation system that made the resources accessible to continental and world markets, they have shaped the pattern of migration and settlement; and the vagaries of demand that typify their markets have been reflected in the varying fortunes of many Shield communities.

At the close of the twentieth century, a further set of influences originating outside the region are helping to mould its future. Changing cultural values in the dominant urbanized societies of the industrialised world (including heartland Canada) are transforming perceptions of the natural environment, albeit frequently in ambiguous ways.[3] The same is true with respect to recognition of the rights and place of aboriginal peoples in modern society. Rather than being viewed exclusively as a cache of marketable resources, the "wild" landscape of the Shield is itself being seen as a resource, with aesthetic and spiritual values that need protection, as well as with recreational potential that can be developed commercially. Localized conflicts between proponents of an "environmentalist" agenda of wilderness preservation and those of a "livelihood" agenda of maintaining a conventional resource-based economy have become fairly common. But the need for "sustainable" resource-use practices within the staple industries has become increasingly accepted. The previous destructiveness of many forest-based industries and mining operations has been curbed, and policies that accommodate multiple land-use objectives at the regional scale are increasingly the norm. Where they have a significant presence, Native peoples are assuming a greater role in the regulation of resource management.

The Shield has contributed significantly to the evolution of a specifically Canadian consciousness and historiography. As the realm of the fur trade — of Native woodlore, the canoe, the self-reliant *coureur de bois* (French trader) and of physical challenge and harsh winters — it gave rise to a frontier mythology with distinctive central images.[4] After Confederation, this evolved into the national mythology of the newly independent Dominion, with the Shield's environment providing the objective locational anchor for Canada — "the true North, strong and free".[5] When the Group of Seven sought to develop an art that was authentically Canadian and not derivative of European styles, they discovered in the Shield's distinctive landscape a natural source of inspiration.[6] As the nation developed an industrial economy, the principal theme became the apparent inexhaustibility of the Shield's resources, which would assure Canadians of a prosperous future. As late as 1958, John Diefenbaker's vision that Canada's destiny lay in embracing the challenges and reaping the rewards of harnessing its northlands resonated with popular sentiment. Academic interpretations of Canadian history and economic development have equally given a central place to the Shield. Donald Creighton and other historians of the "Laurentian School" explained the evolution of the nation in terms of the strategic importance of the St. Lawrence waterway and the resources, notably the furs and forests of the Shield, to which it gave access. Harold Innis' staples theory recognized that one of the central features of Canada's economy, based on resource exploitation in regions such as the Shield, was that it entrenched relationships of metropolitan dominance over remote hinterlands.

THE PHYSICAL ENVIRONMENT

The enormous significance of the Shield in shaping Canada's national consciousness owes as much to its location as its size. Certainly, with an area of some 4.6 million square kilometres, it is the country's largest physiographic region, and Canadians have, of necessity, been forced to deal imaginatively with both the distinctive opportunities and the limitations associated with its physical environment. But the fact that the rugged terrain of the Shield cuts the Canadian ecumene in two, separating the heartland communities of central Canada from the population of the Prairies and the West, is what has made its influence inescapable. In a country built on the vision of uniting people "from sea to sea," the Shield of northern Ontario has invariably served more as a barrier than as a link. The high construction costs and sparse local earnings of domestic transportation corridors joining southern Ontario and the Prairies have, in the past, been seen as a threat to Canada's nationhood, for east–west routes south of Lake Michigan, through Chicago, have always been more attractive financially. In the 1870s, the issue was whether an all-Canadian railway to the Pacific was feasible; in the 1950s, the pros and cons of routing oil and gas pipelines from Alberta to southern Ontario north of

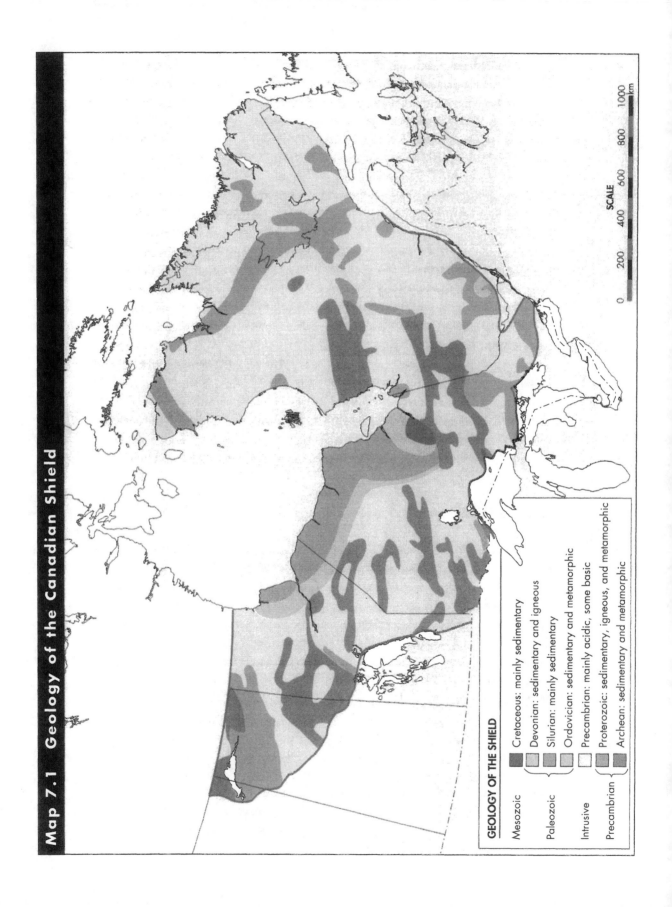

Map 7.1 Geology of the Canadian Shield

GEOLOGY OF THE SHIELD

Mesozoic
- Cretaceous: mainly sedimentary

Paleozoic
- Devonian: sedimentary and igneous
- Silurian: mainly sedimentary
- Ordovician: sedimentary and metamorphic

Intrusive
- Precambrian: mainly acidic, some basic

Precambrian
- Proterozoic: sedimentary, igneous, and metamorphic
- Archean: sedimentary and metamorphic

SCALE

0   200   400   600   800   1000 km

Lake Superior were passionately debated. The prominence of the Shield's environment within the imagination of cultural elites has, additionally, been secured by the close proximity of its southern edge to the metropolitan centres of central Canada.

The geology and glaciation record of the Shield are fundamental to the character of the region (Map 7.1). Its rocks are among the oldest and most stable elements of the earth's crust. The ancient Precambrian structures had been worn down, overlaid with limestone, and re-exposed by erosion long before the Pleistocene ice age moulded the current terrain. Over three-quarters of the surface is made of granite gneiss: "the sameness of Shield scenery over vast areas must be explained by this constancy of rock type."[7] The predominant landscape of the Shield south of latitude 60° N is that of a generally rolling, lake-dotted upland with elevations averaging about 650 m above sea level. The presence of greenstone belts (metamorphic basic volcanics) within the southern Shield, containing the region's principal gold and base-metal deposits, has determined the location of mining settlements.

The bare rock, thin soils, and muskeg are telling signs of the widespread imprint of glaciation (Map 7.1). So too is the fragmented drainage pattern. The vast Laurentide ice sheet that covered most of Canada in the last stage of the

"Flowerpots" from northern Ontario

Pleistocene ice age reached its maximum extent about 20 000 years ago. It gradually retreated northeastwards, leaving most of the southern Shield 8000 years ago. At times during this retreat, glacial lakes formed between heights of land and the ice margin. In particular, Lake Agassiz, the forerunner of lakes Winnipeg and Winnipegosis, extended into parts of northwestern Ontario; and Lake Barlow–Ojibway formed south of James Bay, leaving the so-called Clay Belts of Ontario and Québec. The sediment of these lakes, and of marine deposits (notably the extension of the Champlain Sea into the Lac St. Jean Lowland), forms the basis of the only extensive arable land in the Shield, but soil acidity and poor drainage have limited the expansion and productivity of farming. A short frost-free season and the modest accumulation of heat units restrict agricultural land use in these areas to pasture and fodder crops. Otherwise, Shield soils are predominantly shallow, permeable, and of low fertility, better suited to forestry than crop cultivation.

Although the homogeneity of the boreal forest (the major vegetation belt in Map 7.2), the frequency of lakes, and the widespread occurrence of rocky outcrops suggest environmental uniformity, there are significant regional variations within this pattern. The faulted escarpment of the Shield overlooking the valleys of the lower Ottawa and St. Lawrence rivers, for example, provides a distinct visual contrast to the gently sloping sedimentary lowlands and thinning tree cover extending toward Hudson Bay in northern Ontario and Manitoba. The barren upland plateau of Québec–Labrador is markedly more rugged than the cottage country of the Shield's fringe in northwestern Ontario and Muskoka. There are important differences in river systems also. The Hudson Bay–Atlantic drainage watershed is so located as to produce only minor southward-flowing rivers in northern Ontario but

Drying capelin, Hopedale, Labrador

Map 7.2 Land Cover and Pulp and Paper Mills in the Canadian Shield

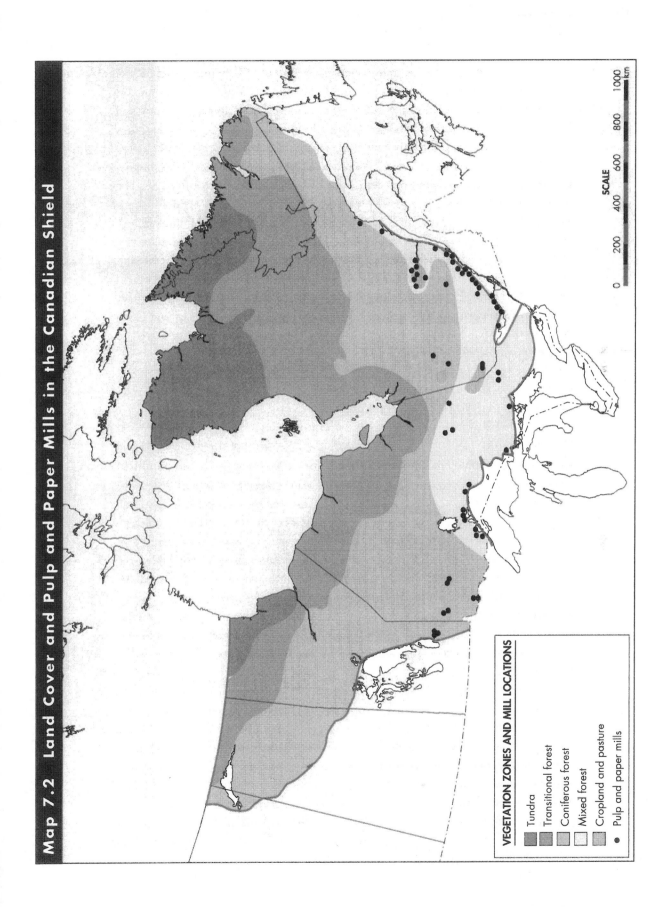

VEGETATION ZONES AND MILL LOCATIONS

Tundra
Transitional forest
Coniferous forest
Mixed forest
Cropland and pasture
• Pulp and paper mills

SCALE

0   200   400   600   800   1000
km

a large number of rivers with large water-power potential draining south off the Shield in Québec (together with the Churchill River in Labrador). In addition, of the rivers draining into Hudson Bay, those in Ontario have their major breaks of slope further upstream than those in Manitoba and Québec, thus reducing their hydroelectric power development potential.

Taken together, these features of the physical environment of the Shield largely account for its sparse population and discontinuous pattern of settlement. The environmental conditions, which Europeans initially found inhospitable because they discouraged permanent agricultural settlement, are ones to which the region's Native peoples adapted by means of their land-extensive subsistence economy. Having long survived by living off the limited resources of the natural environment, they have in recent years increasingly contested forms of economic development that would imperil its fragile ecological balance. But having had their territory steadily encroached upon by the expansion of Euro-Canadian society, the Native people of the Shield have experienced an uneasy transition from traditional to modern forms of life — one whose tensions are still unresolved.[8]

HISTORICAL DEVELOPMENT

The development of the Canadian Shield as a resource frontier of the Western European heartland economy began with the fur trade. Pelts were compact, high-value commodities, well suited to a transportation system dependent on the canoe and demanding no expensive infrastructure. The lumber trade, handling much more bulky materials of low unit value, was initially confined to locations where logs could be readily assembled (a task made easier by snowy winter ground) for floating downriver. This too required little infrastructural investment. But as the Industrial Revolution spread from Britain to the northeastern United States, and with it demands for a greatly expanded range of natural resources, so too did the revolutionary technology of the railway, giving unprecedented access to remote inland locations. Hence the geography of settlement in the Shield closely reflects the sequence of railway construction and of trends in the international economy that provided the impetus for it.

Beyond the widely scattered small villages or reserves where Native peoples form the majority of residents, the population of the Shield is concentrated in relatively isolated resource-based settlements. The spatial pattern reflects three causative processes. First was the widespread encroachment of settlement at points along the Shield's southern boundary from more productive agricultural areas in the St. Lawrence Lowlands. Here, farming communities occupied the land in the wake of the advancing commercial lumbering frontier as it progressed westward during the nineteenth century. The Ottawa Valley was the principal axis of this movement. Similarly, the Lac St. Jean Lowlands were opened up by colonists from

Native settlement
in the Shield

the St. Lawrence Valley, although here, in a more subsistence-oriented economy, the chronological distinction between the movement of the lumbering frontier and the subsequent advance of agriculture was less clear cut. Throughout the Shield, however, the agricultural frontier was in retreat by the 1940s.

The second process was the construction of transcontinental railways during the years 1880–1915. These stimulated settlement and resource developments within the Shield quite incidental to their primary purpose of linking east and west. The crowded sequence of station names that appeared on railway maps across northern Ontario and along the Québec section of the former National Transcontinental Railway belied the sparsity of residents in the strings of small settlements along these corridors. On the other hand, Sudbury, Thunder Bay, and many smaller, single-industry towns came into being or acquired their modern significance during this phase of development, which greatly increased knowledge of the mineral resources of the Shield.

The third formative element in the evolution of the regional settlement pattern has been the growing demand for resources to fuel industrial growth and financial accumulation in the heartland economies of North America and the wider world. Beginning with construction of the Temiskaming and Northern Ontario Railway (now the Ontario Northland) in the early 1900s, which brought the silver reserves of Cobalt into play, numerous rail lines and mineral-based communities expanded the resource frontier. Especially during the period of rapid global economic growth during the 1950s and 1960s, from Lynn Lake in the west to Schefferville in the east, major investments were made to tap the Shield's mineral resource potential.

New areas of forest exploitation, such as around St. Félicien, Québec, also expanded the ecumene during this period. But since the mid 1970s, the global comparative advantage of the Shield's resources has declined. The changing structure of the world economy (the declining resource-intensity of production), its shifting geographical focus (away from eastern North America, toward the Asia-Pacific region), and the stronger competitive position of such resource-rich nations as Chile and Indonesia (more inviting than in the past to international investors) has dulled, but far from eliminated, the appeal of the region's resource base.

THE FOREST INDUSTRIES

Forest dominates the landscape of the Canadian Shield, and the industries it supports have long been the foremost component of the regional economy. The first period of commercial exploitation extended from the early nineteenth century until World War I, and was based on lumber production. The second period began in the early twentieth century and was devoted almost entirely to pulp and paper manufacturing. The third period, in so far as recent developments justify recognizing one, is distinguished by more intensive competitive pressures in global markets, heightened attention to environmental impacts and forest sustainability, and efforts to maximize the efficiency of resource utilization through closer integration between the pulp and paper sector and a revived commercial lumber sector. The initial period was characterized spatially by the westward movement of exploitation, as the forest was selectively and wastefully mined rather than harvested. The second period, in contrast, was based on the establishment of pulp and paper mills at strategic locations, each one fed with timber from within defined catchment areas. Had forest regeneration been pursued more rigorously in these catchments over the past eighty years, there would have been less pressure in the 1970s to extend the frontier of exploitation, and less concern currently about localized woodfibre deficits that may appear in the early years of the twenty-first century.

British demand for Canadian lumber grew rapidly at the beginning of the nineteenth century, when access to Scandinavian forests was severed by Napoleon's blockade of the Baltic. The cutting of squared timber from virgin stands of pine, which, together with hemlock and birch, make up the mixed forest of the Great Lakes–St. Lawrence region (Map 7.2), was encouraged by preferential tariffs and financed by British merchants based in the city of Québec. Operations in the woods, however, were brought progressively under the control of Canadian entrepreneurs based in Ottawa–Hull, where, as the frontier advanced up the Ottawa Valley, log rafts were assembled for dispatch downstream. Impoverished French Canadian and Irish workers constituted the bulk of the labour force and bequeathed to the Ottawa Valley its distinctive ethnic mix.

By the early 1840s, changing external conditions began to reorient the Shield's forest industries to a new market. With Britain's access to the Baltic restored, and with its growing commitment to free trade, the tariff protecting Canadian shipments was removed. As a result, squared timber exports began a prolonged decline. Simultaneously, the rapidly growing industrial economy of the northeastern United States was emerging as a new heartland, whose demand for constructional sawn timber was outstripping its domestic supply. The composition and destination of shipments from the Shield's forests thus gradually switched at the same time as the frontier of exploitation maintained its westerly momentum. From the 1850s, sawmills increased in size and in levels of mechanization, and the largest concentrations emerged at strategic locations such as Ottawa and Parry Sound. By the close of the nineteenth century, the focus of timber production had moved beyond the Ottawa River catchment, through the Georgian Bay region, and on to the North Channel shore of Lake Huron.

This geographical shift was matched by significant changes in the industry's structure. The long-established firms of the Ottawa Valley made little attempt to extend their control over production further west, where newer and smaller firms predominated and competition from United States interests was more direct. Eventually these western firms, particularly vulnerable to adverse developments in trade relations between Canada and the United States, felt constrained to seek government help. Beginning in 1890, a period of relatively free trade in forest products between Canada and the United States saw a sudden rise in the export of unprocessed logs from the Ontario shore of the upper Great Lakes, towed in rafts across to sawmills in Michigan. This threat to the prosperity of the Ontario mills intensified in 1897, when the United States reimposed duties on Canadian timber, but not on saw logs. In appealing to the provincial government for support, the mill owners were able to capitalize both on the significance of forest revenues to the Ontario treasury and on growing popular sentiment in favour of strengthening the province's expanding industrial economy. The time was ripe for "Empire Ontario" to curtail the export of unprocessed primary products typical of a resource hinterland, and to increase the amount of domestic manufacturing. An amendment to the Crown Timber Act, "the manufacturing condition," specified that all pine cut on Crown land in Ontario was to be sawn into lumber in Canada. This move, ensuring that the revenues and employment generated by the upgrading of a provincial resource remained within the province, immediately stimulated sawmill construction all along the North Channel and Georgian Bay shores, as well as in the Lake of the Woods area.[9]

The second major phase of forest exploitation was already underway by the turn of the century. Technological advances had established the viability of wood-based paper production, and a growing demand for pulpwood soon developed to offset the declining availability of quality saw logs. Spruce replaced pine as the species attracting most attention, as the forest industries, aided by the growing railway network, expanded northwards toward the boreal forest belt. Early pulp and paper mills were concentrated in the Ottawa and St. Lawrence valleys, close to the metropolitan newspaper markets of the northeastern United States. By 1910 the distribution of mills extended from Sault Ste. Marie to Chicoutimi.

As was the case in the sawn timber trade, the earliest newsprint markets centred on New York, to be joined later by those centred on Chicago. Ontario's attempt to enforce "the manufacturing condition" against the export of pulpwood logs was unsuccessful in the absence of parallel legislation in Québec, then the chief source of shipments to American markets. But by 1911, foreseeing an inadequate domestic wood supply, leading newspaper publishers in the United States successfully persuaded their government to permit the duty-free entry of Canadian newsprint. This move immediately resulted in investment by both American and Canadian interests in pulp and paper mills north of the border and laid the foundations for the industry's dramatic growth in the 1920s. By that time, too, the eclipse of the Shield's lumber industry was being accelerated by the growth of West Coast lumber producers.

The geographical distribution of pulp and paper mills has been remarkably stable since the early 1920s, when a brief period of extremely high newsprint prices stimulated massive overinvestment in production capacity. Between 1925 and 1930, for example, newsprint consumption increased by one-fifth, but production capacity doubled in response to the momentum of earlier capital commitments. Naturally, this excess capacity brought about a disastrous fall in prices, which merged into the broader economic dislocation of the Depression. From a high of $137 in 1921, the price of a ton of newsprint fell to $57 in 1931, and to a low of $40 in 1934. These developments created severe problems for the financially overextended pulp and paper manufacturers, for the Canadian banks whose loans were at risk, for the many single-industry communities whose livelihoods were totally dependent on employment generated by the local mill, and for the provincial governments who faced demands from each of these other interests. Attempts were made, first by Montréal-based banks and subsequently by the premiers of Ontario and Québec, to enforce a cartel on the industry in an effort to prorate production and eliminate price competition.

These measures by Canadian heartland institutions to bring stability to the hinterland resource economy met with limited success. One obstacle was the divergent interests of those American firms, notably the International Paper Company, whose corporate linkages gave their Canadian mills guaranteed newsprint markets and hence relative immunity from the dislocations suffered by Canadian-owned producers. Stronger markets eventually reappeared in the late 1930s, but not before towns such as Espanola and Pine Falls had endured heavy unemployment and attendant communal stress. In general, institutional responses to the industry's plight did more to protect heartland stock- and bondholders than hinterland woods and mill workers from financial insecurity.[10]

These pulp and paper mills using the forest resources of the Shield (note that many mills in the valleys of the St. Lawrence and the lower Ottawa are physically beyond the Shield's edge) are concentrated in the Lac St. Jean area, along the St. Lawrence Valley between Trois-Rivières and Québec, in the lower Ottawa Valley, and at Thunder Bay. Elsewhere, the location of isolated mills reflects the alignment of transcontinental railway lines — in the case of northeastern Ontario, at the points where they cross northward flowing rivers. Three mills were constructed during the 1940s along the northern shore of Lake Superior, representing a new production subregion made viable by changes in pulping technology that permitted extensive use of jack pine. Precisely these technological advances, however, helped to reduce the overall appeal of the Shield as a locus of postwar investment by pulp and paper companies serving the American market. The pine forests of the southeastern United States emerged as a competitive source of raw material, giving producers there several advantages (including lower wood, labour, and transportation costs) over competing Canadian suppliers.

New pulp and paper mills were built in the Shield in the late 1960s and early 1970s, but their locations were more peripheral than those of earlier mills, and their financing involved considerable amounts of public money in the form of regional development incentives. Taken together, these two factors go a long way toward explaining why the projects became commercial liabilities. Large mills at The Pas, Manitoba; Port-Cartier, Québec; and Stephenville, Newfoundland, faced uncompetitively high woodfibre costs in regions of slow and low-density tree growth. The environmental limitations of these production sites were discounted by an excessive political commitment to encourage investment and create jobs.[11] The mill at Port-Cartier closed in 1979 after operating for six years. Only after substantial restructuring of its production and wood supply, and with improved labour relations, did it reopen (at the second attempt) in 1995 with an apparently secure future.[12] A new small pulp mill at Prince Albert, Saskatchewan, fared better than the grandiose schemes, as did a paper mill opened at Amos, Québec, in 1982.

Even in long-established locations, however, the pulp and paper industry in the Shield has faced a tougher competitive environment since the 1980s. Many mills in Ontario and Québec have lagged in upgrading their equipment, leaving their machines older and smaller (and so higher-cost) than those in western Canada, and even more so than in competing mills in the United States. Moreover, especially in Ontario, the region's traditional electricity cost advantage has been eroded by rate increases stemming from overinvestment by Ontario Hydro in nuclear generation. Finally, public concern about the environmental impact of the paper industry has become much more influential. Legal requirements to reduce the water pollution associated with pulp production have become more stringent, although a federal-provincial program of grants in the early 1980s assisted firms to upgrade mill technology to reduce chemical discharges. The strong growth in demand for newspaper recyling has been much more difficult to adjust to, however.

Driven by growing problems of solid-waste disposal in metropolitan regions in the United States and Canada, newspaper recycling has significantly altered the geography of locational advantage for paper mills, favouring those close to supplies of used newsprint rather than those in the forest hinterland.[13] De-inking equipment has been installed at a number of mills in the Shield, including ones at Thunder Bay, Gatineau, and Shawinigan, but all three have to ship in recycled material from distant cities.

Offsetting these disadvantageous trends in the Shield's newsprint sector have been technological advances in timber processing and the introduction of new wood-based products, such as oriented strandboard. These have disproportionately benefitted firms utilizing the species and smaller trees of the boreal forest. They have also made the integration of lumber and pulp production more logistically feasible and economically attractive and have brought considerable cost savings to many paper mills. Reinforcing this development has been the effective closing of the frontier of unexploited pulpwood supplies, prompting paper companies to purchase lumber firms to secure their cutting rights. With significant economies of scale in lumber production, the large modern sawmills in the Shield have become increasingly competitive in the major metropolitan markets of central Canada and the northeastern United States.

In the more competitive global newsprint markets of the 1990s, firms have continuously had to seek productivity improvements and/or move to higher value-added products. In the process, a number of mills in the Shield have been threatened with closure or put up for sale by their multinational parents. Some American firms have pulled out of the Shield to concentrate their output in newer and larger mills in the southern United States, or because they are abandoning low-profit commodities (such as pulp) to concentrate on more specialized and lucrative end-products. One result has been that a number of mill-dependent towns in the Shield have been faced with the loss of their major employer. They have increasingly responded by attempting to continue production under local control, following the successful example of Tembec. This firm was formed when an obsolescent American-controlled mill at Temiskaming, Québec, slated for closure in 1972, was returned to profitability by locally based owners (employees, managers, the municipality) with provincial government assistance. Some of these recent buy-outs have succeeded commercially (as at Spruce Falls, Ontario, aided by a provincial energy subsidy); a few have not.

Both the forest-based industries and their landlords, the provincial governments, have been negligent in the past with respect to forest regeneration. Inadequacies in collecting basic data and in implementing effective reforestation policies have created a situation in some areas in which the recent levels of annual cut cannot be sustained. The problem is most acute in northwestern Ontario, where three-quarters of manufacturing employment is dependent on the forest sector. Fewer difficulties are anticipated in northeastern Ontario and the Abitibi region of

Québec, although the transportation costs of obtaining pulpwood from more northerly regions will rise. In the 1990s, concerted efforts by industry and government to implement effective forest regeneration practices offer better prospects for the future, but the legacy of previous shortsightedness will not disappear overnight.

By the mid 1990s, almost sixty pulp and paper mills, representing 40 per cent of the national total, were drawing upon the forest resources of the Shield (see Map 7.2). These mills account for almost 70 per cent of Canadian newsprint capacity and over half of national woodpulp capacity. The United States continues to consume over 80 per cent of the Shield's pulp and paper production, although Canada's share of the American newsprint market has declined (to just over 60 per cent) and is further threatened by recycling in United States markets. The Shield's growing lumber exports, from Québec in particular, have been caught up in the long-running Canada–United States softwood lumber trade dispute, which has forced the federal government to impose quotas on cross-border shipments.

MINING

Mining, no less than the forest-based industries, has been developed in the Shield primarily in response to the demands of non-Canadian markets. As a permanent element of the regional economy, the industry dates only from the last decade of the nineteenth century. The first major investments in mineral resources accompanied the initial construction of railways across the Shield. These routes often hit upon rich deposits more by accident than design, as the Canadian Pacific Railway did at Sudbury (1883), and the Temiskaming and Northern Ontario Railway did at Cobalt (1903). However, with the exception of precious metals and of nickel (in which Sudbury quickly acquired a near-monopoly of the world's supply), the minerals of the Shield faced weak markets: at that time, the United States, in particular, had no shortage of accessible domestic supplies. After World War II, however, this situation changed, and the mining industry of the Shield underwent a prolonged period of sustained expansion. The production capacity of established mining regions was enlarged, and new metals from new regions, notably iron ore from Ungava, were added to the spectrum of mineral output (Map 7.3). Foreign competition, the high cost of borrowing capital, and the drop in the price of gold brought this growth phase to a halt in the 1970s. Similarly, the severe recession of the early 1980s left the mining industry of the Shield, with the exception of gold producers, smaller in size and facing stronger global competition.

Limited demand was thus one factor that delayed the emergence of a mining industry in the Shield; a second was the complex composition of many of its mineral deposits. Whereas the nineteenth-century gold rushes in British Columbia and the Yukon were based on placer (alluvial) material, accessible to the individual panner, the gold of Porcupine and Kirkland Lake required hardrock mining, and

Map 7.3   Minerals in the Canadian Shield

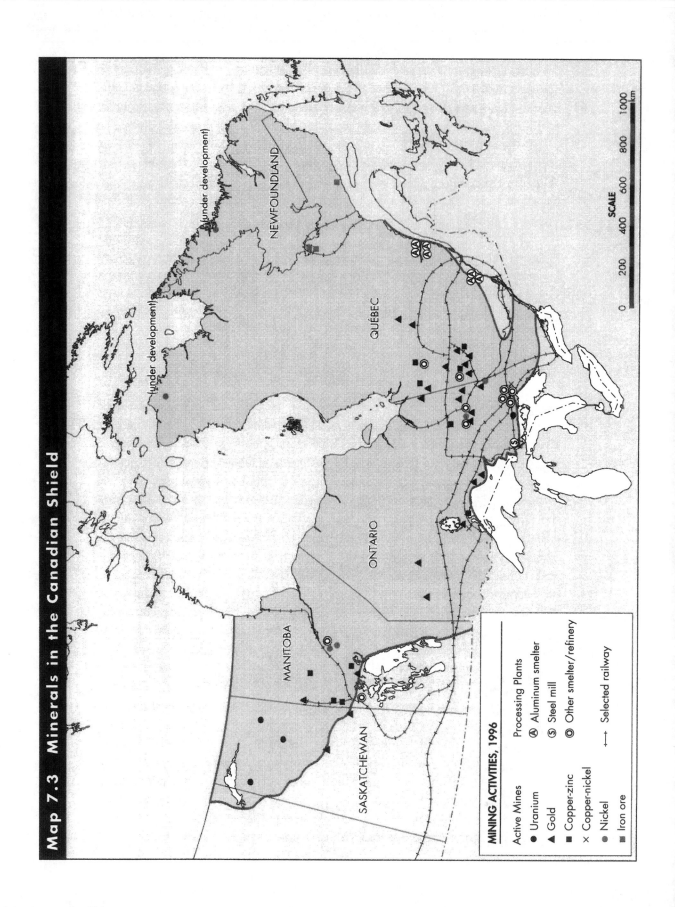

(under development)

NEWFOUNDLAND

QUÉBEC

(under development)

ONTARIO

MANITOBA

SASKATCHEWAN

MINING ACTIVITIES, 1996

Active Mines		Processing Plants	
●	Uranium	Ⓐ	Aluminum smelter
▲	Gold	Ⓢ	Steel mill
■	Copper-zinc	◎	Other smelter/refinery
✕	Copper-nickel	┼┼	Selected railway
●	Nickel		
■	Iron ore		

SCALE

0   200   400   600   800   1000
km

could only be separated by a capital-intensive chemical process such as cyanidation. Even more challenging were the nickel-copper ores of the Sudbury basin. The technological breakthrough that finally made possible the separation of the two metals was achieved by an American company (a forerunner of Inco) with the expertise, the financial backing, and a secure market in its domestic armaments industry — all factors absent in the hinterland Canadian economy. Armed with this technological monopoly, the company was able, quite easily, to frustrate Ontario's attempt in 1900 to apply "the manufacturing condition" to the export of ores. Refining facilities remained firmly situated in the United States until changes in refining technology (from a coal-based to an electrolytic process), and the prospect of having its properties expropriated during World War I, prompted the company to establish a Canadian refinery at Port Colborne in southern Ontario, using electric power from Niagara.[14]

The third factor to influence the pace of mineral development in the Shield is accessibility. With few exceptions, mineral products are of low value in relation to their bulk. In inland areas they are therefore dependent upon cheap rail transportation for commercially viable shipment. The spread of mining was thus directly linked to that of an expanding railway network. Prior to the 1920s, railways were built through the Shield to serve a variety of national and provincial purposes unrelated to mining, although significant mineral discoveries were made in the process. Thereafter, as the development of bush flying gave prospectors a new-found mobility, mineral deposits could be evaluated well in advance of railway construction. Subsequent decisions to develop a mine then identified the specific transportation infrastructure required. Mining activity in the Shield is therefore most intense where a regional core of active mining developed, which could then be expanded by small incremental investments in the necessary transportation network. The history of development in the Cobalt–Timmins–Noranda area prior to 1930, and in northwestern Québec and northern Manitoba in the decade prior to 1965, demonstrated this pattern.[15]

In contrast, the most distinctive feature of the dramatic expansion in the Shield's mineral economy beginning in the early 1950s was the massive investment in railway construction undertaken to bring distant and isolated deposits into production. Development of the Québec–Labrador iron fields required laying 1155 km of railway, as well as building related port facilities at Sept Îles, Port-Cartier, and Havre St. Pierre. Altogether, over 5000 km of railway carrying significant mineral traffic were built in Canada between 1950 and 1975, nearly half serving mines in the Shield.[16] Only a number of gold mines in northwestern Ontario and northwestern Québec and uranium mines in northern Saskatchewan have been able to sustain production solely on the basis of road (including winter road) and air transport.

The Shield is one of the largest and most diversified mining regions in the world. After a century of production, it still contains extensive reserves of a variety of minerals and continues to attract exploration dollars and investment in new mines. The rich Hemlo gold mines, discovered in the 1980s, and the massive Voisey's Bay nickel property, discovered in the 1990s, confirm that the region's geology still offers a lot of hidden potential. Yet the prosperity, and hence the expansion, of the mining industry in the Shield has become increasingly sensitive to changes in its external environment, both in Canada and the world. Canada's share of world mineral markets has declined during the past fifty years as a result of the tremendous growth of output in developing nations in Latin America and southeast Asia, and also in Australia. Within the global perspective of the world's leading transnational mining companies, including those that are Canadian-owned, the Shield is now only one resource hinterland among many. The comparative attraction of investment in the Shield, as opposed to elsewhere, is influenced by such factors as the relative richness of particular deposits, comparative operating and infrastructure costs, Canadian taxation and environmental regulations (both primarily a provincial matter), the strength of the dollar, and the geographically differential impacts of changes in technology and markets.

Nickel provides a singular illustration of the significance of these factors for the Shield's mining industry. In the 1950s, the Sudbury basin produced over 80 per cent of world supply and Inco monopolized the world market. Yet despite expanded output at Sudbury and the development of a large deposit at Thompson, Canada's share of global output shrank to less than one-third in the late 1970s and dipped to only 14 per cent in 1982. This dramatic shift in the fortunes of what was still, in 1980, the world's lowest-cost nickel producer by a comfortable margin, reflected a number of developments.[17] First, continuous growth in world demand had come to a sudden end, and as a result there was a global surplus of production capacity. Many of the new foreign mines, planned before the slowdown, maintained or even increased their output (which helped to lower world nickel prices), either to provide their governments with hard currency or to maintain a cash flow to service their high debt load. Canada's private-sector producers found it increasingly difficult to compete with these predominantly state-controlled or state-subsidized enterprises.

The second factor was that Canadian firms (Inco and Falconbridge) had invested in nickel production overseas to protect their access to specific markets, partly at the expense of their Shield operations. Certainly, the simultaneous start-up of an Inco property in Indonesia and major redundancies at its Sudbury and Thompson mines in 1977 led to accusations that the company was exporting jobs. But the Sudbury layoffs, which triggered a major strike in 1978–79, were primarily driven by technological change. The adoption of bulk mining methods in place of traditional rock-face drilling brought major improvements in labour productivity; and the application of robot technology to some of the more hazardous tasks underground further reduced the need for miners.[18] Mining-based employment in the Sudbury area, which peaked at 25 700 in 1971, had by the early 1990s stabilised at just over 10 000.[19] When the very low-cost mine at Voisey's Bay, Labrador, opens, it will become the focus of expanded nickel output in the Shield.

Coppercliff
smelter at
Subury

The Shield's iron ore deposits have never been as commercially attractive internationally as its nickel. Output has declined from its peak in 1979 as a result of major restructuring in the North American steel industry. It was demand for new sources of raw materials by the United States steel industry that drove the large-scale investment which opened up the iron fields of the Labrador Trough in the 1950s. The original mine and settlement at Schefferville, at the northern end of the Québec North Shore and Labrador Railway, was closed down in 1982–83. Mines were also closed at Fire Lake and Gagnon on the Québec-Cartier Railway in 1984–85.[20] Overall, ore production in the Québec–Labrador region dropped by almost half (to 28 million tonnes) between 1979 and 1983. Output had increased, to 36 million tonnes, by the mid 1990s. Smaller mines in Ontario (at Atikokan, Capreol, and Red Lake) also closed in the 1980s, leaving only that at Wawa, supplying Algoma Steel.

The Shield's copper mines, which are distributed in a continuous arc from Lynn Lake, Manitoba, to Chibougamau, Québec, have also faced more severe competition since the 1970s. In the early 1980s, copper prices were as low in real terms as during the Depression of the 1930s. By the 1990s, Canadian copper producers were as likely to invest in new mines in Chile as explore prospects in the Shield. Only the fact that the Shield's copper is invariably found in complex ores — with zinc, nickel, or precious metals — has saved the region from widespread mine closures. The Sudbury basin accounts for over one-fifth of Canadian copper production and supports two local smelters. Smelters at Noranda and Flin Flon also process ore from a number of mines in their respective regions, whereas a third, at Timmins, draws primarily on the output of the nearby Kidd Creek mine. This zinc-

Iron ore carrier on the St. Lawrence

copper deposit was discovered only in 1964, but was quickly developed into one of the largest and most lucrative base-metal properties in Canada. The Shield's zinc smelters are located near Timmins and at Flin Flon, although there is also a substantial flow of concentrates to the smelter at Valleyfield, near Montréal.

Gold and uranium have been the metals primarily responsible for new mine developments in the Shield since the recession of the early 1980s. Despite the increasingly uncertain future of the nuclear power industry in many parts of the world, some power corporations have remained interested in securing long-term access to uranium supplies. Intensive exploration in northern Saskatchewan in the 1970s, involving much European capital, revealed high-grade deposits which have begun to be exploited. Saskatchewan has replaced Ontario as the centre of the Canadian uranium industry, although the original mine at Uranium City, dating from the 1950s, closed in 1982. The low-grade deposits at Elliot Lake, whose exploitation was subsidized for many years by Ontario Hydro, ceased to be mined in 1996. A nearby uranium refinery, at Blind River, was opened in 1983: its output is further processed at Port Hope, in southern Ontario. However, gold was the most active sector of the Shield's mining industry in the 1980s. In particular, the discovery of rich and easily accessible deposits at Hemlo, east of Marathon, triggered the most recent example of a Canadian gold rush. Three mines there now

Eagle Point uranium mine at Cameco's Rabbit Lake operation

produce almost half of Ontario's gold output. The long-established gold mining area of northeastern Ontario and northwestern Québec continues to see new developments coming into production, whereas mining activity in northwestern Ontario has declined in recent years.

HYDROELECTRICITY

Had it not been for the widespread distribution of substantial hydroelectric power potential, the pulp and paper and mining and smelting industries of the Shield would have developed later and less extensively than they did. With the exception of low-grade lignite deposits near Onakawana, in the James Bay Lowlands, the Shield is devoid of fossil fuels. The delivered cost of coal (which was significant prior to 1950) and of petroleum products to industries in the Shield rises rapidly with increasing distance from the distribution network of shipping on the St. Lawrence and Great Lakes. Natural gas became available to most of the industrial communities in northern Ontario (and Noranda) on completion of the TransCanada Pipeline in 1958, but cheap hydroelectricity has been basic to any comparative advantage enjoyed by resource-processing industries in the Shield.

Since the 1960s, however, the stimulus to harnessing the region's power potential has passed from these internal markets to the steady growth of electricity demand in the urban-industrial heartlands of southern Canada and adjacent regions of the United States. Technological advances in long-distance, high-voltage transmission have made increasingly remote hydroelectric power generating sites economically competitive with market-oriented thermal power plants. However, in the 1990s, changing attitudes to the environmental and social impacts of dammed rivers and flooded terrain, and more flexible technological options in thermal electricity generation have cooled much of the enthusiasm for further large-scale hydroelectric projects.

Despite the common environmental factors underlying hydroelectricity development throughout the Shield, the evolution of the industry within each province has been quite different. Québec, Manitoba, and Newfoundland share the characteristic that 80 per cent or more of their total provincial electricity generating capacity consists of hydroelectric stations located in the Shield. In contrast, the equivalent figure in Ontario is 15 per cent. The relatively limited hydroelectric power generating potential of the Ontario Shield, as compared to that of the other provinces, stems primarily from regional variations in physical geography, described above (Map 7.4). As a result, Ontario, which has the highest provincial demand for electricity, has the most restricted opportunity for generating hydroelectric power in the Shield. Under prevailing cost conditions, the commercially attractive potential has been fully developed; and with its southern hydroelectric power resources, notably at Niagara Falls, already exploited, the province has been forced since 1960 to rely increasingly on thermal generation. Ontario Hydro's substantial commitment to nuclear power in the 1970s and 1980s was made in this context.

Interprovincial variations in the institutional framework of hydroelectric power development have also been important. Electricity generation in all four provinces is currently dominated by public utilities, although in Québec and Newfoundland this situation dates only from the 1960s. Throughout the Shield, many resource industries developed (and retain) private hydroelectric power plants because they needed electricity at locations remote from existing distribution grids. Ownership of this generating capacity, even if it is now supplemented by power purchases from the provincial utilities, can provide important cost savings. Links between pulp and paper and power production have been most numerous (e.g., the Maclaren group, and International Paper-Gatineau Power, both on tributaries of the Ottawa), but by far the largest industrial producer of electricity is Alcan, whose aluminum smelting and refining operations in the Lac St. Jean region depend on the output of 2350 megawatts of captive capacity (about 10 per cent of the Québec total).

The Shield became a source of hydroelectricity for metropolitan markets in the central Canadian heartland much earlier in Québec than in Ontario. Early power development in Ontario was concentrated at Niagara Falls, close to major urban industrial markets. The inherent monopoly position enjoyed by the owners of generating capacity there prompted the provincial government to pioneer a public electrical utility in 1907. In contrast, the first major hydroelectric developments in Québec took place at Shawinigan, where there was no local market of captive consumers. Rather, the private utility had to create its markets, which it did by attract-

Map 7.4 Main Hydroelectric Power Developments in the Canadian Shield

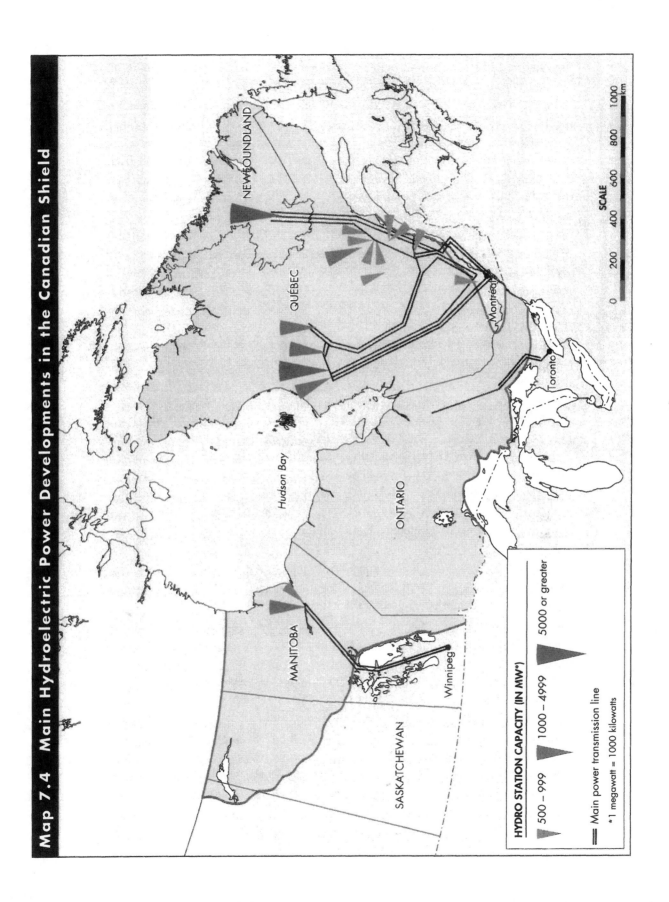

NEWFOUNDLAND

QUÉBEC

Hudson Bay

ONTARIO

MANITOBA

SASKATCHEWAN

Montréal

Toronto

Winnipeg

HYDRO STATION CAPACITY (IN MW*)

500 – 999

1000 – 4999

5000 or greater

Main power transmission line

*1 megawatt = 1000 kilowatts

SCALE

0   200   400   600   800   1000
km

ing power-intensive industries and by pioneering long-distance transmission of electricity to distant Montréal, 145 km away, by 1903. In fact, the first move by Ontario Hydro to tap the hydroelectric power generating potential of the Shield for the Toronto region was through a power purchase contract from a Québec utility, the Gatineau Power Company. But thereafter, Ontario Hydro began to develop power sites in the north of the province, primarily to meet the growing industrial demand in the Sudbury region. Construction of power plants on the Ottawa River was delayed until the 1950s by intergovernmental disputes concerning jurisdiction over the river.[21]

Canada's rapid economic growth after World War II fuelled a massive increase in electricity demand, especially in the metropolitan heartland of southern Ontario and Québec. Despite some water diversion schemes to increase hydroelectric power generating capacity north of Lake Superior, Ontario soon found itself without further major power sites in the Shield, and was forced to turn to market-oriented thermal generation. Québec and Manitoba, however, were in a position to embark on ambitious schemes to harness remote sites. In Manitoba these schemes have involved sequential development of the Nelson River on a scale well beyond that necessitated by domestic provincial demands. High-voltage transmission corridors bring power not only to southern Manitoba (with links to Saskatchewan and northwestern Ontario), but also to American utilities in the upper Midwest. In Québec, the systematic exploitation of river basins increasingly distant from the major markets of the Montréal region became both a major engine of provincial economic growth and a symbolic statement of the technological expertise of French Canadian society following the "Quiet Revolution." Successive dam construction on the Betsiamites, Outardes, and Manicouagan rivers culminated in the massive $15 billion James Bay development, with a total capacity of 12 280 megawatts (or one-third of the provincial total).

Controversy surrounded the James Bay project from its inception in 1971. The provincial government established a separate organization, the James Bay Development Corporation, to undertake construction, but also gave it such comprehensive powers over such an extensive territory (350 000 square kilometres) that it was likened to an autonomous republic. The native inhabitants of the area, the Cree, whose lands and interests were fundamentally threatened, were initially ignored. It took a court battle for them to establish their rights in the region, and only after protracted negotiations with the provincial government were they able to reach an acceptable land-claim settlement.[22] From the start, the huge size of the project raised questions about which markets it was designed to serve. That exports to New York State were clearly intended strengthened the opposition of groups who saw the scheme, not only as a betrayal of the Cree, but as another example of the degradation of the Canadian environment to satisfy American consumers. The technical and economic aspects of the project underwent numerous changes in the course of its development. The final version involved doubling the flow of the

La Grande River by diversions from four other rivers, and increasing the peak power generating capability at the expense of some of the firm (base-load) power capability initially planned for. This indicated that electricity demand in the 1980s was growing more slowly than anticipated in Hydro Québec's projections.

When Hydro Québec unveiled plans to press ahead with its second major James Bay project, the Grande Baleine (Great Whale), in 1990, it encountered much stronger and well-prepared opposition. The importance of New York financial institutions in raising capital for the scheme was not lost on the Cree and their supporters, both Canadian and American. They were able to invoke the uncertainties created by Cree repudiation of the Parti Québécois assumption that an independent Québec would include Cree territories of the north. They also engaged in astute public relations to draw attention to the environmental and human rights issues put at risk by the project. These concerns, together with significant improvement in the economics of alternative power provision in New York and New England, have removed the Grande Baleine development from Hydro Québec's agenda for the time being.

Although located in Labrador, the large Churchill Falls power plant feeds entirely into the Hydro Québec network. Indeed, in 1992 it provided 17 per cent of the electricity consumed in Québec.[23] This situation has long been a source of contention between the governments of Newfoundland and Québec. Rights to harness the Churchill River were part of a substantial bundle of concessions granted in 1953 by the Smallwood government of Newfoundland to Brinco, a consortium of European financial interests. Construction of the 5225 megawatt generating station waited, however, upon the identification of markets for its power. Advances in long-distance transmission technology made New York State a potential customer, but necessarily one that could be reached overland only via the province of Québec, and thus only on terms acceptable to that province. Québec's dissatisfaction with the 1927 ruling on its border with Newfoundland compounded the problems associated with complex financial arrangements. With its greater strategic leverage, Hydro Québec was able to obtain a 65-year contract for 90 per cent of the power generated, at fixed rates. But upheavals in world energy markets in the 1970s meant that, by the 1980s, Newfoundland was receiving only about one-tenth of the prevailing market value of Churchill Falls' output.

The government of Newfoundland nationalized Brinco's water power rights in 1974 and made determined efforts to obtain a more equitable share of the electricity revenues. However, the Supreme Court of Canada ruled in 1984 that it had no legal basis for breaking the existing agreement, and this cheap power allowed Hydro Québec to expand lucrative electricity exports to the United States.[24] Events south of the border, rather than political accommodation between the two provinces, have finally (in 1997) brought the prospect of resolution to this dispute. Deregulation of the United States electricity market requires Hydro Québec to open its transmission grid to outside power suppliers if it wants to continue making cross-border sales. This has greatly increased Newfoundland's leverage, for it can now consider further power developments in Labrador which would not have to rely on Hydro Québec's cooperation, and might even compete with it.

RESOURCES OF LAND AND LAKE

The wealth to be gained from exploiting its natural resources has been the Shield's principal attraction to those who have been drawn to the region from elsewhere. However, there have long been other elements in its appeal to heartland interests. Although severely limited, the Shield's agricultural potential was viewed favourably in the early decades of the twentieth century. Recreational and tourist activity, based on the appeal of the region's terrain, was established in some areas as early as the 1860s. This is now a major source of livelihood throughout most of the Shield's southern edge, though a markedly seasonal one except in some favoured locations. In recent years, the wilderness character of many parts of the Shield has become more highly valued in its own right, even by people who have never seen it. These different public perceptions of what the Shield offers have in common that they are, or were, formed primarily in the minds of outsiders. They also, at times and in particular places, result in conflicting assessments of what policies or human practices are appropriate. In general, the views of those living in the resource-based communities of the region have not achieved prominence through the political process, nor have they been communicated through the arts or the media. Since the 1970s, however, the views of Native peoples living in the Shield have been articulated more strongly, and their perceptions of the region have sometimes challenged the construction placed on it both by local non-Native people and by the metropolitan society to the south.

AGRICULTURE

The opening of the Prairies to settlement in the late nineteenth century presented the Ontario government with an uncomfortable prospect. The transcontinental railways that were opening up the southern Shield to economic activity were also inviting "the sons of Ontario ... [to] seek a home in another Province." But surveyors' reports of "excellent agricultural land" in the Ontario Clay Belts suggested that these potential emigrants could, with some assistance, be retained to create a prosperous "New Ontario" in the north.[25] The government's decision, in 1902, to undertake construction of the Temiskaming and Northern Ontario Railway as a colonization line was a direct result of this concern. Settlers quickly discovered the constraints of northern agriculture, but official publications nevertheless conveyed a promising image of Clay Belt farms throughout the 1920s. In Québec, church and state collaborated closely, and acted more directly than did the Ontario government, to promote rural colonization as a culturally idealized way of life. Demographic pressure in the farming communities of southern Québec had already prompted substantial out-migration to the industrial, commercialized, and secularized society of Montréal and to the mill towns of northern New England. Rural settlement in the Abitibi region appeared to the authorities to offer security to Québecers from the evils of cultural and linguistic assimmilation.

Interprovincial differences in the degree of support for, and perceived role of, agricultural colonization in the Shield are reflected in contrasting spatial patterns of settlement. These are particularly evident in the northern Clay Belt, which is bisected by the provincial boundary. The systematic creation of contiguous parishes linked by a regular and dense pattern of rural roads in the Québec portion contrasts with the more sporadic and lower-density settlement on the Ontario side, where colonization proceeded under an essentially laissez-faire regime.

The final thrust to promote farming in the Shield took the form of a desperate but ill-conceived response to the massive urban unemployment of the 1930s. Various federal and provincial measures aimed to assist northern settlement so that the unemployed might at least feed themselves; but the measures did little more than transfer the burden from municipal welfare rolls in the heartland to ill-equipped individuals in an unpromising hinterland. "[A] class of pseudo-farmers was introduced on to the land without capital and real knowledge of farming, usually in units too small to allow capital formation and subsequent farm enlargement and specialization."[26] These essentially subsistence holdings were rapidly abandoned after World War II, initiating a more general shrinkage of the farming population, which has since continued at a slower rate.

The agricultural ecumene of the Shield comprises, in addition to the continuous belt of farmland on its southern margins, three principal regional concentrations. In 1991, the largest of these three was the Clay Belts, containing 304 000 hectares north of Lake Temiskaming and along the axis of the railway from Cochrane to Senneterre. In the other two regions, the Lac St. Jean area contained 208 000 hectares, and in northern Ontario three distinct subregions, focussed on Thunder Bay, Rainy River, and Dryden, made up a further 117 000 hectares. Agricultural activity is similar in all three regions, with differences primarily reflecting variations in accessibility to metropolitan markets. Thus, about 45 per cent of farmland is improved in northwestern Ontario, 55 per cent in the Clay Belts, and 65 per cent around Lac St. Jean. Hay is the dominant crop throughout, followed far behind by oats and barley. Dairying is the principal farm enterprise around the urban centres of the Lac St. Jean region, the Clay Belts, and Thunder Bay, but beef production is dominant in more remote areas. In the early 1980s, rising land values in southern Ontario prompted a number of established livestock farmers to move north into the Clay Belts, where forage could be produced more cheaply. But the marked shift of Canadian beef production from Ontario to Alberta in recent years has shrunk the market for cattle raised in the Shield. The only significant specialty crop raised on farms in the Shield is regionally concentrated in the Lac St. Jean area, which is the source of 98 per cent of Québec's blueberry crop.

Farms in the 160–240 hectare-size range are the modal class in census divisions where beef is dominant. Dairy farms tend to be slightly smaller. Despite the continuing trend to improve the viability of farming in the Shield by increasing the size of remaining holdings, the majority of farmers in Ontario census divisions are part-timers, deriving other income from a wide variety of occupations. Levels of off-farm work are somewhat lower in Québec, especially in the Lac St. Jean region.

THE SHIELD AS "ENVIRONMENT"

Proximity to heartland cities underlies the spatial pattern of tourist and recreational activities in the Shield. From the Laurentian Hills to Muskoka, its southern boundary is within day-trip distance of the major metropolitan centres between Québec and Toronto and, except in provincial parks, most of its many lakeshores have been incorporated into vacation properties. The metropolitan populations of Winnipeg, Minneapolis–St. Paul, and other American cities in the upper Midwest have created a similar concentration of seasonal homes in northwestern Ontario and southeastern Manitoba. Camping, canoeing, sport fishing, and, in the fall, hunting are the major attractions for those who come to the Shield to do more than simply enjoy the scenery of lake and woods. Accessibility by road is a major factor in determining the detailed pattern of these activities, but hunting and sport fishing also attract a substantial clientele willing to pay for fly-in tourist facilities at remote sites. Especially in northwestern Ontario, these lodges and camps provide a significant source of employment for Native peoples, as guides.

Seasonal employment of Native peoples in the tourist, forest, and other industries of the Shield does not, however, destroy the distinction, made by Mr. Justice Hartt, that "Indian people live in the land, while non-natives live on it."[27] Especially in the north, traditional activities such as trapping, hunting, and fishing yield an important part of Native peoples food consumption, and provide the means of earning supplementary income. In the Kenora area, wild-rice harvesting represents a similar combination of traditional lifestyle and source of economic independence, one which the Ontario government acted to protect in 1978 against demands by non-Natives peoples that the crop be more fully exploited commercially, with the use of mechanized harvesting.

The use and enjoyment of land and lake by Native peoples and non-Natives has not escaped the negative environmental impacts of the Shield's industrial economy. For many decades, the pollution of watercourses by pulp mill effluent or mine tailings, or of the atmosphere by smelting operations, was simply allowed to go on. The resolve of governments to enforce even minimal control measures tended to erode at the threat that such action would destroy the economic basis of single-industry communities. Today, as a result of increased public awareness and more stringent environmental legislation, adverse industrial impacts have been significantly reduced. On the other hand, the increased importance attached by many people in recent years to the preservation of wilderness and old-growth forests has resulted in the recognition of the threats posed by logging to the integrity of the Shield's environment.

One issue where considerable success has been achieved is that of acid rain and related environmental impacts of the Shield's metal smelting industry. Acidic precipitation, by reducing or destroying the biological productivity of lakes and forests, directly threatens the viability of the region's tourist and forest product industries. In addition, the particulate content of atmospheric emissions, resulting in metal contamination of the soil, has an equally negative, if more localized, impact on vegetation. Nickel smelting at Sudbury has a particularly notorious history of environmental contamination. In the early 1970s, when public concern had begun to result in policies to combat the pollution, the city's smelters were the largest single source of sulphur dioxide emissions on the continent. Construction of Inco's 381 metre-high "superstack" in 1972 did improve the air quality around Sudbury, but at the expense of more distant areas downwind. Government regulations that demanded a steady reduction in sulphur dioxide emissions through the 1970s and 1980s, and thereby stimulated major innovations in the smelting process, have contributed to noticeable environmental improvement. By 1995, annual sulphur dioxide emissions were less than one-fifth, and particulate emissions approximately one-tenth, of those in the early 1970s.[28] Remedial measures, such as the addition of limestone to selected acidified lakes, have accelerated the restoration of ecosystems. Many parts of the barren and denuded landscape close to Sudbury have begun to sprout greenery. The negative environmental impacts of smelting have in no sense been totally eliminated, nor can they be, but it is indicative that Sudbury is now internationally recognized as a model of landscape restoration. Measures to reduce smelter damage have been pursued less vigorously in Québec (Noranda) and Manitoba (Flin Flon), but environmental improvements have been noted.

URBANIZATION

The nature and pattern of human settlement in the Shield reflects the opportunities offered by its resource economy. A limited agricultural land base and the predominance of basic employment in mines, the woods, and paper mills combine to make industrial towns the most characteristic centres of population. These industrial towns are markedly different from the urban centres of the heartland.[29] Most hinterland communities exhibit limited economic diversification, owing to their dependence on a single major employer. This relationship similarly constrains occupational diversity, which is in most cases further reduced by the truncated managerial hierarchy of the local (branch) plant, and by the limited variety of service sector employment in a small town. Except in northeastern Ontario and the Lac St. Jean area, individual urban centres are widely separated. Such isolation contributes to the strength of community organizations, which many residents regard as an attractive feature of these settlements; but it nevertheless limits the scope of social and employment opportunities, usually forcing young people to migrate. The Shield's resource towns often displayed dramatic growth in their early years, but thereafter they have been prone to stagnate, both in size and socio-economic composition. This in turn reduces their subsequent chances of achieving industrial diversification and self-sustaining growth.

Single-industry communities, varying in size from less than 1000 to more than 10 000 inhabitants, are found throughout the region. Larger urban centres, where employment is more diversified but still notably dependent on the resource industries, are concentrated close to the Shield's southern margin. (Along the lower Ottawa and St. Lawrence river valleys the definition of a town as a Shield community is unavoidably arbitrary. In Map 7.5, for instance, the pulp and paper town of Gatineau is excluded, as part of the Ottawa–Hull metropolitan area, but Trois Rivières is included.) The main rail routes by which heartland interests traversed and penetrated the Shield determined the precise location of most towns, and the external metropolitan centres of Montréal, Toronto, and Winnipeg continue to exert a pervasive influence. Sudbury, Chicoutimi–Jonquière, and Thunder Bay, with 1996 populations of 160 000, 160 000, and 126 000 respectively, are the foci of the three most populated subregions.

Construction of the Canadian Pacific Railway consolidated Montréal's position as the dominant metropolis of the Shield in the nineteenth century. Sudbury, Sault Ste. Marie, Thunder Bay, and Winnipeg were connected to Montréal by rail for almost twenty years before railway links from Toronto offered any substantial competition. That city's emergence as a rival metropolis to Montréal for control of the Shield economy was directly associated with the rise of the mining industry and thus (albeit fortuitously) with the construction of the Temiskaming and Northern Ontario Railway. From the 1880s, financiers in Toronto had shown an interest, not shared by their Montréal counterparts, in promoting mining properties, and in 1898 two small mining exchanges amalgamated to form the Standard Stock and Mining Exchange. Its stature as the centre for mining industry financing in Canada was solidly entrenched in the aftermath of the Cobalt silver discoveries (1903) and the Porcupine and Kirkland Lake gold discoveries (1911–12) in northeastern Ontario. Nearly all subsequent expansion of the non-ferrous mining industry in the Shield has been organized from or through (in the case of foreign-controlled firms) Toronto.[30] The weak presence of Québec-based interests in the mining industry of their own province was symbolically confirmed in 1928 when the Ontario-owned,

Map 7.5 The Urban Centres of the Shield

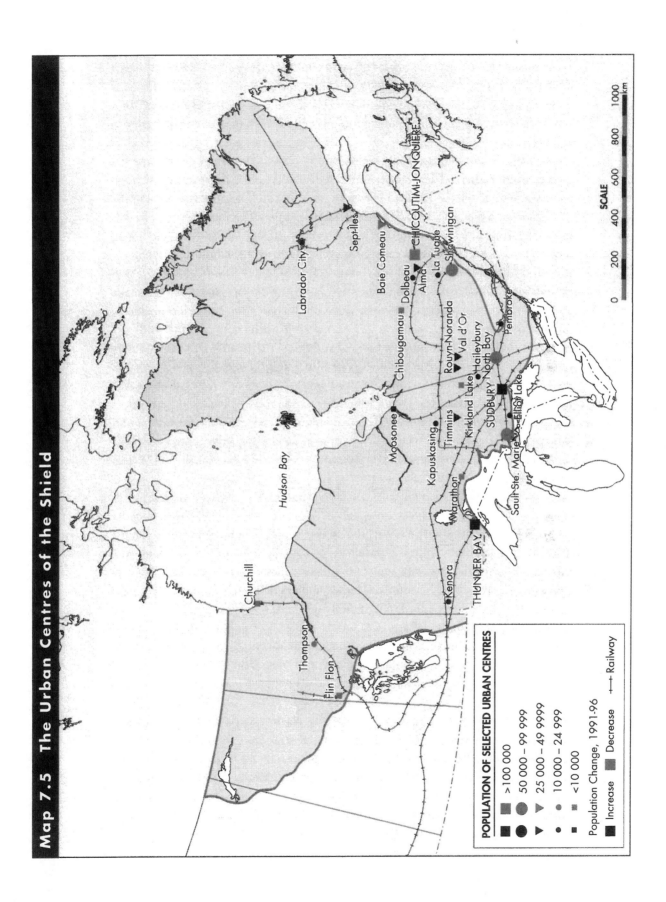

Hudson Bay

Churchill

Thompson

Flin Flon

Kenora

THUNDER BAY

Marathon

Kapuskasing

Moosonee

Timmins

Kirkland Lake

Sault-Ste. Marie · Elliot Lake

SUDBURY · North Bay

Pembroke

Rouyn-Noranda · Haileybury

Val d'Or

Chibougamau

Dolbeau
Alma
La Tuque
Shawinigan

CHICOUTIMI-JONQUIÈRE

Baie Comeau

Sept-Îles

Labrador City

POPULATION OF SELECTED URBAN CENTRES

- ■ >100 000
- ● 50 000 – 99 9999
- ▼ 25 000 – 49 9999
- ● 10 000 – 24 999
- ■ <10 000

Population Change, 1991-96

- ■ Increase   ■ Decrease

— Railway

SCALE

0   200   400   600   800   1000
km

Toronto-controlled Temiskaming and Northern Ontario Railway linked the Rouyn mines to its system, despite legal opposition from the Québec government. The creation of SOQUEM (the Québec government's mineral exploration agency) in 1965 was an attempt to reduce Toronto's control over the non-ferrous mining industry in Québec.

It would be misleading, however, to view the Shield as a region locked into an unchangeable pattern of heartland–hinterland relationships. These are evolving in response to trends in the national economy. As a result of corporate mergers in the 1970s, for example, Toronto reduced Montréal's dominant control over the Shield's forest industry, but the continuing attractiveness of Montréal's infrastructure of professional services in this sector was confirmed in 1997 when the merged corporation, Abitibi-Consolidated, the largest forest products firm in Canada, chose the city over Toronto. The same westward shift of economic power has seen the Vancouver Stock Exchange take over some of Toronto's role in the promotion of junior Canadian mining companies. The increased willingness of provincial governments in the 1970s to enforce their constitutional rights in the resource sectors and to participate actively in production resulted in further challenges to Toronto's hold over the Shield's economy. For example, the significance of Saskatoon was enhanced by the Saskatchewan government's substantial involvement (now sold to the private sector) in the province's growing uranium industry. In contrast to other provinces, Manitoba has seen little investment in the resource industries of its Shield hinterland by either public or private sector actors based locally, other than Manitoba Hydro.

Nevertheless, the fact that the critical decisions made by Alcan, Inco, Abitibi-Consolidated or Hydro Québec are made in Toronto or Montréal clearly limits the power of hinterland cities to direct the course of the Shield's economy. Sudbury, Thunder Bay, and Chicoutimi act primarily as major subregional service centres, a role which gives their urban economies some resilience to the ups and downs of resource industry cycles. At Sudbury, for example, the downturn in the nickel market in 1977–78, followed in turn by a prolonged strike at Inco, proved two points. The municipality's fortunes were not totally dependent upon the mining industry payroll; at the same time, its economy was not immune to fluctuations in the mineral economy. Its population shrank by 6500, or 4.5 per cent, in just two years, and this experience prompted both the federal and provincial governments to expand public service sector employment in the city in the 1980s. The location of Ontario Lottery Corporation's headquarters in Sault Ste. Marie, when the future of the city's steel plant was in question in 1989, was prompted by similar concerns.

About half the Shield's larger urban centres, and most of its smaller ones (below 10 000 inhabitants) lost population between 1991 and 1996, reflecting the impact of recession and economic restructuring on resource-based communities (Map 7.5). Mining-dependent Manitouwadge (3400 inhabitants), having grown 13 per cent between 1986 and 1991, recorded the highest loss (14 per cent) among smaller centres. Among larger centres, Labrador City lost 8 per cent, Kirkland Lake 5 per cent, and Thompson 4 per cent. The continuing layoffs leading to the ultimate closure of Elliot Lake's uranium mines left a community built to accommodate up to 30 000 people with only 13 600. At the other end of the spectrum, there was rapid growth in a few small centres (e.g., 41 per cent, to 3250, at Chisasibi, centre for the James Bay hydro project), but expanding larger urban centres showed only weak growth (under 4 per cent).

The larger urban centres of the Shield are no longer as dependent as they once were on the natural resource industries. The three Census Metropolitan Areas (CMAs) of Sudbury, Chicoutimi–Jonquière, and Thunder Bay have established themselves as subregional capitals, providing a wide range of services to their respective hinterlands, but all three felt the impact of declining employment in resource–related sectors. Technological change has reduced labour demand in Sudbury's mining industry, in the aluminum processing plants of the Chicoutimi area, and in Thumder Bay's grain handling sector. But these trends have been offset until recently by growth in public-sector employment, notably in the health and higher education sectors (now in the throes of downsizing). There is limited manufacturing that is not related to the resource industries or to the provision of regional needs (such as food processing), but Bombardier employs 800 people in its Thunder Bay transit equipment plant.

Each of the three metropolitan areas reflects its strategic position in broad cultural and economic patterns that reach beyond the boundary of the Shield. Thunder Bay, despite its setting and its paper mills, could lay claim to be a city of the Prairies. Ethnically, only half its population is of English or French origin, and it boasts the largest concentration of Finns in North America. Functionally, its storage and transshipment role in the grain economy of western Canada, though diminished since the 1970s, is the most highly specialized element in an otherwise diversified employment base. Chicoutimi–Jonquière is the most exclusively French by ethnic origin of all Canadian CMAs, and shares with the city of Québec the distinction of experiencing the highest linguistic assimilation of born anglophones. Its spatial structure is that of a multicentred linear conurbation, in which Chicoutimi serves as the regional service centre, and various former company towns (Kenogami and Port Alfred with their paper mills, Arvida with its aluminum smelter) retain their manufacturing orientation. An equal mix of English and French ethnic groups makes up three-quarters of Sudbury's population, reflecting both its location and its period of initial settlement. Although the city's growth has been historically tied to that of the nickel industry, from the beginning Sudbury developed as a service centre for a region of scattered mining communities (subsequently consolidated in company towns such as Copper Cliff, Coniston, and Levack) rather than as a mining town itself.[31]

Smaller centres are less resilient, especially if they are remote and dependent on an inevitably finite mineral deposit. The cessation of mining at Uranium City

(1982) and Schefferville (1983) illustrated the problems. The workers and their families directly affected by the closure of a mine face major upheavals, but in many respects they are the least disadvantaged group in the community. At Uranium City, for instance, the United Steelworkers of America, representing the miners at the Eldorado uranium mine, was able to secure full relocation expenses, severance pay, and the cancellation of various household debts owed to the company. Within a few months of the closure announcement, over 60 per cent of the mine employees had been recruited for other jobs. In contrast, the owners of private businesses, such as drug stores or restaurants, were faced with the sudden loss of their livelihood and of the capital (often their life savings) tied up in it. Neither the mining company nor the provincial government had specific obligations to protect their interests, although some assistance was provided. Within a year, the population of Uranium City shrank from 3000 to 700, and about 90 per cent of its buildings fell vacant. The closure of the Iron Ore Company of Canada mine at Schefferville, where over one-third of the permanent work force had accumulated twenty-five years of seniority, had similar consequences.[32] The community survives only because the local Native peoples, most of whom did not work in the mine but on the land (hunting and fishing), stayed on.

That these problems will inevitably arise at some stage in the life of many hinterland resource communities has forced both governments and corporations to reconsider their role in planning resource towns. Generally, it was the pulp and paper manufacturers who first found themselves in the position of having to establish new communities. Unlike many of the early, small-scale mining enterprises, whose activities sponsored unplanned and ephemeral collections of makeshift structures, pulp mills represented large-scale capital investments employing substantial numbers of people on a continuing basis to process a renewable resource. As a result, a firm had to undertake at least a minimum of community development. Unimaginative urban layouts, and housing which entrenched the social hierarchy of the workplace, were at first the norm, but after World War I, experimentation with British Garden City planning concepts took place at Temiskaming, Kapuskasing, and Dolbeau. In the interwar period, the large-scale mining and smelting operations of the Sudbury basin, together with smaller concentrations at Flin Flon and Noranda, gave rise to planned company towns. Later, during the major postwar expansion of the Shield's mining industry, many new communities, such as Labrador City and Thompson, were developed in remote areas on the basis of comprehensive planning principles devised and used in southern metropolitan centres.[33] Nevertheless, the restricted range of social and employment opportunities for women, together with limited educational and occupational facilities for older children, still prompt some families to leave the communities after a few years.[34] Both the community and the major employer suffer if the rate of population and employee turnover becomes excessive, and cooperative attempts to improve the quality of life in resource towns are widespread.

Some mining towns have acquired functional importance as regional service centres (Thompson for northern Manitoba and Matagami for the James Bay developments). Others, such as Val d'Or, can look to a relatively secure future of continuing mineral production. But given the uncertainties of contemporary markets, the mining industry no longer wishes to create new communities around mines which may have an active life of thirty years or less. Increasingly, therefore, it has become the practice, wherever possible, to designate an existing community as the home base for miners employed at a new remote mine site. This may involve considerable daily commuting, as between Mattabi Mine and Ignace (160 km round trip), or else "fly-in" commuting, on a seven-days in, seven-days out basis, as at the Rabbit Lake uranium mine in northern Saskatchewan. There, an air shuttle service moves personnel from as far away as Saskatoon (690 km) to live in motel-type units at the mine site. Although this arrangement is estimated to cost more than construction of a townsite, it is more appealing to the majority of miners and their families, and thus reduces the costs and social disruption associated with the high rates of labour turnover typical of remote resource communities.[35]

Communications flows and migration movements link the Shield's CMAs to heartland metropolitan centres more directly than to each other. Chicoutimi–Jonquière has had the least favourable net migration ratio (within Canada) and the lowest external immigration ratio of all Canadian CMAs. Like Thunder Bay, it has gained migrants only from non-metropolitan areas of its own province, which suggests that these CMAs act as intermediary points for those moving outwards and (socio-economically) upwards from the hinterland to more dynamic heartland cities. Migration and communications flows link Chicoutimi most strongly to Québec and Montréal, just as they do Thunder Bay to Toronto and, to a lesser degree, Winnipeg. Sudbury's interaction with the central Canadian city system is more complex, but links with Toronto are most prominent. The main scheduled air services mirror these patterns, fostering north–south contacts between Shield and heartland, rather than east–west links between hinterland centres. Networks of local air services in northeastern and northwestern Ontario are linked together by a route connecting Sudbury, Sault Ste. Marie, Thunder Bay, and Fort Frances.

THE HINTERLAND EXPERIENCE

Having examined heartland interest in the resources of the Shield and the nature of the urbanization to which it has given rise, we now consider the consequences for the hinterland population. A useful analytical framework is provided by G. R. Weller's classic study of northwestern Ontario.[36] According to Weller, resource-based economies are organized by, and principally serve, metropolitan interests. The myriad effects of this *economics of extraction* (subregional fragmentation, selective out-migration, the alienation of local control, etc.) cumulatively weaken the capability of the hinterland to define and pursue its own goals. This process results in the *politics of extraction*, an unequal struggle between heartland and hinterland. Hinterland interests attempt to bring about fundamental change in their relationship with the heartland (*the politics of futility*), whereas heartland interests seek to maintain hinterland acquiescence in the status quo through the *politics of handouts*. Repeated failure of the hinterland to change its status within wider economic and political systems affects the character of intraregional public life. Hence the *politics of frustration* expresses itself in the appeal of politically radical groups and of fringe movements, such as the regional separatists of the New Province Committee of Northern Ontario. Alternatively, this frustration is sublimated into the degenerate *politics of parochialism*, best exemplified by the rivalry which kept Port Arthur and Fort William independent (and uncooperative) municipalities for almost a century.

Local initiative can combat hinterland conditions but it cannot fundamentally change the region's dependent status. Hinterland problems are invariably interpreted from a heartland perspective. Around 1970, for example, much publicity was given to the Mid-Canada Corridor, a concept perpetuating the myth of the north's "inexhaustible resources." As a band across a map, linking towns from Prince Rupert, British Columbia, to Schefferville, Québec, it was suggestive, but the inferred processes of massive population growth and industrialization ran directly counter to the prevailing dynamics of demographic and economic change within the region.[37] The "growth centre" strategies favoured by federal and provincial regional development planners in the 1970s and early 1980s did little more than promote the orderly provision of services in a region of static or declining population.

Hinterland alienation has in some instances fostered a radical reaction. The effectiveness of such a response within the broader political and economic system has been limited, however, by the numerical weakness of the hinterland population and the fragmentation of its protest. Parliamentary protest has been uncoordinated. At both the federal and provincial levels, there has been as much tendency for Shield residents to support governing parties as those in opposition. Early unionization of resource industry employees was hindered by ethnic diversity, with each group conscious of its separate identity and often ideologically at odds with others. Active management opposition meant that unionism took root more slowly in paternalistic company towns, such as those of the Nickel Belt, than amongst Cobalt and Timmins miners, who generally lived away from the company's premises. Having finally established itself in Sudbury in the 1940s, union activity was hampered for the next twenty years by internal rivalry, leaving vital issues such as occupational health and safety relatively neglected. Significant improvements in working conditions were secured subsequently, and since the early 1980s there has been a substantial improvement in labour relations in the nickel industry.

Evidence of heartland control over the hinterland economy is pervasive. The prices of metals and forest products originating in the Shield are set in continental or global markets that give no incentive to industries that use them to locate within the region. Similarly, gasoline sold in Red Lake, Ontario, retails at the Sarnia price plus freight to Red Lake (2050 km), although the town is only 500 km by road from the Winnipeg refineries. Hinterland frustrations are also manifest in other ways. For example, the residents of Matagami told Ontario's Royal Commission on Electric Power Planning in 1976 that they were "still puzzled that although their community had to be relocated in 1921 in order that a dam for hydroelectric power generation could be built, it was only five years ago [in 1971] that electricity was installed in their community."[38]

The Shield's future will undoubtedly be marked by continued dependence on an economy of forest products, mining, and tourism. Employment will remain cyclical, in varying degrees of severity, in all these industries. Increasing capital intensity in the mining and forest sectors means that greater output is achieved by a smaller work force. The occupational structure and limited backward linkages of the tourist industry will do little to weaken the pressures on young people to migrate to regions where career prospects are more diversified and promising. The regional industrial incentive programs of the 1960s and 1970s demonstrated, by their lack of success, that the Shield is not an attractive location for most manufacturing plants, other than a few directly linked to the region's resource-based sectors. A more diversified employment base will come into being only through concerted efforts that improve the educational opportunities available in the region; encourage local entrepreneurs to identify and exploit business ventures capitalizing on regional advantages; and that make effective use of the research capabilities of the region's colleges and universities. This combination is most readily delivered in the major metropolitan centres. Sudbury, for instance, has seen the development of firms manufacturing equipment for the mining and forest industries, as well as, less conventionally, one growing tomatoes in greenhouses warmed by exhaust air from mineshafts.

Native peoples entrepreneurship still faces many obstacles, particularly in gaining access to capital and management training. Businesses are typically small and serve local markets. A survey in northern Ontario found 526 Indian-owned or controlled businesses in 1986, roughly evenly divided between locations on main highways and those in remote areas. Retailing (mainly general stores), tourism and recreation (especially related to hunting and fishing), and personal services were the dominant enterprises. Most firms were individually or family owned, but band-owned businesses are becoming more common. This is especially the case where land-claim settlements have provided substantial financial resources, as exemplified by the agencies established in the wake of the James Bay and Northern Québec Agreement (1975). The $225 million payment (spread over twenty years) which the Native peoples received in exchange for relinquishing land claims to over half the area of Québec has enabled them to gain control over significant elements of their regional economy. In the transportation sector, the local airlines, Air Creebec and Air Inuit, are Native-owned, and a variety of cooperative enterprises pursue Cree initiatives in forestry, tourism, and mining exploration.[39] Following the 1987 sale by the Hudson's Bay Company of its network of northern stores, many Indian bands took over the operation of the local outlet in their community as a joint venture with the successor company, the North West Co.[40]

The process of persuading heartland governments to grant Native peoples of the north greater self-government has been protracted and ridden by federal-provincial conflict. Only in the 1960s did the government of Québec begin to take an active interest in the administration of its "provincial north," where the Native peoples were serviced by a strong federal (and anglophone) presence.[41] Much of the tension and acrimony associated with the early years of the James Bay agreement was caused by the failure of the federal and provincial governments to live up to their undertakings, frequently placing responsibility for problems on each another. Native peoples in northern Ontario voiced their frustration with government attitudes to the members of the provincial Royal Commission on the Northern Environment, established in 1977. Its initial mandate, to study a proposal by the Ontario government to release the last large tract of uncommitted Crown forest to the paper company whose Dryden mill had been responsible for mercury pollution of the English-Wabigoon river system, was broadened to cover the whole range of social and economic problems of the "provincial north." But the work of the Commission dragged on into the mid 1980s, to little effect, leading Native peoples, who had initially welcomed the enquiry, to see it as a government substitute for effective action.

Intensified federal attention in the 1990s to unresolved land claims, and to the wider agenda of bringing an appropriate end to the internal colonialism of Canada's aboriginal peoples, has highlighted the relationship between Native peoples and their natural environment. This has multiple threads, encompassing socio-cultural, spiritual, and economic dimensions, and embodies the comprehensive conception of "sustainability" that has increasingly become a motif (often more in rhetoric than action) of non-Native society also. The Native peoples' vision (too often mocked by the local reality of an alienated and disoriented culture, scarred by discrimination and welfare-dependence) fosters a diversified and balanced approach to the utilization of the Shield's resources, an outlook often shared by non-Natives who have made the Shield their home and who derive their livelihood from it.[42] This hinterland vision contrasts with the singleness of purpose with which heartland interests have tended to approach the region. The southern vision of a wealth-generating "resource frontier" has been expressed in hydroelectric power-generating megaprojects, extensive forest-cutting rights, and mining, smelting and paper-making operations with toxic environmental "externalities," the negative impacts of which have been borne by the Shield's residents. But that contemporary vision of urbanized environmentalists, which sees the Shield as a place of threatened remnants of "wilderness," to be preserved untouched for its intrinsic value and for the good of humanity, can equally define a narrowly conceived, externally imposed future for the region, the costs of which are localized in the hinterland. The complex currents of these disparate visions have found expression in specific land-use conflicts, such as those in the Temagami region.[43]

An ethical evaluation of the resource-based economy that has been created across the rugged expanse of the Shield hinterland might judge it to be, simultaneously, both a major achievement and an indictment of heartland society. Risk-taking and technological ingenuity have been required to discover and harness the natural resources of the region, and the wealth derived from their exploitation has been widely diffused through the economy and society of Canada. But in the process, the sustainabilty of the environment and the well-being of the residents of the Shield have too often taken second place, which is a succinct statement of the imbalances inherent in heartland–hinterland relations.

NOTES

[1] *The Financial Post 500* (1995) identifies two firms headquartered in Sault Ste. Marie (Algoma Steel and Algoma Central, a transportation company), one in North Bay (Boart Longyear, a mining equipment manufacturer), and one in Val d'Or (Le Groupe Forex, a timber company).

[2] G. R. Weller, "Hinterland Politics: The Case of Northwestern Ontario," in C. Southcott, ed. *Provincial Hinterland: Social Inequality in Northwestern Ontario* (Halifax: Fernwood Publishing, 1993), chapter 1, reprinted from *Canadian Journal of Political Science*, 10 (1977).

[3] I. Wallace and R. Shields, "Contested Terrains: Social Space and the Canadian Environment," in W. Clement, ed. *Understanding Canada: Building on the New Canadian Political Economy* (Montréal, McGill-Queen's University Press, 1997), chapter 17.

[4] R. C. Harris and J. Warkentin, *Canada Before Confederation: A Study in Historical Geography* (Ottawa, Carleton University Press, 1991), 8–15.

[5] R. Shields, "The True North Strong and Free," *Places on the Margin* (London, Routledge, 1990), chapter 4.

[6] D. Cole, "Artists, Patrons and Public: An Enquiry into the Success of the Group of Seven," *Journal of Canadian Studies*, 13 (1978), 69–78.

[7] J. B. Bird, *The Natural Landscapes of Canada*, 2d ed. (Toronto: John Wiley and Sons Canada, 1980), p. 184.

[8] Canada, *Report of the Royal Commission on Aboriginal Peoples*, Vol. 2 *Restructuring the Relationship*, Part 2 (Ottawa: Supply and Services Canada, 1996).

[9] H. V. Nelles, *The Politics of Development: Forests, Mines and Hydro-electric Power in Ontario, 1849–1941* (Toronto: Macmillan, 1974), chapter 2.

[10] Nelles, *Politics of Development*, pp. 443–64; Gilles Piédalue, "Les groupes financiers et la guerre du papier au Canada, 1920–1930," *Revue d'histoire de l'Amérique française*, 30 (1976), 223–58.

11 P. Mathias, *Forced Growth: Five Studies of Government Involvement in the Development of Canada* (Toronto: James Lewis and Samuel, 1971), chapter 6.

12 J. B. Nadeau, "Miracle on the North Shore," *Report on Business Magazine*, November 1996, 133–146.

13 *The Role of Wastepaper in the Canadian Pulp and Paper Industry* (North Vancouver: Temanex Consulting Inc., 1993), prepared for Forestry Canada.

14 Nelles, *Politics of Development*, pp. 326–35, 349–61; Jamie Swift, *The Big Nickel: Inco at Home and Abroad* (Kitchener, Ont.: Between the Lines, 1977), pp. 21–8.

15 J. Lewis Robinson, *Resources of the Canadian Shield* (Toronto: Methuen, 1969), pp. 23–7 and Fig. 3.1.

16 Iain Wallace, *The Transportation Impact of the Canadian Mining Industry* (Kingston, Ont.: Centre for Resource Studies, 1977), pp. 22–8. Calculations are based on a revised form of Table 3.

17 *Mineral Policy: A Discussion Paper* (Ottawa: Energy, Mines, and Resources Canada, 1981), p. 43.

18 Wallace Clement, *Hardrock Mining: Industrial Relations and Technological Changes at Inco* (Toronto: McClelland and Stewart, 1981).

19 Oiva Saarinen, "Creating a Sustainable Community: The Sudbury Case Study," in M. Bray and A. Thomson, eds. *At the End of the Shift: Mines and Single-Industry Towns in Northern Ontario* (Toronto: Dundurn Press, 1992), chapter 11.

20 John H. Bradbury and Isabelle St.-Martin, "Winding Down in a Québec Mining Town: A Case Study of Schefferville," *Canadian Geographer*, 27 (1983), 128–144.

21 John H. Dales, *Hydroelectricity and Industrial Development: Québec 1898–1940* (Cambridge, Mass.: Harvard University Press, 1957), chapters 3, 4, 7; Nelles, *Politics of Development*, pp. 464–87.

22 Richard F. Salisbury, *A Homeland for the Cree: Regional Development in James Bay, 1971–1981* (Montréal, McGill-Queen's University Press, 1986).

23 *Le Québec Statistique 1995*, p. 377.

24 R. C. Zuker and G. P. Jenkins, *Blue Gold: Hydro-Electric Rent in Canada* (Ottawa: Economic Council of Canada, 1984). The authors estimated that Québec derived over half a billion dollars in economic rent in 1979 from the power it purchased from Churchill Falls.

25 Albert Tucker, *Steam into Wilderness: Ontario Northland Railway, 1902–1962* (Toronto: Fitzhenry and Whiteside, 1978), p. 7, quoting Ontario Premier George Ross.

26 Ivor G. Davies, "Agriculture in the Northern Forest: The Case of Northwestern Ontario," *Lakehead University Review*, 1 (1968), 129–53. Quotation on p. 133.

27 Justice E. P. Hartt, *Royal Commission on the Northern Environment: Issues Report* (Toronto: Queen's Printer, 1978), p. 54.

28 Raymond R. Potvin and John J. Negusanti, "Declining Industrial Emissions, Improving Air Quality, and Reduced Damage to Vegetation," in J. M. Gunn, ed. *Restoration and Recovery of an Industrial Region: Progress in Restoring the Smelter-Damaged Landscape Near Sudbury, Canada* (New York: Springer-Verlag, 1995), chapter 4.

29 James E. Randall and R. Geoff Ironside, "Communities on the Edge: An Economic Geography of Resource-Dependent Communities in Canada," *Canadian Geographer*, 40 (1996), 17–35.

[30] The Québec–Labrador iron mines were initially developed as predominantly vertically integrated subsidiaries of United States steel corporations, and so were largely independent of the Toronto mining and financial communities. The Iron Ore Company of Canada, for instance, with its executive office in Cleveland, Ohio, and head office in Wilmington, Delaware, retains its Canadian office in Montréal. In the 1990s, Canadian and foreign (non-US) steel companies have acquired greater control over the iron mines.

[31] C. M. Wallace and Ashley Thomson, eds. *Sudbury: Rail Town to Regional Capital* (Toronto: Dundurn Press, 1993).

[32] Bradbury and St.-Martin, "Winding Down."

[33] L. D. McCann, "The Changing Internal Structure of Canadian Resource Towns," *Plan Canada*, 18 (1978), 46–59.

[34] Miriam Wall, "Women and Development in Northwestern Ontario," in C. Southcott, ed. *Provincial Hinterland: Social Inequality in Northwestern Ontario* (Halifax: Fernwood Publishing, 1993), chapter 3.

[35] Mark Shrimpton and Keith Storey, "Fly-In Mining and the Future of the Canadian North," in M. Bray and A. Thomson, eds. *At the End of the Shift: Mines and Single-Industry Towns in Northern Ontario* (Toronto: Dundurn Press, 1992), chapter 12.

[36] Weller, "Hinterland Politics."

[37] Mid-Canada Development Foundation, *Essays on Mid-Canada* (Toronto: MacLean-Hunter, 1970); Ivor G. Davies, "The Emergence of Mid-Canada," *Lakehead University Review*, 3 (1970), 75–97.

[38] Ontario Royal Commission on Electric Power Planning, *The Meetings in the North* (Toronto: Queen's Printer, 1976), p. 14.

[39] Peter George, "The TASO Research Program — Retrospect and Prospect: Native Peoples and Community Economic Development in Northern Ontario," Lakehead University Centre for Northern Studies Research Report #15 (n.d.).

[40] *The Globe and Mail*, February 1, 1994.

[41] G. R. Weller, "Local Government in the Canadian Provincial North," *Canadian Public Administration*, 24 (1981), 44–72.

[42] Neither Native nor non-Native society within the Shield is homogeneous, and there are conflicting interests with respect to resource use within both communities. They are well illustrated in a case study of Ignace, Ontario, by Maureen G. Reed, "Governance of Resources in the Hinterland: the Struggle for Local Control," *Geoforum*, 24 (1993), 243–262.

[43] Issues at stake in the Temagami region are illustrated in Jeremy J. Shute and David B. Knight, "Obtaining an Understanding of Environmental Knowledge: Wendaban Stewardship Authority," *Canadian Geographer*, 39 (1995), 101–111; Ron Prefasi, "Temagami," in M. Bray and A. Thomson, eds. *At the End of the Shift: Mines and Single-Industry Towns in Northern Ontario* (Toronto: Dundurn Press, 1992), chapter 10; "Temagami remains a tinderbox," *The Globe and Mail*, September 2, 1996; "Class conflict in Temagami," *The Globe and Mail*, September 19, 1996.

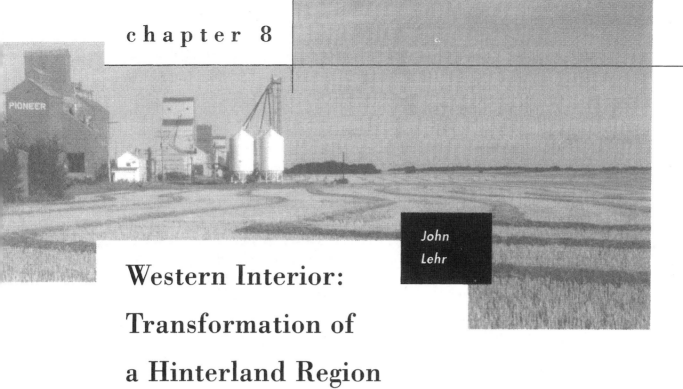

chapter 8

John
Lehr

Western Interior:

Transformation of

a Hinterland Region

The Western Interior has traditionally been viewed as a region dominated by the economic force and political power of central Canada which, in turn, has been guided by similar and equally strong forces in Britain and the United States. The chief commodities of the region, including primary agricultural and mineral products, have been subject to burdensome transportation costs, inequitable tariffs, and alienating market prices determined far beyond its boundaries. Moreover, the regional economy has advanced by using labour and capital imported from developed regions. Even though the influx of immigrant farm labour diminished during the Depression of the 1930s, the development of energy and mineral resources after World War II continued the tradition of relying on metropolitan sources of capital and entrepreneurship. Many of the advanced skills and the technology necessary to develop highly productive petroleum reserves still come, for instance, from the Texas Gulf region. Borrowed capital, technical expertise, and entrepreneurial skills have also brought organizational dependence on, and subordination to, multilocational and multinational firms, whose chief interests focus on the spatial integration of a hinterland's resources with its own industrial and management systems in other regions and nations. From a heartland–hinterland perspective, therefore, the Western Interior has shown many characteristics of hinterland dependency. In this world order, change and innovation are diffused from developed regions into areas struggling with remoteness, environmental obstacles, and limited social and economic opportunities.

Historically, the most direct lines of dependency have been to central Canada. The West, most of which was settled during the period of intensive immigration between 1896 and 1914, was initially agricultural in character, and supplied primary products to distant industry in central Canada and overseas markets. For decades after Confederation, efforts to develop an even limited industrial base were unsuccessful, mainly because of inequitable freight rates, isolation, and remoteness from markets. Federal government policies tended to reflect the realities of Canadian demography — the power of the electorate lay in Ontario and Québec — and concentrated on nurturing the growing financial and industrial interests of the heartland, often to the disadvantage of the West. In the debate before World War I over reciprocity between Canada and the United States, for example, Ontario and Québec favoured a protectionist stance, defending what they viewed as their fragile industrial structure. By contrast, the West favoured reciprocity and access to cheaper American goods, especially farm machinery. Reciprocity was defeated: the heartland's gain was the West's loss.

Even in the late 1990s there is still a polarization of interests between western and central Canada. During the debate over the Free Trade Agreement (FTA) with the United States, the western provinces were generally more in favour of the move, whereas Ontario and Québec were more circumspect. Within the region, enthusiasm for the North American Free Trade Agreement (NAFTA) was generally stronger in Alberta where the National Energy Policy was unpopular and NAFTA offered the prospect of riding on the energy boom once again. In contrast, in Manitoba opinion was tempered by the uncertainty of free trade's impact on Winnipeg's manufacturing sector and fear of job losses in other sectors of the economy.

The full impact of free trade has yet to be measured. Indeed, one could argue that true free trade has yet to be achieved, and it will be some time yet before its economic effects become truly apparent. What is clear is that the West is in the process of transformation as it adjusts and restructures in response to a plethora of forces, most of which emanate from beyond the region, suggesting that the heartland–hinterland paradigm still aptly describes many of the forces binding the region to its markets and suppliers.

In the 1990s the region has emerged from the uncertainties of the 1980s with a new sense of confidence in its destiny. This is especially true of Alberta, which benefitted from the collapse of the National Energy Policy, high prices for oil and gas, and technological developments in the energy sector. Saskatchewan and Manitoba look enviously westward at the buoyant Albertan economy and a ____ince exuding political confidence as it assumes a more confident stance in its ____tionship with the heartland provinces.

From a metropolitan perspective Calgary has displaced Winnipeg as the largest Prairie city. Calgary's rapid growth in the late 1980s and 1990s has enabled it to outstrip both Edmonton and Winnipeg, although both retain an impressive array of metropolitan functions. There is a danger of discounting Winnipeg's importance as a metropolitan centre in the Prairies. It lacks the glamour of Calgary and perhaps Edmonton. Many of its own inhabitants tend to be somewhat pessimistic about its future. Nonetheless, Winnipeg ranks fifth in Canada in total revenues from non-financial corporations, well ahead of Edmonton in tenth place but behind Calgary in third place. In terms of the total assets held by Canadian financial corporations Winnipeg lags behind Calgary but easily outstrips Edmonton.[1] Winnipeg's strength lies in the diversity of its economy and its dependence on the service and manufacturing sectors, which give it some immunity from the wild swings of the energy sector at the expense of a less exciting economic future.

Calgary's rapid growth and the increasing significance of the energy sector in Alberta and Saskatchewan can obscure the fact that agriculture is still a crucial element of the Prairie economy. Rural depopulation in response to a global restructuring of agriculture, a consequent increase in farm size and the erosion of many small Prairie service centres, sometimes masks the areal importance of agriculture to the West. Whereas in 1941 half of all Prairie residents lived on farms, less than ten per cent now do so, for over half the region's population is housed in the five metropolitan areas. As mining, including oil and gas production, and the tertiary sector have expanded, agriculture has declined in relative importance. This is no regional anomaly but a reflection of national trends. The Prairies still remain Canada's agricultural heartland, and wheat is still paramount, creating over 30 per cent of prairie farm income.[2]

Despite considerable restructuring in the Prairie economy it still retains many of the characteristics of a hinterland region. The heartland–hinterland paradigm still remains a valid framework for assessing the development of the Western Interior. In this chapter the historical perspective is emphasized in a consideration of those factors that have shaped the geographical character of the region. While it is acknowledged that the three provinces have very distinct characteristics — even personalities — the Western Interior is treated here as a conceptual entity. The region's common past and general world status are emphasized in the context of the heartland–hinterland pattern. This is prefaced by a description of the physical geography of the region, which has played a significant role in influencing the region's economic and cultural development.

THE PHYSICAL MOULD OF THE WESTERN INTERIOR

The Western Interior of Canada is the northern extension of the central lowlands of the United States. It is a part of a massive physiographic unit bounded by the Rocky Mountains on the west and by the Appalachian and Shield systems on the east. The plains between sweep northwards from the Texas Gulf Coast to the arctic margins of the Mackenzie Delta. The international boundary, which bisects the region along the

Alberta landscape

line of the 49th parallel, divides the land politically, but physically it remains a cohesive unit. Differences of nomenclature — Great Plains and Prairies — cannot obscure the natural linkages wrought by geology and topography.

Geologically, the Western Interior of Canada is a massive, crescentic sedimentary basin located between the crystalline rocks of the Shield and the sedimentary folds of the Rocky Mountains.[3] This basin is composed of gently warped limestones, shales, sandstones, and evaporites, ranging in age from the Cambrian to the Tertiary. The deposits deepen westwards. In Alberta, where deposits of over 3000 m thick are found, oil is present in vast amounts in Devonian reef systems. In the Fort McMurray–Athabasca area, thousands of hectares of oil-bearing sands are found at shallow depths, sometimes outcropping on river banks and seldom overlain by a glacial overburden of more than 100 m thick. Natural gas is found in conjunction with conventional oil reservoirs and independently at shallow depths in sand beds in the southern part of Alberta. These oil and gas reserves are the heart of Alberta's rich resource base. The wealth generated from their development has sparked the economic growth of the region, particularly through the world energy crisis of the 1970s and into the 1980s.

Topographically, the settled area of the Western Interior is not one landscape but three, for the region consists of three great plains each separated by gentle escarpments (see Figure 8.1). The first level runs from the Shield to the rise of the Manitoba Escarpment. Here the cuesta of resistant Cretaceous sediments has been dissected by eastward flowing rivers into a series of picturesque hills: the Tiger Hills, Riding Mountain, and Duck, Porcupine, and Pasquia hills. The terrain rises by 250 m to the second level, a rolling plateau of over 500 m above sea level cut by deeply incised rivers and levelled by many postglacial lake sediments. The western limit of this level is marked by the less defined rise of the Missouri Coteau, a low line of dirt hills running from Weyburn to Moose Jaw, then following the line of the Saskatchewan–Alberta boundary. Westward the third prairie level rises from 670 m above sea level in the northeast to 1200 m above sea level at the base of the Rocky Mountain foothills.

The topography of each of these "prairie steps" owes much to glacial and fluvioglacial processes. In the Quaternary period continental ice sheets invaded from the northeast, covered virtually all the interior, and at their maximum size joined with the alpine glaciers probing eastward from the Rocky Mountain heights. This ice left a discontinuous cover of stony and stratified till to depths of 320 m and deposited thick accumulations of end moraine. Stagnant ice moraines dotted with "myriad kettle lakes" and rolling till plains are common in the southern part of all three prairie provinces, as are the wide, steep-sided, flat-bottomed glacial spillways now occupied by misfit streams, sinuous lakes, and swamps.[4] These offer some of the most magnificent scenery in the Canadian West.

The most characteristic prairie scenery — the level plain — was formed by damming or glacial meltwaters (glacial lakes) in preglacial troughs. The largest of these lakes was Lake Agassiz, which, at its maximum extent, covered parts of the Lake of the Woods region in Ontario, most of southern Manitoba, and extended into east central Saskatchewan. It left a series of gravelly beach ridges — strandlines — at the foot of the Manitoba Escarpment and left deep deposits of lacustrine clays

FIGURE 8.1   East-West Profile of the Interior Plains of the Western Interior

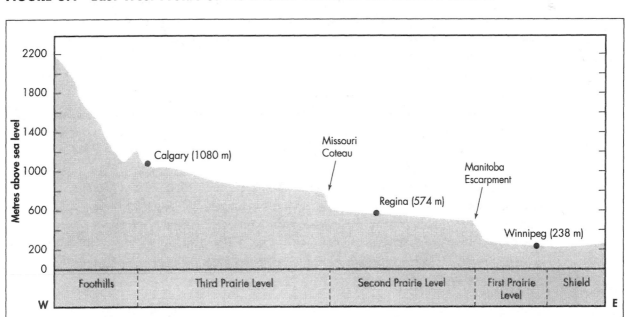

across southern Manitoba. The old lakebed constitutes one of the most fertile, and to some unsympathetic eyes, most monotonous regions of the prairie.

In postglacial times the Western Interior has been drained by two major river systems. The Nelson–Saskatchewan basin extends from the international boundary to latitude 54° N, while the basin of the Mackenzie–Athabasca–Peace system drains the northwestern plains. Though the escarpment and hills were important to the course of occupation by Native peoples and Europeans, the river systems were the single most important geographical feature of the region. They dictated the flow of communication, the path of trade, and the tide of exploration and early settlement.

Since the rivers flow north and east, delayed ice break-up in the north makes many river valleys prone to spring flooding when meltwaters hit icebound territory or ice-jammed rivers. In sparsely settled areas this flooding merely disrupts agricultural activity, but in the densely settled Red River Valley spring flooding causes considerable social and economic disruption. Ironically, the phenomenon that has awarded the valley its great fertility constitutes its most serious natural hazard, one which has led Winnipeg to construct a massive flood diversion channel to deflect the waters around the city.

The vegetation and soils of the Western Interior reflect both the climate and topography of the region (see Map 8.1 and Map 8.2). Short grasses form most of the cover on the less fertile brown chernozems of Palliser's Triangle. In the extreme south of the region sage and common cactus are also found. Bordering this short-grass zone is an arc of mixed grass prairie developed on more fertile dark brown chernozems. Here precipitation levels are higher and more reliable. At the eastern extent of this arc of vegetation, on the Agassiz plain, lies the true prairie — a lush cover of tall grasses some 1.5 m in height developed on black chernozems and gleysols. Between the mixed grass prairie and the boreal forest sits the parkland belt, where aspen bluffs and prairie clearings developed on dark brown and black chernozems, and on gleysols. These gradually merge northward into continuous deciduous woodlands and coniferous boreal forest developed mostly on dark grey chernozems and grey luvisols.

The climate of the Western Interior is cool continental, marked by hot summers and extremely cold winters. Climate has been a major obstacle to the successful agricultural settlement of the region. Late and early frosts restrict the growing season from 120 days in the south to 100 days or less in the north, a period that was insufficient to ripen wheat in the early years of settlement, even in the best regions.[5] Today, despite the development of early maturing varieties, the growing season on northern margins of the region is barely adequate for successful cereal cultivation. Other climatic hazards include hail from severe convectional summer storms, the occasional tornado, which wreaks havoc in the southern margins, and, of course, drought, which is usually most serious in the water deficient areas of the prairies north and south of the Cyprus Hills.

Map 8.1 Vegetation of the Western Interior Main Agricultural Lands

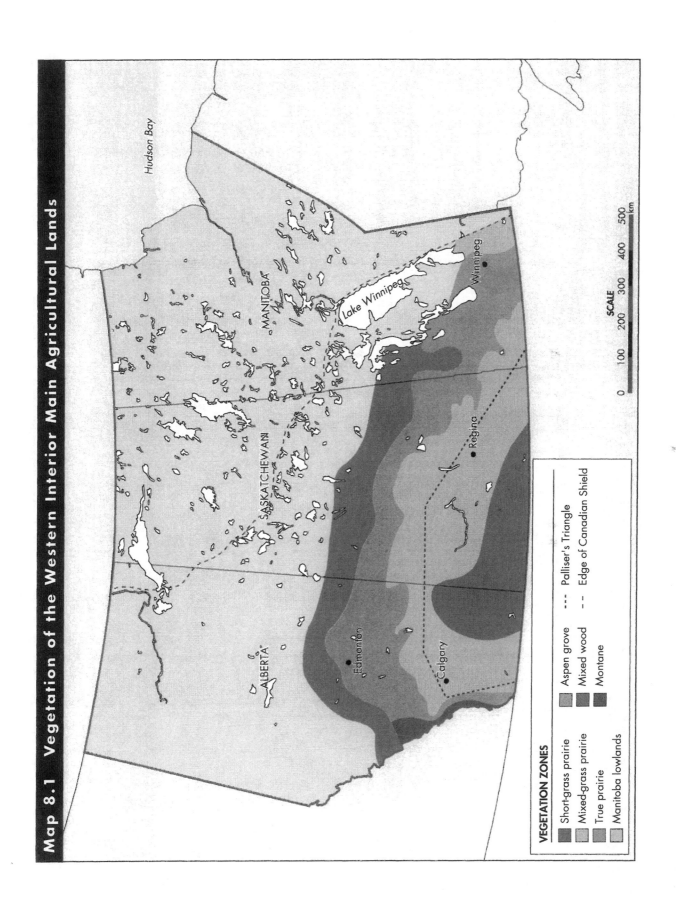

VEGETATION ZONES

Short-grass prairie
Mixed-grass prairie
True prairie
Manitoba lowlands

Aspen grove
Mixed wood
Montane

--- Palliser's Triangle
-- Edge of Canadian Shield

Hudson Bay

MANITOBA

Lake Winnipeg

Winnipeg

Regina

SASKATCHEWAN

ALBERTA

Edmonton

Calgary

SCALE

0   100   200   300   400   500
km

Map 8.2 Soils of the Western Interior Main Agricultural Lands

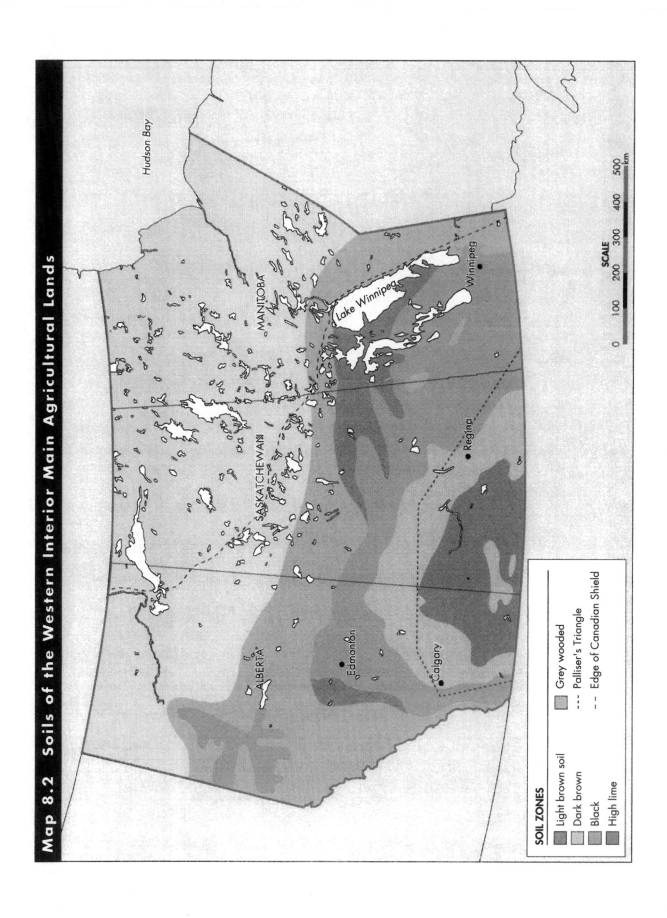

Hudson Bay

MANITOBA

Lake Winnipeg

Winnipeg

SASKATCHEWAN

Regina

ALBERTA

Edmonton

Calgary

SCALE

0   100   200   300   400   500
                              km

SOIL ZONES

- Light brown soil
- Dark brown
- Black
- High lime
- Grey wooded
- - - Palliser's Triangle
- -- Edge of Canadian Shield

The severity of the climate is tempered in the extreme west by chinooks that sweep down from the Rocky Mountains and raise temperatures dramatically over short periods, clearing the snow cover from the range lands east of the mountains. Microclimatic variations assume importance in local agricultural endeavours, as, for example, at Morden, Manitoba, where apple orchards survive in the shelter of the escarpment.

The Western Interior is, as many Prairie authors have observed, a stark land of haunting beauty, a land of contrasts and extremes in climate, vegetation, and topography; one that places a vertical man on a horizontal land, and pits the puny skills of mankind against the immensity of the plain and the unpredictable vagaries of an unforgiving continental climate.[6] Many perceptive writers of the prairie provinces have attributed the attitudes and opinions held by the region's inhabitants to the environment in which they live. This may be facile determinism, but it is clear that the historical evolution of the Western Interior as a hinterland region has done little to dissipate the westerners' perception of themselves as a people beset by a harsh environment.

CASTING THE PATTERN

The Hudson's Bay Company and the Canadian Pacific Railway cast the initial geographical patterns of the Canadian West. Indeed, as Harold Innis has pointed out in *The Fur Trade in Canada*, the present political boundaries of the region are a reflection of the efficiency of the Canadian fur trade.[7] When the Hudson's Bay Company was awarded a royal charter to trade through Hudson Bay in 1670, it received proprietary title to Rupert's Land, a vaguely defined territory that came to be equated with the drainage basin of Hudson Bay. Within this area, the company enjoyed the privilege of monopoly trading and, apart from the small area of colonial Canada, it was for two hundred years the sole agent of British imperial authority. The Hudson's Bay Company focussed its activities upon the prosecution of the fur trade. It reluctantly established a small colony on the Red River near its confluence with the Assiniboine River at the present site of Winnipeg, but only to secure an agricultural and labour supply base for its wilderness trade.

The greatest impact of the company was its permanent alteration and reorganization of patterns of aboriginal occupance over much of the northern half of the continent. This European geography imposed by the company was characterized by three distinct patterns of occupance and trade. In the first place, from 1670 until 1774, settlement was confined to fortified entrepôts located on the estuaries of the major rivers flowing into Hudson Bay. The company remained on the coast, depending upon Native peoples acting as middlemen for the extension of company trade to the outer limits of the fur trade hinterland. The placement of these forts eventually drew the Native peoples north and west in search of access to the central routeways to the Hudson Bay shores. This movement in turn provoked inter-tribal conflict. The Cree and Assiniboine tried to retain their supremacy in the north by maintaining their role as fur-trade middlemen, but in doing so, they denied the Gros Ventre, Blackfoot, and Chipewyan access to the Hudson Bay forts. The diffusion of trade goods and manufactures from a European core wrought havoc among the aboriginal population, transforming their economy and altering the human geography of the region.[8]

The second phase of Hudson's Bay Company occupance was marked by the movement of trading posts inland, away from Hudson Bay shores, in an attempt to counter rival fur trade companies seeking to divert the flow of furs eastward to the St. Lawrence. Compelled to move into the interior, the company did so selectively, establishing forts and posts at strategic locations to control the rivers, the arteries of the fur trade. Finally, the amalgamation of the North West Company and the Hudson's Bay Company in 1821 terminated rivalry and ushered in an era of fur trade consolidation. Many of the strategic post locations identified in the heat of competition consolidated their positions. Some, such as Fort Garry and Fort Edmonton, served as springboards for urban growth when Rupert's Land was acquired by the Dominion of Canada in 1870.

During the two centuries of company jurisdiction, Native peoples' society and economy were transformed. Often within a generation, the Native peoples' inde-pendent way of life was changed by the fur trade to a life of dependency upon European mercantile interests. Thus, to the uncertainties of the fisheries and the hunt were added those of supply and demand in Europe. The vagaries of the European market were felt keenly in the emerging West.

Early Settlement, 1870–1895

Three years after Confederation, the government of Canada acquired Rupert's Land. It received a territory where the fur trade reigned supreme, where agricul-ture was either subsistence or designed to provision the fur trade, and where, along the banks of the Red River, the few Selkirk settlers and the Métis maintained a ten-uous hold on the easternmost fringes of the prairie. The most pressing demand on the Dominion government was to settle this vast virgin territory, because settlement would check the aspirations of the United States for expansion into the Northwest. In addition, settlement of the West promised to stimulate the sluggish economy of central Canada.

Any settlement of the Western Interior depended on three prerequisites: estab-lishment of a land survey, creation of reservations for Native peoples, and con-struction of a rail link to eastern Canada. Before 1869, agricultural settlement followed a river lot survey system not unlike that used in Québec. Although it offered social and economic advantages for the early settler, it was nevertheless deemed unsuitable for the territory as a whole. Instead, the township system, modelled upon the American survey system, was chosen to facilitate "the rapid and accurate division of the prairie region into farm holdings."[9] (Figure 8.2) The township survey usually preceded settlement and superimposed a rigid, stereotyped pattern across the landscape without any regard to topography. Only in rare instances did the system deviate as it did to accommodate certain Métis demands for local river-lot surveys. With the completion of this survey, one major obstacle to settlement was removed. The conclusion of a series of treaties, which placed Native peoples on reservations as wards of the Crown, removed another. The third, most difficult, obstacle still remained: Manitoba and the NorthWest Territories were isolated from central Canada by tracts of forest, swamp, and rock. None of the existing routes into the West was easily traversed. The two all-Canadian routes — the Hudson Bay–Lake Winnipeg route and the Dawson route via Lake of the Woods and the Great Lakes — were both arduous. The alternative route, by rail to Moorhead, Minnesota, and then by boat north along the Red River to Emerson or Winnipeg, was easier; but it was not until Winnipeg obtained a direct rail link with the United States railhead at Emerson in 1878 that the communications barrier was truly broken.

Settlers did enter Manitoba and the Northwest before the completion of this rail link, but they were few in number and settled mainly in the Red River Valley and on the prairie margins. The government had stimulated settlement by encouraging the immigration of ethnic and religious groups, for whom special reserves of land were set aside, creating a mix of individual and colony settlement in the first phases of regional occupation. In 1874, Mennonite settlers from Ukraine were attracted to

FIGURE 8.2   The Township System of the Western Interior

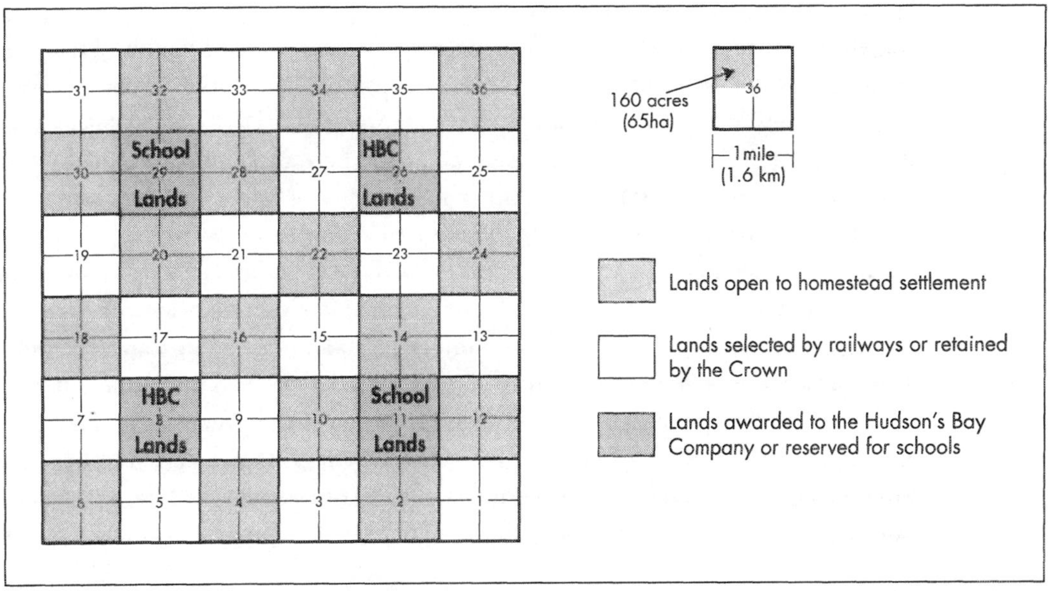

Western Canada by the Dominion government's guarantee of freedom of conscience, exemption from military service, and the exclusive use of an eight-township block of land in the Red River Valley. In the same year, French Canadians from New England were given land reservations to facilitate their settlement in blocks; and in subsequent years similar reservations were granted to other ethnic and religious groups that the Dominion government was anxious to attract into the West, including peoples from Britain, the United States, northwest Europe and, later, eastern Europe, England, Scotland, Iceland, Belgium, and Russia.

Although the government experienced some success with its group colonization policy (for example, over six thousand Mennonites entered Manitoba between 1874 and 1878), the general rate of settlement remained disappointingly slow. In the 1870s, the limitations of Red River farming had yet to be overcome. New settlers were either reluctant or unable to leave the security of the rivers or woodland fringe to challenge the open prairies. The fact that the government granted the Mennonites eight townships of land east of the Red River well illustrates the conventional attitudes to land evaluation. The area chosen was only 50 km southeast of Winnipeg and well endowed with wood and water, but it possessed large areas of swamp and gravelly soils. It was a good location for a typical settler of the time — a settler who lacked experience in farming the open prairie, who depended on wood for fuel, building, and fencing, and who relied on surface streams and ponds for watering stock. But for the Mennonites from Ukrainian steppes it was an inferior environment of limited potential and opportunity for sustained economic progress. More suited to their experience and needs were the twenty-two townships of the Mennonite West Reserve, opened for their exclusive settlement in 1875. There, on rich chernozem soils, they used adaptive strategies evolved in Ukraine. The use of dung for fuel, the employment of temporary sod buildings, and the communal herding of stock reduced their dependence on timber for fuel, building, and fencing.

To other potential settlers, lacking not only the necessary agricultural technology and adaptive strategies but also access to eastern Canadian markets, the West held little attraction. The situation became worse in 1875 when plagues of grasshoppers destroyed the standing crops of southern Manitoba and greatly discouraged further immigration. Further uncertainty was added by the malaise of economic depression then afflicting the industrial world.

The completion of the rail link between St. Paul, Minnesota, and Winnipeg in 1878, together with the easing of the world economic situation, soon renewed the flow of immigration into the Western Interior. By 1881, most of highland Manitoba south of the Riding Mountains had been occupied. So, too, had the Canadian base of the parkland crescent lying in the western section of the province. Lands to the west remained generally void of activity. Settlement was still confined to areas where wood, water, and hay were readily available; to districts easily accessible by rail; or to those areas with good prospects for the imminent development of rail

Clearing land, early settlement in Manitoba

communications. Although some 1 092 000 ha of land had been occupied by 1881, only a fraction of that, a mere 113 000 ha, was actually improved for agriculture.

The linking of Winnipeg with eastern Canada by the Canadian Pacific Railway (CPR) in 1883, and the completion of the transcontinental route in 1885, initially failed to meet settlement expectations. Many of the homestead entries in Manitoba and in adjoining parts of the North-West Territories between 1883 and 1890 were speculative ventures. From 1874 to 1896, homestead entries averaged under three thousand a year. In some years there were as many cancellations as there were new entries, partly because the Dominion Lands Act allowed relocation if the initial homestead proved disappointing. Of more serious concern was the contrast in settlement rates with the Dakotas to the south, which were being settled rapidly, in large part by emigrant Canadians. The Winnipeg *Times* complained that the trails from Manitoba to the States were ". . . worn bare and barren by the footprints of departing settlers."[10]

The paucity of settlers in the 1880s was understandable. The memory of the grasshopper plagues was still fresh, and the uncertainty of cereal production in the region began to diminish only in 1885 with the introduction of early-maturing Red Fife wheat. The price of manufactured goods from central Canada was high, and cheap American imports were denied access by tariff walls created to protect Canadian manufacturing. Transportation costs were excessive and wheat prices depressed. Credit costs were steep, as were farm mortgages: both inhibited settlement, especially the highly capitalized settlement of the prairie drylands. Thus, settlement clung to the Red River Valley, the park belt within Manitoba, and the Qu'Appelle region of the Assiniboia district. It advanced only slowly toward the prairie lands, clinging to the lifeline of the CPR as far as Moose Jaw, where the move into the arid conditions and rougher lands west of the Missouri plateau was halting and tentative. The prevailing official view was that most of Palliser's Triangle was fit for settlement, but few were willing to risk farming in this semi-arid environment.

Other factors besides natural and economic conditions were responsible for the slow rate of settlement. Contemporary critics, admittedly partisan, pointed to the lacklustre promotion of the West by the Department of the Interior, which they claimed did little to promote Western settlement. Clifford Sifton offered a stinging critique: "[It is] a department of delay, a department of circumlocution, a department in which people could not get business done, a department which tired men to death who undertook to get any business transacted with it."[11] The CPR also promoted Western settlement, but despite an apparently energetic and imaginative campaign, it, too, experienced little success before 1896. It seems, therefore, that the failure of the government to promote settlement was due more to economic conditions beyond its immediate control than to departmental inaction.

In order to finance the building of the transcontinental link between central Canada and the Pacific, in 1881 the federal government gave the CPR a subsidy of $25 million and 25 million acres (10 117 500 ha) of land, in addition to the grant for roadway and station sites. This land was assigned in alternative, odd-numbered sections of 259 hectares from territory extending 39 kilometres from each side of the railway. The railway company was free to select its lands from those "fairly fit for settlement" at a rate of 4856 hectares per kilometre for the first 1449 kilometres of track, 6744 hectares for the next 724 kilometres, and 3894 hectares for a final 1300 kilometres. Although the CPR built 1050 kilometres of its total track through Ontario and 431 kilometres in British Columbia, it selected its lands primarily in Manitoba, Saskatchewan, and Alberta: 883 494, 2 515 932, and 3 968 272 hectares respectively. It was a heavy burden for these provinces to bear. The system of securing alternate sections retarded the introduction of irrigation, dispersed settlement, and removed a substantial portion of potential taxation revenue from government sources. Alberta and Saskatchewan were hit particularly hard, because the railway companies selected land which promised the greatest appreciation in value. Six railroad companies, running no track in Saskatchewan, chose their grant from that province. Alberta relinquished over 3 966 060 hectares to the CPR alone, yet only 540 kilometres of track actually crossed its territory.[12]

To westerners, the CPR enjoyed the unenviable reputation of a rapacious agent of eastern Canadian capitalism. Not only had the company alienated millions of acres of potential homestead land, but its real estate offices often manipulated station locations to ensure maximum profits.[13] Even more galling to westerners was the fact that much of the CPR's landholdings in the burgeoning urban centres were tax exempt. The CPR's land-disposal policy, based on delaying the sale of the land in the more northerly areas until the improvements of homesteaders had caused a sharp appreciation in the value of railway sections, was, of course, in the best interests of the corporation but not of western progress. Indeed, the CPR even delayed selecting lands "fairly fit for settlement" to avoid paying taxes on entitled but unpatented lands. By doing this, settlement was effectively prevented in some of the most promising farming areas because the Department of the Interior could not open for homesteading those lands which the railway might eventually select as part of its grant. Exacerbating this situation, the CPR's promotional schemes centred on lands located adjacent to its transcontinental line, where accessibility had obviously increased land values. These sites could only be developed by settlers with considerable capital and experience, the most elusive type of immigrant in the 1880s and 1890s.

The CPR's freight rates were also a controversial issue. Westerners called them discriminatory because they forced settlers to bear the brunt of transportation costs on both exports and imports. The CPR was viewed as an expensive alternative to American railways. Certainly the rates were burdensome, but this was, perhaps, a simple reflection of the hinterland status of the West. Relying on exports of raw materials and facing strong competition from other established suppliers, the western farmer had to bear the cost of transporting his goods to distant markets. The absence of manufacturing industry in the Western Interior created a relatively inelastic demand curve for imported manufactures, so freight costs were borne by the consumer — the western farmer.[14]

The laggardly pace of early western settlement and economic growth is attributable in large measure to such economic circumstances. To what extent governments of the day can be blamed is more difficult to assess. Nevertheless, by the mid 1890s progress had been made: over 3 million ha had been occupied, of which a little over one-fifth had been agriculturally improved. In Manitoba, the base of the parkland crescent land had been settled, and in the shadow of the Rocky Mountains, a small group of Mormon settlers had established a tenuous bridgehead on the plains in the Cardston area. Still, it was a disappointingly low level of achievement for fifteen years of effort. One disillusioned westerner described the region in the early 1890s as having " . . . a small population, a scanty immigration and a north west empty still."[15]

The Sifton Years, 1896–1905

The mid 1890s were watershed years in the settlement of the Western Interior. At a time when the international economic climate was improving, political change in Canada thrust Clifford Sifton forward as the Minister of the Interior in the newly elected Liberal government of Sir Wilfred Laurier. More than ever before, or in the years to follow, one man's personality and policies dramatically shaped the social geography of a region. To Laurier, the twentieth century belonged to Canada, but to Sifton it belonged to the Canadian West. Soon after joining the new government in November, 1896, Sifton reappraised his department's objectives and policies in the area of immigration. He was determined to create a settled and prosperous West as a foundation for Canadian economic prosperity.

Previous administrations had attempted to secure immigrants from Great Britain and northwestern Europe, emphasizing qualities such as loyalty to the Crown, cultural affinity, and assimilative potential while generally overlooking both the financial and occupational factors, which ultimately determined the success of an immigrant. Indeed, attempts to secure immigrants with agricultural experience and capital were largely misdirected when focussed on the highly urbanized British Isles. Many Britons came to Canada as artisans with no farming experience; they subsequently failed in agriculture and drifted to urban centres. When Sifton took office, Canada's immigration policy was, therefore, general in scope and very lenient to those of European background. Entry was prohibited to many others, including the diseased, the criminal or vicious, and those likely to become public charges. Sifton did not change this greatly. He simply recast the net for immigrants and pursued the agricultural immigrant with single-minded determination. To Sifton, the ideal settlers were Canadian or American farmers. They were usually blessed with capital, were familiar with North American agricultural practices, were independently minded, and posed no problems of assimilation. Even though the Americans may have been tainted by republicanism, Sifton thought them to be ". . . of the finest quality and the most desirable settlers."[16] But because they could not be acquired in sufficient numbers, Sifton looked toward the peasant heartland of central Europe for "immigrants of quality." His policies in promoting Western settlement focussed on the basic need for agricultural immigrants.

The most notable, and controversial immigrants to settle in the West during Sifton's tenure were the Slavs, mostly Ukrainians and Doukhobors, but Hungarians, Germans, Poles, and Romanians came as well. Convinced that a ". . . stalwart peasant in a sheepskin's coat, born on the soil, whose forefathers have been farmers for ten generations, with a stout wife and half a dozen children is good quality,"[17] Sifton encouraged Slavic immigration and gave his blessing for the formation of a clandestine organization of steamship agents, known as the North Atlantic Trading Company, to promote Canadian immigration in areas where, for political reasons, the Canadian government could not openly conduct business. Thus, the peasant heartland of Europe was scoured by agents of the North Atlantic Trading Company, all propagandizing Western Canada as the place of free land, equality, and opportunity. Sifton even paid a bonus on the head of each agricultural immigrant booked to Canada. The amount reflected Sifton's perception of their potential as pioneers: $5.00 for a Ukrainian male adult but only $1.75 for a British agriculturalist!

The social geography and ethnic character of the Prairie West were forged during the nine crucial years of Sifton's tenure as Minister of the Interior. His policies gave unity and presence to the region. Although the foreign immigrants entering

Canada in this period never threatened to overwhelm those of British extraction, in certain years Slav immigrants comprised almost half of those who passed through the Winnipeg gateway. But their foreign ways and peasant dress, combined with their arrival by train in large groups, often on successive days, exaggerated their impact to English-speaking westerners. The question of Slavic immigration, indeed the future direction of western settlement, became a national issue. The Conservative press, and even some elements of the Liberal press, furiously attacked Sifton's importation of Slavic "peasants." Even the pro-Liberal elements, while stoutly defending the Slav immigrants against racist invective, voiced the opinion that the newcomers should be assimilated, and demanded that they reject their cultural heritage as proof of loyalty to their adopted country.[18]

The great majority of Slav immigrants were Ukrainian peasants from the then Austrian provinces of Galicia and Bukovina.[19] The social background and poor economic status of the typical Ukrainian peasant immigrant led to a remarkably uniform pattern of settlement decision making. Coming almost entirely from the hills and forests of the Western Ukraine, they lacked experience in steppeland farming and craved the security of parkland habitats. In these areas, the penurious settler could find the essential resource base of wood, water, and meadowland that was vital for successful subsistence farming in the early years of settlement. They also sought the social advantages of settling together in order to preserve a milieu of familiar religious, cultural, and linguistic characteristics.

The spectre of "little Ukraines and Polands" arising in the West caused alarm in the anglophone society, both in the West and Ontario, because the British character of the new lands seemed to be in danger. The early years of Sifton's tenure

Ukrainian immigrants bring old world home designs to the new

were accordingly marked by frantic efforts to prevent the growth of large blocks of ethnically homogenous settlement immune to assimilative pressures. Unless checked, this would be the inevitable result of the common aspirations of Ukrainian immigrants and their tendency to follow their friends and kin into established blocks of settlement. From 1896 until 1900, therefore, the government worked diligently, and generally with success, to accommodate the wishes of the immigrants and to appease the demands of an anglophile Canada. Slavic, Scandinavian, German, and Anglo-Celtic blocks were intermixed across the northern fringes of the parkland belt. Unfortunately, the singular determination to settle adjacent to friends and kin led later Ukrainian arrivals to disregard the environmental quality of the land they were settling, and to move on to progressively poorer land in order to maintain contact with their compatriots. Sifton's detractors subsequently charged that the Slavs were deliberately placed on submarginal territory in order to develop western resources and western markets for Ontario manufacturers. In fact, Sifton's immigration and settlement policies played a key role in establishing the macro patterns of the Western Interior's social geography.

Sifton's foreign campaigns to attract immigrants were matched in Canada by his manoeuvers to free as much land as possible for settlement. By the mid 1880s, millions of hectares of potential homestead land remained in the grip of railway companies and land speculators. The railways were granted 11.5 million hectares of land, but in 1896 had selected and patented only some 0.8 million. Huge reserves of land "fairly fit for settlement" remained unsettled. Realizing that this unsettled land was detrimental to the progress of settlement, Sifton attacked the legal foundation that allowed the situation to exist. He soon forced the railways to complete their selection and patenting of lands in 1900, thereby opening up vast areas for homesteading. He also cancelled time sales, which jointly freed further areas for settlement and discouraged speculation in unsettled land. There was little to be done about underdeveloped patented lands — the results of unchecked speculation in the 1880s — and this situation continued to prevent the settlement of some of the better soils in Manitoba.

Sifton's policies altered little of the general pattern of economic development in the Western Interior. Although population grew rapidly, and the area of land brought into agricultural production expanded dramatically (see Map 8.3), the West remained a producer of primary products and an importer of manufactured goods. Urban development was based on service to the primary sector, and what little industrial growth took place was similarly oriented. Winnipeg, for example, experienced rapid growth in warehousing and transportation functions, but there was little processing of the materials shipped to eastern Canada. Sifton's conception of a prosperous West did not encompass vigorous industrial growth; the West's destiny lay in the achievement of prosperity through fulfilling the needs of the Industrial Heartland. By 1905, when Sifton resigned as Minister of the Interior, the position of the West had been consolidated and the social fabric established.

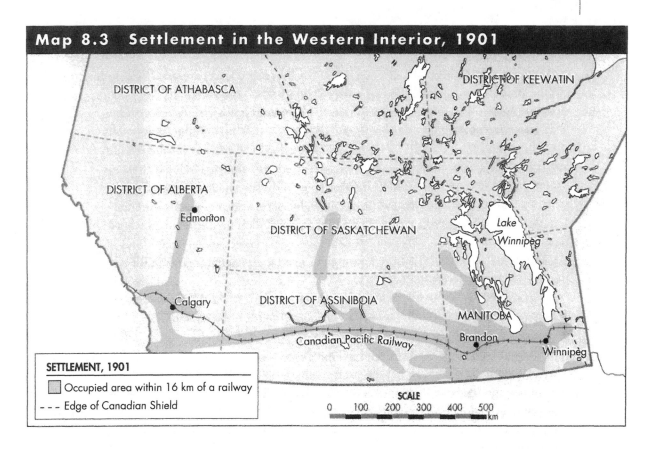

Map 8.3   Settlement in the Western Interior, 1901

DISTRICT OF ATHABASCA

DISTRICT OF KEEWATIN

DISTRICT OF ALBERTA

Edmonton

DISTRICT OF SASKATCHEWAN

Lake Winnipeg

Calgary

DISTRICT OF ASSINIBOIA

MANITOBA

Brandon

Canadian Pacific Railway

Winnipeg

SETTLEMENT, 1901

☐ Occupied area within 16 km of a railway
- - - Edge of Canadian Shield

SCALE
0   100   200   300   400   500
km

Within a few years, settlement would reach the last frontiers and people would strive to exploit the semi-arid plains in the heart of the Palliser's Triangle, enter the boreal forest, and penetrate the grasslands of the Peace River District.

The Last Frontier

As more attractive lands became increasingly scarce, homesteaders moved boldly toward the open prairie lands of southern Alberta and Saskatchewan. Two important innovations contributed significantly to this movement: the adoption of Red Fife wheat and the employment of black summer fallow. American technology, including the chilled steel plow, steel windmill, barbed wire, and self-binding reaper, facilitated the costly move into these drier and drought-prone regions. Capital costs of settlement were high. The peasant immigrants from Europe were excluded by financial constraints, although some second generation Slavs did move into such areas — around Prelate in Saskatchewan, for example. As with most movements onto new lands, this westward thrust was undertaken in a spirit of enthusiasm and optimism. Technological adaptations were made, but the old institutional framework based on the quarter-section farm unit, more suitable for humid regions, was maintained. In the drier areas, even the half section was not economic as a farm unit, and from the start it was clear that many of the homesteaded areas of the short-grass prairie were marginal. Droughts, frosts, and grasshopper infestation triggered farm abandonment even in the early years of settlement, but it was not until the Depression of 1930s that many finally realized, or admitted, that the arable frontier had extended beyond wise environmental limits.

But not all settlement was forced to retreat. Irrigation in southwestern Alberta was introduced by Mormon settlers in the 1890s. Based at first on the small-scale flooding of river flats, a series of large-scale irrigation ventures was established in 1897 by the Mormon church in cooperation with both the Alberta Coal and Railway Company and several Utah-based sugar producers. In the shadow of the Rockies, irrigation was undertaken largely to ensure fodder crops in the event of drought, but east of Cardston and south of Lethbridge it was undertaken to cultivate sugar beets. The significance of Mormon irrigation lies not in its areal component, but in its precedent, for it paved the way for the later and larger schemes of the CPR and the Prairie Farm Rehabilitation Act.

The Peace River Block — 8000 square miles (20 500 square kilometres) in extent — was the last sizable part of the Western Interior to be colonized. Settlers began moving into the area in earnest about 1910, just ahead of the railway, which did not reach High Prairie until 1914 and the Peace River townsite until 1915. The Peace River district approached the prairie microcosm in both temporal and spatial senses. Its remoteness, climate, and soils encouraged dependence on wheat as a staple crop, but mixed and subsistence farming prevailed in the northern fringe areas, where the risks attendant on early frosts curtailed crop yields. A short growing season forced Peace district farmers to face problems already overcome in more southerly areas by the introduction of Red Fife wheat. Indeed, it is by no means clear whether settlement was *pulled* into the Peace by the perceived attractiveness of the land, or was *forced* into the area by the shortage of favourable land elsewhere in the prairies. Certainly some second generation Ukrainian farmers, who left the marginal bush country of southeastern Manitoba just before World War I, chose the challenge of the Peace, bypassing the intervening opportunities of dryland farming in southwestern Saskatchewan.

Like the prairie region to the south, the Peace River district has been characterized by four well-defined stages of settlement. Until the second decade of the new century, the region was locked in the outpost stage. Agriculture was insignificant, though not totally absent. In the next phase of agricultural expansion, immediately before the outbreak of World War I, settlement was beset with problems such as distance to markets and lack of winter feed for livestock. With the entry of the railway, the region entered a stage of expansion and integration. Security of tenure increased for most farmers, and many expanded their operations until the malaise of the Depression years necessitated retrenchment. The final stage, which in some measure is still in process, has been marked by regional centralization, focussing on the growth of regional service centres. However, it is a measure of the region that even in the late 1990s, there is not one clearly dominant service centre. The broken topography and variety of soil types in the Peace River district have created widely separated settlements. Low rainfall and climatic uncertainty continue to prevail against the viability of the small farmstead. Off-farm jobs — insurance against the vagaries of nature — are commonplace. These elements have all done much to conserve the region's sense of being an agricultural, as well as a resource, frontier.

AGRICULTURAL PROGRESS IN THE INTERWAR PERIOD

The boom that accompanied the settlement of the Western Interior ended in 1912, but two years later the outbreak of World War I boosted demand for wheat. Prices rose and remained relatively strong through most of the 1920s. The resumption of immigration after the hiatus of the war stimulated a last burst of land settlement and railway building, and by the end of the decade there was very little settled territory that was not within 16 km of a railway (see Map 8.4). But this flowering also carried the seeds of disaster. Encouraged by high wheat prices, marginal land in the dry belt was broken, and wheat production continued to increase. At the same time, overseas markets, which previously had absorbed a large proportion of the wheat crop, began to soften as Europe recovered its agricultural productivity and established protectionist policies. The stock market crash of 1929 further reduced foreign and domestic markets for Canadian grain: prices fell dramatically (wheat dropped from $1.02 to 35 cents a bushel between 1929 and 1931), and farmers suffered accordingly. A series of dry years heightened the tension. The dry belt was hardest hit; here wheat farmers were faced with the erosion of both land and income. Ironically, certain prosperous areas were hit worst by the drought and ensuing depression. Cash flow dependence affected the commercial farming population more than the marginal farmer who, located on the northern fringes of settlement, could easily fall back on a subsistence way of life. This fact prompted some farmers to migrate northwards from the dried-out areas to the fringe of settlement.

By 1935, the agricultural calamity had assumed such dramatic proportions that the provincial governments were unable to cope effectively with the host of escalating problems. The federal response was the creation of the Prairie Farm Rehabilitation Administration (PFRA), which was designed to reconstruct the social and economic fabric of the drought-stricken and soil-drifting areas of the prairies. The PFRA encouraged dryland farmers to adopt conservationist techniques such as strip farming and plowless fallow (trash farming); to switch from grain to grass; and in the worst hit areas, to resettle on PFRA irrigation projects. Within eight years, over 60 700 ha (ultimately some 1 112 910 ha) of submarginal land were recovered from grain farming and turned into community pastures man-

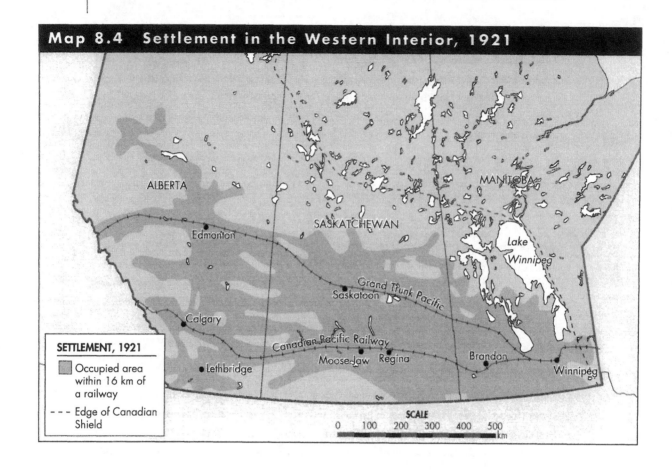

Map 8.4   Settlement in the Western Interior, 1921

ALBERTA

SASKATCHEWAN

MANITOBA

Edmonton

Lake Winnipeg

Grand Trunk Pacific

Saskatoon

Calgary

Canadian Pacific Railway

Moose-Jaw   Regina

Brandon

Lethbridge

Winnipeg

SETTLEMENT, 1921

Occupied area within 16 km of a railway

- - - Edge of Canadian Shield

SCALE

0   100   200   300   400   500
km

Dust storm near Lethbridge, Alberta

aged by the PFRA. Initial PFRA action was mostly consultative, based on the dif-fusion of information from experimental farms to various agricultural improvement associations which initiated and coordinated community action. In keeping with this stance, early PFRA efforts were typically small in scale and directed at the individual farmer or community; but subsequent projects became more complex when, for example, the PFRA cooperated with the Alberta Eastern Irrigation District to provide resettlement opportunities for displaced farmers. Eventually, PFRA projects increased in size to the scale of the St. Mary's River Project in southern Alberta, where over 89 000 ha of land were irrigated in an attempt to diversify crops and add value to local production.[20]

The years of drought and depression were also years of rural population loss. From 1931 to 1951, Saskatchewan experienced an absolute decline in population, and natural increases in Manitoba and Alberta barely balanced losses to out-migration. Before the 1930s, some rural-urban drift had been evident (the per-centage of the prairie population classed as rural fell from 75 in 1901 to 63 in 1931) as a consequence of farm consolidation and the gradual elimination of the inefficient quarter-section. But the Depression left a deeper, more lasting scar on the psyche of the regional farmer. The abandoned farmhouse on the dry prairie bore witness to the vagaries of nature; to the suffering farmers it stood as a symbol of the indifference of banks and mortgage houses of central Canada. From this sort of discontent in Saskatchewan and Alberta sprang the Western populist protest movements of the Co-operative Commonwealth Federation (C.C.F.), Social Credit, and later, in the 1980s, the Reform Party, three radically different responses to the same sense of western alienation.

THE URBANIZING WEST

For the most part, widespread urban development in Western Canada was a prod-uct of the railway age. Although the Hudson's Bay Company had established a del-icate tracery of routes and a scatter of posts across the West, this structure had little impact upon initial urban development. The railway companies, in fact, vir-tually dictated the geography of urban growth. Station halts or shipping points for grain exports — the life blood of a railway — were placed at 16 km intervals and became the nuclei of rural service centres. Major centres emerged at divisional points where railway maintenance facilities were located.

For obvious financial reasons, the railway companies preferred to establish townsites on their own lands. The CPR, for example, steered clear of the Hudson's Bay Company sites. Even the choice of the southern Kicking Horse route placed the CPR well away from established HBC posts in the parkland belt. As a result, the CPR established many new towns in most areas: Brandon, Moose Jaw, and Regina were all its creations. When a self-interested railway company located a station upon its own land a few kilometres outside an existing settlement, as was case both at Calgary and Vegreville in Alberta, then the existing settlement was oved to the railway site. Winnipeg narrowly escaped eclipse only by offering the CPR lucrative concessions, including a cash bonus of $200 000, free land for a station site, and exemptions from municipal taxes in perpetuity, as an inducement to route the line through its borders, rather than have it pass 32 km north, through Selkirk, as was originally intended.[21] Not until the building of the Canadian Northern and the Grand Trunk did the northern tier of settlements based upon fur trade posts rise in importance, and only then was the Hudson's Bay Company able to capitalize on its entitlement to a land reserve surrounding each post.

Only four prairie settlements had more than 1000 inhabitants in 1891: Winnipeg (30 000), Calgary (4000), Brandon (3800) and Portage la Prairie (3400). Winnipeg and Calgary benefitted from their strategic locations at the extremities of the prairies, and assumed rudimentary metropolitan functions shortly after the turn of the century. Winnipeg, especially, capitalized on its location at the eastern edge of the prairies and its initial advantage as an established urban centre. The gateway function of the city was bolstered by an initial and discriminatory freight-rate structure which made it cheaper to ship goods west from eastern Canada through Winnipeg, rather than directly to other prairie centres. The city's entrepôt function would undoubtedly have arisen without such schemes, but they certainly encouraged the rapid growth of Winnipeg's wholesaling and warehousing function.

Grain elevator, southern Saskatchewan

At the same time, the city exploited its natural advantages in the financial and commercial sphere. At the conclusion of the major rush of immigration in 1912, Winnipeg boasted the Winnipeg Grain Exchange (1887) and the Winnipeg Stock Exchange (1903), as well as an array of other financial and insurance companies. Although most railway privileges had been removed by 1914, weakening a monopoly on transshipment functions, Winnipeg's position as the leading city of the West was firmly established. But, even as the major city of the Western Interior, with a population of 128 000 in 1911, it was still firmly linked to the economy of central Canada. West of Winnipeg, the emerging urban centres of Saskatchewan and Alberta were equally in the shadow of the Industrial Heartland's economic strength. Their growth was linked closely to servicing the local hinterland, and little else. The benefits of western agricultural expansion were reaped largely by the manufacturers, suppliers, and financial institutions of Ontario and Québec.

The outbreak of war in Europe in 1914 failed to stimulate urban growth in the Western Interior as it did in the manufacturing cities of eastern Canada. Not only did war terminate the injection of immigrant capital into the supply centres of the West, but the region also failed to benefit from military expenditures. In the early stages of World War I, military contracts were let to civilian contractors without tender, the emphasis being placed on rapid delivery. Western cities suffered under the double disadvantage of having a poorly developed industrial capacity and a paucity of political friends in the high councils of the Militia Department. The federal government explained its wartime economic policies as the rational allocation of resources; in its view, the Western Interior should concentrate on production of wheat and eastern Canada on munitions manufactures. It did "not consider it to be in the interest of the Empire to encourage the erection of plants for forging shells in Calgary when there are sufficient plants already in the Dominion."[22] This continuation of the subservient hinterland status of the Western Interior was rational economic planning to the Ottawa bureaucrat, but to the westerner it was simply protection of established economic interests, and a clear reaffirmation of Sir John A. Macdonald's discriminatory National Policy.

During the interwar period, urban growth advanced little. Winnipeg remained the dominant centre despite slow growth, increasing its population by only 2 per cent between 1931 and 1941. The city used its strategic position to diversify a manufacturing industry that strongly reflected the hinterland status of the region it served: processing of agricultural products (meat packing and flour milling), the manufacture and repair of railway equipment, the production of a variety of consumer goods (especially clothing), and the manufacture of farm equipment. Edmonton and Calgary achieved moderate size at this time. Edmonton had been selected as provincial capital and used this position to secure the University of Alberta. Calgary, which coveted both, did not fare badly, however. It was a CPR divisional point, and in 1911 the company located its major western repair shops in the city. Soon after, oil was discovered 40 km to the southwest in the Turner Valley, and in 1921 western Canada's first oil refinery was set up in the city. Still, little diversification followed, and

Calgary remained, like other western cities, a transportation and distribution centre and the limited processor of primary products. In marked contrast stood Saskatchewan, the most rural of the three Western Interior provinces. The province had the largest total population but the smallest urban population. Regina and Saskatoon were overshadowed by Winnipeg, so their growth was determined by the demands of the limited agricultural hinterlands they served. Until the 1950s, they remained only small centres in a "have-not" province.

ECONOMY AND SOCIETY IN TRANSITION

From the beginning of the pioneer era, the settled landscape of the Western Interior was a landscape of institutional efficiency. The system of survey, the selection of the quarter-section as the optimum size for the homestead unit, the rigid geometry of communications, and the regular spacing of villages and towns along railway lines all reflected corporate and governmental concern for the efficient management of a new colonial territory. Just as the economic and topographic framework was set by metropolitan forces so was the religious landscape.[23] Protestant churches, headquartered in Toronto, divided the prairies into spheres of interest for missionary work. Methodists took the territory along the CPR main line, Presbyterians the territory along the CNR, while the Congregationalists were given a free hand in the north. The framework of the landscape was clearly of metropolitan origin.

The material culture of the region, in contrast, was derived from the placement of a polyglot population on this institutionalized landscape. Since most immigrants preferred to settle with others of their own nationality, early settlement was marked by the transference of a good deal of material culture from the Old World to the New. For the most part, ephemeral or transient elements distinguished the pioneer landscape: domestic vernacular architecture, formal religious architecture, farm building arrangements, and fence types. With few exceptions, traditional village forms disintegrated under the rigid system of land survey. Only those who secured the privilege of officially sanctioned group settlement, such as the Doukhobors and Mennonites, or those who arrived independently but were united by strong religious bonds, as were the Mormons, were able to overcome the constraints of the sectional survey and retain their old-world village systems. But even the ephemeral signs of ethnic settlement began to decay within a decade or two, as pattern book and even prefabricated houses became commonplace, replacing traditional vernacular forms.

However, the social fabric was more enduring (Table 8.1). The stability of rural society has resulted from a slower rate of economic change in the countryside. Out-migration has only heightened social stagnation, enabling rural-urban migrants in the burgeoning cities to idealize rural life as a repository of basic moral and social values. A measure of continuity in the rural Western Interior, therefore, is the extent to which the social geography of ethnic occupance has remained intact. After World War II, the boundaries between ethnic groups in rural areas blurred a little and social environments became homogenized by the assimilation of the second and third generation into the mainstream of Canadian culture. But southern Alberta is still "Mormon country;" southwestern Manitoba is Ontario-British; Manitoba's Interlake district is Icelandic and Ukrainian; and Mennonites are predominant on the reserves set aside for their settlement in the 1870s. Even in the cities, patterns of ethnic segregation have shown surprising longevity. In the pioneer era, a proportion of immigrants from all groups preferred to stay in the cities, to seek work rather than land. For example, Winnipeg's "North End," the Slavic and Jewish immigrant section of the city, was well defined by 1914. There was some mobility out of these ghettos in the interwar period, but social patterns remained little changed, as Jews and central Europeans again formed areas of concentration in the suburbs. Their old neighbourhoods have been taken over by rural in-migrants, particularly the Métis and Native peoples fleeing the poverty of northern reserves and seeking new opportunities in the city.

Despite the resilience of neighbourhood communities, the Western Interior has shed many of its former characteristics. It is now a highly urbanized society. The majority of its population are newcomers who know little of the pioneering phase. With the advance of transportation and communication, isolation has diminished in importance as an integrating force of regional consciousness. A new phase of economic development now pervades the region. Indeed, social and cultural change have followed economic change. Agriculture shaped the initial social and economic character of the Western Interior, and was still of fundamental importance through the Depression and war years, but in the postwar period, energy resources — and particularly oil — have transformed the region. An urban way of life has superseded the rural past as energy resources take on new dimensions of national and international significance.

Before the 1960s, economic growth in the Western Interior proceeded largely in response to external factors originating in Europe, the United States, and central Canada. Much has changed since then. The region's position of subordination to national and international heartlands has been transformed to one of growing maturity, responsibility, and participation in the form and function of the region's interaction with the larger world-economy. Strong political involvement in national councils and in numerous facets of the region's international activity now influences the ownership, sale, transportation, and export of the region's primary commodities. None of these resources can be discovered, explored, developed, or sold without adherence to explicit provincial government policies. While a federal presence still exists, this recent shift in political power is a fundamental factor affecting the economic relationship of the region with the developed world-economy.

Table 8.1   Birthplace and Ethnic Origin Characteristics of the Western Interior, 1911–1991 (by percentage of total population)

Birthplace	1911	1951	1991
Canadian-born	*51.6*	*77.1*	*86.5*
Maritimes	1.7	0.8	1.4
Québec	2.5	1.4	1.3
Ontario	17.1	5.1	4.8
Manitoba	14.9	22.2	20.0
Saskatchewan	7.9	25.3	22.9
Alberta	5.7	21.4	32.7
British Columbia	0.2	0.9	2.7
Yukon and N. W. T.	0.1	0.1	0.2
Foreign-born	*48.4*	*22.9*	*13.5*
Great Britain	17.5	7.4	2.2
United States	12.6	4.2	1.0
Scandinavia	3.2	1.2	0.2
Germany	1.4	0.7	0.8
Russia	9.6	3.5	0.5
Poland	n.a.	2.4	0.8
Italy	0.2	0.2	0.3
Asia	0.4	0.4	3.4

Ethnic Origin	1911	1951	1991
Asian	0.3	0.5	4.7
British	53.5	45.8	18.3
French	5.6	6.8	3.5
German	10.5	11.7	8.8
Italian	0.2	0.3	0.8
Jewish	1.1	1.0	0.5
Netherlands	0.6	4.0	2.0
Polish	1.4	3.7	1.5
Russian	11.1	1.7	0.3
Scandinavian	5.9	6.5	1.8
Ukrainian	n.a.	10.4	5.1
Native Peoples	2.3	2.8	4.5
Total population	1 328 121	2 547 770	4 575 000

Source: Calculated from data in the *Census of Canada.*

Traditional supply and demand relationships are also changing. Developments since the 1960s refute an earlier observation that "the self-sustaining development of many prairie industries, based on independent private capital, is being superseded by government financial support for so-called 'hothouse industries' in which total benefits appear to be less than actual costs, . . . and that the demand for many prairie products is not growing quickly enough to ensure sufficient provision of new industrial employment."[24] While some companies (for example, the J. M. Schneider log-processing plant in Winnipeg) have received promotional support to draw them into specific localities, the dramatic increase in the demand for the region's primary commodities has so altered the region's role in world markets that it must increasingly place priorities on satisfying external demand. The apparent shortage of commodities needed by heartland economies has greatly reduced the penalties of distance and peripherality which have existed for so long.

The assertion of provincial jurisdiction in external economic and political affairs has affected the relationships of different parts of the region to each other. Despite some attempts at coordinating economic development among the three provincial governments, for the most part the provinces strike individual courses. Provincial insistence on controlling resource ownership and associated revenues and taxes, which forces multinational corporations to deal primarily with provincial rather than federal authorities, is common to the region, but ultimately the gains accrue to provinces on an individual basis.

Thus, provincial governments now influence their internal economic development by investigating the environmental, economic, and social impact of resource development, by creating special agencies with concerns for indigenous populations and peripheral areas, and by promoting or supervising economic opportunities throughout the municipal and subregional districts of their jurisdiction. With few exceptions, the initiative and responsibility for developing natural resources and population lie with individual provincial governments, rather than with a unified regional agency or the federal government.

A significant shift in strength between the national political heartland and the provincial governments has thus taken place, but many sub-areas in the Western Interior remain subordinated to political jurisdictions. Provincial agencies have been reluctant to delegate power or jurisdiction to their constituent regions and have, thereby, enhanced core–periphery relationships at an intraprovincial level. Although the physical distance which separates internal problems from the centres of decision making has been greatly reduced, and the number of competing demands has diminished, conflict still separates centre and periphery in individual provinces.

The Manitoba government, for example, has attempted to decentralize government, dispersing departments to rural communities throughout the province. At the centre of this internal pattern, the region's metropolitan centres show positive signs of completing the transition from weakly developed administrative or branch-office station upon its own land a few kilometres outside an existing settlement, as was the case both at Calgary and Vegreville in Alberta, then the existing settlement was moved to the railway site. Winnipeg narrowly escaped eclipse only by offering the CPR lucrative concessions, including a cash bonus of $200 000, free land for a station site, and exemptions from municipal taxes in perpetuity, as an inducement to route the line through its borders, rather than have it pass 32 km north, through Selkirk, as was originally intended.[21] Not until the building of the Canadian Northern and the Grand Trunk did the northern tier of settlements based upon fur trade posts rise in importance, and only then was the Hudson's Bay Company able to capitalize on its entitlement to a land reserve surrounding each post.

Only four prairie settlements had more than 1000 inhabitants in 1891: Winnipeg (30 000), Calgary (4000), Brandon (3800) and Portage la Prairie (3400). Winnipeg and Calgary benefitted from their strategic locations at the extremities of the prairies, and assumed rudimentary metropolitan functions shortly after the turn of the century. Winnipeg, especially, capitalized on its location at the eastern edge of the prairies and its initial advantage as an established urban centre. The gateway function of the city was bolstered by an initial and discriminatory freight-rate structure which made it cheaper to ship goods west from eastern Canada through Winnipeg, rather than directly to other prairie centres. The city's entrepôt function would undoubtedly have arisen without such schemes, but they certainly encouraged the rapid growth of Winnipeg's wholesaling and warehousing function.

Grain elevator, southern Saskatchewan

Map 8.5   Areas of Population Gain, 1981–1991

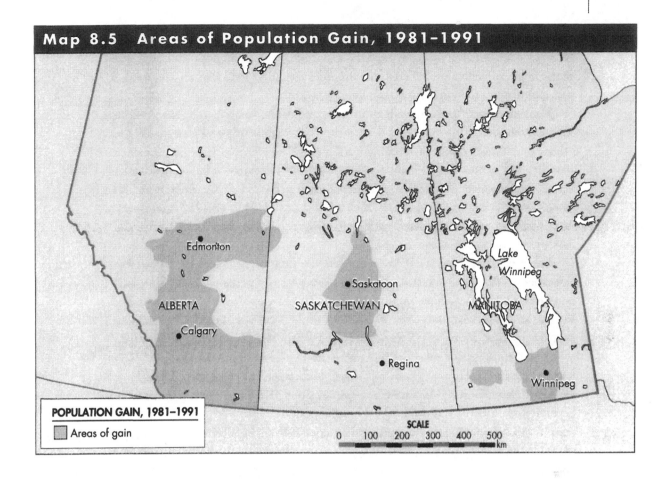

POPULATION GAIN, 1981–1991

☐ Areas of gain

SCALE
0   100   200   300   400   500 km

palities are those surrounding the region's major cities, many of which have gained in population as urban residents have drifted to land in exurbia.

According to Simmons, the urban system of the Prairies comprises five levels based on size, situation, and economic role.[25] There are three major fourth-order metropolitan centres: Calgary, Edmonton, and Winnipeg. Saskatoon and Regina serve as third-order urban places subordinate to Winnipeg. In addition to these five metropolitan areas, the system has three second-order and thirteen first-order urban centres. There are twenty-one centres in the region's urban system, including one in the Northwest Territories and Kenora in northwestern Ontario. Thus the Western Interior appears to have three urban subsystems, each transcending the familiar provincial political units. These subsystems are ultimately linked not to each other but to Toronto, affirmation of the lingering hinterland status of the Western Interior.

Within the region's urban system, dominance has shifted away from Winnipeg, the major city in the region in the 1960s, toward Edmonton and Calgary. With growth predicated on in-migration in response to the petroleum boom of the 1970s, both cities grew rapidly to overtake Winnipeg in the early 1980s and had moved ahead by 1986. From 1986–1991 Edmonton grew by 8.5 per cent, Calgary by 12.3 per cent, but Winnipeg by only 4.3 per cent. By 1997 Calgary had emerged as the region's economic dynamo, maintaining a high growth rate and showing a new confidence in its destiny as a national growth pole.

Most of Calgary's spectacular growth in the 1970s was squarely based on its regional function as the centre of the Alberta oil and gas industries. In the 1990s, while the energy sector is still crucial for Calgary's continued prosperity, the city's economy has diversified to embrace furniture and chocolate manufacture and a strong service sector. By 1996 it was second only to Toronto in the number of corporate head offices and is home to 92 of the *Financial Post's* top 750 companies. It still maintains its long-standing rivalry with Edmonton, but increasingly will look to Toronto for its economic challenges.[26]

The urban system of the Western Interior has long been dominated by the five largest cities that now comprise about three-quarters of the region's urban population and over half of the total population. In Saskatchewan, the most rural province, Regina and Saskatoon hold over 40 per cent of the total population and comprise over 64 per cent of the urban population of the province. In Manitoba, Winnipeg comprises over 82 per cent of the province's urban population; Alberta's two CMAs make up over 78 per cent of the urban population. In all cases the CMAs have increased their share of the urban population since 1970, and each province has become more highly urbanized over the same period.

Growth has not been confined to the major CMAs, for the number of urban places in the region has increased since the early 1950s, largely through the development of the region's non-agricultural resources.[27] Prior to this growth most urban places were small, their size determined by their basic function of service centres to a surrounding agricultural population. Other small urban places performed additional functions such as mining (Flin Flon), military services (Prince Albert), food processing (Lethbridge and Red Deer), oil and natural gas (Lloydminster and Medicine Hat–Redcliffe), and manufacturing (Brandon). These activities were not enough to stimulate widespread urbanization. Since the 1960s, however, the urban system has changed considerably. Many new centres have been established and some existing centres have grown considerably, while many small non-urban places have ceased to exist. Many urban places based on the fortunes of a single resource have experienced uncertainty as the companies that sustained them rode the roller coaster of international commodity prices. By and large the variety of functions performed by most urban places has increased considerably. They now offer more residential and specialized support service functions including health, recreational, and educational functions. This growth has resulted largely from improvements in the transportation and communications systems and a rapid trend toward centralization.

Resource distribution still plays a role in determining the development of urban places throughout the region. Apart from those centres owing part of their development to the processing of agricultural products from their immediate hinterland, the most notable urban development has been associated with large capital intensive resource-based projects, such as pulp mills, potash mines, and oil sands projects, which use regional raw materials but employ relatively modest numbers of people. This resource-based urban system is dependent upon an extensive rail and pipeline transportation system for the movement of commodities and equipment, and upon comprehensive road and air links for personal support, access to extra-regional amenities, and maintenance of supplies of food and replacement equipment. All of these specialized economic activities and related smaller urban places are subordinate to, and dependent on, the major metropolitan centres of the region.

All indications are that these metropolitan centres will maintain their dominance in the decades ahead. If Thompson's assertion that ". . . the higher-order centres of Winnipeg, Edmonton, Calgary and to a lesser degree Regina and Saskatoon are increasing their functional dominance through centralization and polarization tendencies within the prairie urban system"[28] is correct, then governmental attempts to decentralize government and direct governmental employment away from the CMAs is probably more of a political gesture than a serious attempt at directing regional growth.

Competition within the provinces produces regional asymmetries, suggesting that dominance corresponds not to an interrelated triumvirate but to an independently functioning troika. Competition between Edmonton, Calgary, and Winnipeg is fierce on both an inter- and intraprovincial basis, as they jockey for economic positions. Despite Winnipeg's slower growth it remains of international importance in the grain trade, while in a similar way the oil and gas industry has propelled Calgary and Edmonton on to the world stage. All three cities have significant functions at the regional and perhaps national level. Winnipeg is a major cultural centre. Edmonton, since the building of the West Edmonton Mall, has become a regional tourist and retailing destination, while its government and university activities have spawned a variety of localized functions such as manufacturing and business services. Calgary has diversified and has created numerous linkages in finance and communications services that mark it as a post-industrial city.

ECONOMIC GROWTH AND ENERGY

Long noted as a primary producer and an exporter of unprocessed materials to manufacturing centres outside the region, the Western Interior has, over the last quarter century, moved away from overwhelming reliance on export of raw materials. Agriculture and energy remain important, but in all three provinces the 1980s and 1990s saw movement toward diversification. In Alberta, for example, manufacturing now challenges the energy sector in terms of the amount of corporate income taxes paid ($395 million from manufacturing compared to $309 million from oil wells and mining). Most of this upsurge actually reflects the strong performance of the petrochemical, lumber, and pulp and paper sectors though, as noted earlier, the communications, finance, and insurance sectors are making considerable contributions to the provincial economy.

Saskatchewan's growth, though far more modest than that of Alberta, is also based on the resource sector, principally potash, oil, and wheat. The number of oil wells drilled in 1996 was increased by 2840, or 34 per cent. In contrast, Manitoba's equally modest gains have been largely unrelated to developments in the resource sectors of the economy although, as in Alberta and Saskatchewan, energy plays a significant role in the provincial economy.

Perhaps because of its global importance and its place in the Albertan economy since the 1940s, there is a tendency to view petroleum and gas as the main energy resources of the Western Interior. Nevertheless from very early days other forms of energy played important roles in all three provincial economies.

Coal is present in all three provinces, though it is no longer mined in Manitoba. Changing technologies in home heating, and conversion from steam to diesel by the railways, destroyed traditional markets for coal across the Prairies, but these markets have been replaced by the demands of thermal power stations in Alberta and Saskatchewan. Saskatchewan now has six open-pit mines in the southeast, which feed thermal power stations generating almost 70 per cent of the province's power. Alberta is the largest coal producer with ten pits, all but one surface mines, which produce both metallurgical coal for export to Pacific Rim countries and thermal coal for the generation of over 80 per cent of the province's

Potash processing plant, Saskatchewan

electrical energy. Manitoba imports some coal from Saskatchewan to power its two small thermal generating stations, though 99.2 per cent of its electricity is generated by hydroelectric installations in the province's Shield area (Table 8.2).

In 1970 the petroleum and gas industries were making an important contribution to the economy of Alberta, but few could have foreseen the coming boom, the result of high energy prices triggered by events far beyond the region. Following the Yom Kippur Arab–Israeli war of 1973, the Organization of Petroleum Exporting Countries (OPEC) flexed its economic muscle and drove up oil prices world wide through production controls. Alberta was a major beneficiary of this artificial oil shortage, for higher crude prices triggered renewed interest in exploration and development of Canadian oil and gas reserves.

The prosperity generated by the 1970s energy boom was always somewhat precarious because it was dependent upon forces lying far beyond the region and over which Alberta producers could exercise little, or no, control. Alberta relied heavily on distant markets in the United States and central Canada to sell its natural gas and crude oil — markets that have been expensive to reach by overland pipeline systems, and have been protected in the United States by quotas and institutional barriers. The attempts by the federal government to establish a National Energy Policy was seen by producers in the Western Interior as unwarranted interference in the fortune of the West, and as a policy promising to secure cheap energy for central Canada at the expense of lower wellhead prices for Western producers. Such federal policies were interpreted in Alberta, and to a certain extent in the region as a whole, as an example of central Canada's self-interest, and its view of the Western Interior as a resource-rich hinterland.

From the years of high energy prices came a renewed viability of processes aimed at extraction of oil from the massive tar-sand deposits of northern Alberta. Small experimental plants had operated since the late 1960s, but it was not until international prices for crude oil rose, and technological innovation reduced the cost

Table 8.2 Changing Demands for Primary Energy in Canada

	1979 (%)	1990 (%)
Oil	54.6	39.6
Natural Gas	19.2	26.6
Coal	10.8	13.2
Hydroelectricity	9.0	12.9
Other	6.4	7.7
Total	100	100

Source: Statistics Canada 1992

of production, that massive investment in large-scale endeavours was warranted. Today, the Syncrude plant at Fort McMurray is one of the largest private sector employers in Alberta, and has a production capacity that meets about 12 per cent of Canada's total petroleum needs, shipping in excess of 73 million barrels per year.[29] Operations are on a massive scale, entailing the removal of thousands of tonnes of overburden before the sands can be mined from leases covering an enormous area of land. Indications are that tar sand operations will continue to expand as increasing efficiency reduces the cost of production below its present level of $13.69 per barrel, and as royalty and tax changes implemented by the federal and Alberta government continue to encourage investment in the industry. Nothing, however, is certain, for the oil industry is sensitive to changes in the world price of crude oil. Overproduction elsewhere could have a devastating effect upon the profitability of Syncrude's extraction process.

Alberta's petrochemical industry has also benefitted from changes in the world hydrocarbons supply and its pricing structure since 1974. When planning began for a network of pipelines and processing plants in south-central Alberta, the economic future was less promising that it appears today. Pipelines now link the natural gas fields to ethane extraction plants at Empress, Cochrane, Waterton, and Edmonton, and provide links from there to the gas ethylene plant at Jaffre, near Red Deer, and on to the vinyl chloride monomer plant at Fort Saskatchewan. Diversification of the province's energy sector into processing will be of increasing importance as Albertans re-evaluate the role of energy in the provincial economy.

Tar sand oil extraction near Fort McMurray, Alberta

Natural gas is increasingly seen as a raw material and coal is undergoing a similar re-evaluation, both as a fuel and an industrial raw material. It will be a true measure of regional maturity when the area achieves the economic sophistication to process the bulk of its commodities, and ceases to derive the bulk of its resource revenues from export of raw materials.

PROBLEMS AND CHALLENGES

It is always tempting to see in the past a time when the Prairies were stable, little influenced by the process of change. Prairie writers in their novels hark back to changeless times, usually those of childhood, when the true essence of prairie life was to be found. The Prairies, however, have always been in flux, changing according to technological advances, adjusting to fluctuations in global markets and adapting to the vagaries of the physical environment. It might seem that, as the region approaches the millennium, the pace of change is accelerating and the region is losing many of the traits that gave it its distinctive social and visual character. This, no doubt, is illusory, though the region is faced with a number of challenges and problems that will shape its development in both the short and long term. To a surprising degree these challenges and problems are intertwined. Agriculture on the prairies, for example, is affected by a host of extra-regional factors, government policies, tariffs, freight rates, and transportation subsidies, as well as by pressures from within the region, such as provincial policies, local environmental concerns, and, of course, the spectre of climatic change. Although there is a high degree of interconnectivity, some of the issues whose resolution will have a significant bearing on the future of the region are treated thematically in the following discussion.

Agriculture: The Changing Heart of Prairie Life

In the early 1990s a joke was circulated on the Prairies: "Did you hear that the banks are going to help get farmers back on their feet? . . . . they're going to repossess their pick-ups." It captured beautifully the westerners' cynicism with regard to the major corporations, which are often regarded as agents of eastern interests. In effect, it was the modern version of a cartoon created some fifty years before, which showed a farmer in front of his hail-damaged field of wheat, shaking his fist at the sky, and shouting, "Goddamn the CPR." Prairie farmers saw — and probably still see — themselves as beset by climatic, economic, and geographic forces beyond their control, vulnerable to exploitation by powerful extra-regional institutions. This perception played a major role in the establishment of farmer cooperatives, credit unions, and wheat pools, much as it undoubtedly did in spawning the protest politics of the Alberta-based Reform Party.

The Prairie region, even if it becomes more urban, is still home to almost half of Canadians who live on farms. In 1991 there were about 401 500 people living on Prairie farms, about 14 per cent less than in 1981 and 60 per cent less than in the 1930s. The region, which has 80 per cent of Canada's agricultural land (see Figure 8.3), produces over half of Canadian agricultural output, and still derives over 7 per cent of its gross domestic product (GDP) from agriculture. In Saskatchewan, agriculture produces 12.5 per cent of the GDP, in Manitoba, 5.1 per cent, and in Alberta, 4 per cent. Primary agriculture and downstream processing account for nearly 9 per cent of all Prairie economic activity and over 11 per cent of total employment in the region. The number of farms is declining but acreage is expanding, so much so that the number of large farms with over $50 000 gross receipts has doubled since 1981.[30] This trend is most noticeable in Saskatchewan, where wheat monoculture is most prevalent. Wheat is still the most significant crop. In 1991 almost 32 per cent of prairie farm income was from wheat (see Map 8.7). No other crop rivalled wheat: barley constituted 6.4 per cent, and canola 8.2 per cent of crop value.

Canada's share of the world wheat market is now almost 20 per cent and predicted to rise to 25 per cent by 2005. Its regional and national importance as a staple crop will continue, though the termination of the Crow rate transportation subsidy means that many wheat farmers are facing doubled transportation costs. Demands to end the marketing monopoly of the Wheat Marketing Board by farmers better positioned to ship grain to United States markets are indicative of the regional stresses playing upon prairie farming in the 1990s. Fluctuation in world agricultural commodity prices also complicates the lot of the prairie farmer. In 1995, for example, responding to stagnant grain prices, many farmers diversified into sunflowers and chick peas, only to see wheat and barley prices soar.

FIGURE 8.3 Cropland in Canada, 1991

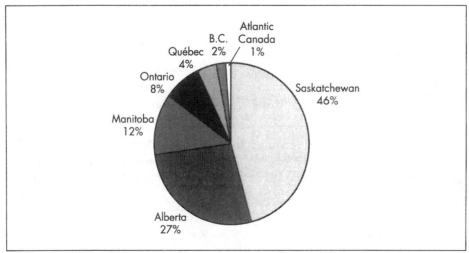

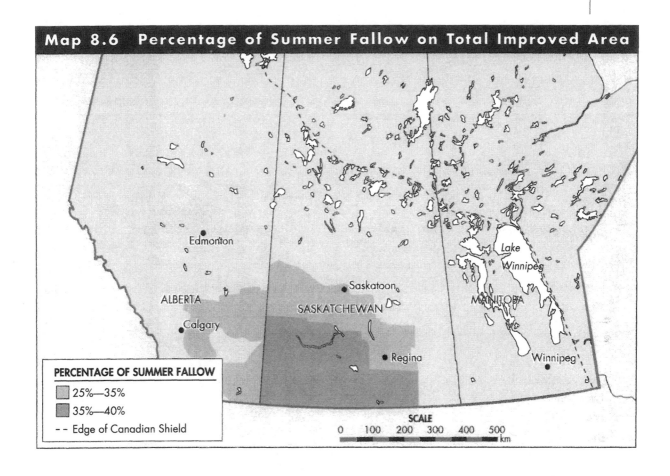

Map 8.6 Percentage of Summer Fallow on Total Improved Area

PERCENTAGE OF SUMMER FALLOW
- 25%—35%
- 35%—40%
- – – Edge of Canadian Shield

SCALE
0 100 200 300 400 500
km

Feed lot,
southern Alberta

Map 8.7 Agricultural Production in the Western Interior

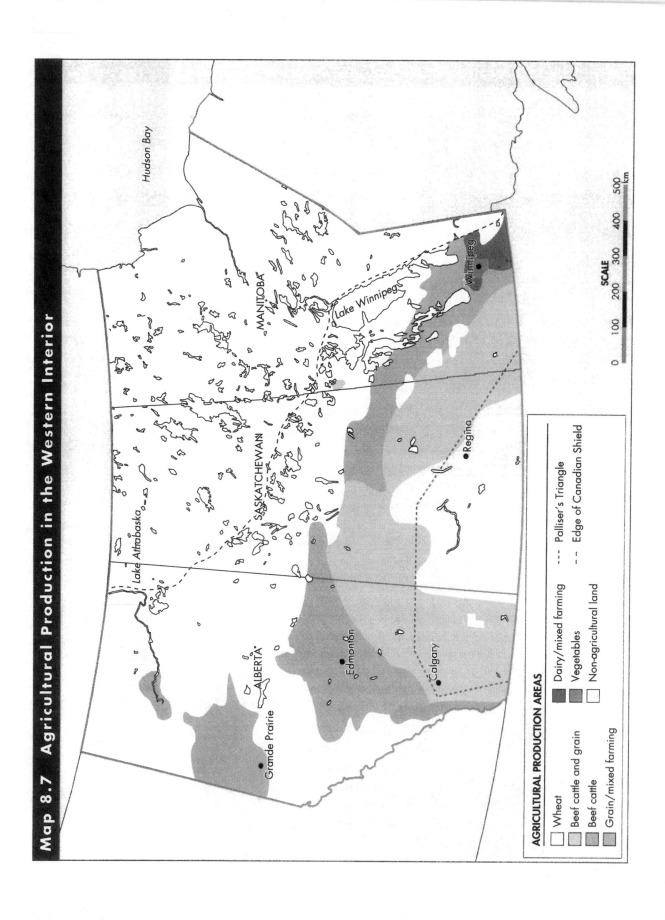

Hudson Bay

Lake Winnipeg

MANITOBA

Winnipeg

SASKATCHEWAN

Lake Athabaska

Regina

ALBERTA

Edmonton

Calgary

Grande Prairie

SCALE

0   100   200   300   400   500
km

AGRICULTURAL PRODUCTION AREAS

Wheat

Beef cattle and grain

Beef cattle

Grain/mixed farming

Dairy/mixed farming

Vegetables

Non-agricultural land

- - - Palliser's Triangle

- - Edge of Canadian Shield

Agriculture's importance to the Prairie economy extends far beyond the farm. Most people in Winnipeg, for example, have little idea of the importance of the grain trade to the city, yet the Wheat Board has annual grain sales of over $5 billion and employs five hundred people. The grain trade is a steadying influence on the local economy, smoothing out the peaks and dips of the normal economic cycle. The trade also supports a massive transportation infrastructure focussed on the shipment of grain for export to open-water terminals.[31]

Transportation on the prairies has always had a symbiotic relationship with the grain trade and the prairie farmer, though not always a happy one. Historically, freight rates have favoured monoculture and the processing of agricultural products outside the region. Farmers today often feel that the railways are content to exploit a captive market without devoting sufficient resources to meeting the needs of their clients. Grain producers, seeing their profits eroded by the inability of the railways to ship sufficient grain during peak season, are not particularly sympathetic to the railways' explanations of the interruptions to west coast deliveries occasioned by Rocky Mountain bottlenecks resulting from avalanches, lack of double tracking, or line wash-outs.

In attempts to rationalize their systems and cut costs, railways have closed branch lines and placed the onus for grain delivery to its shipping points on to the farmer. Since the 1950s the number of grain delivery points has declined by over 50 per cent, but the number of elevators has fallen dramatically from more than 5000 in 1953 to about a thousand in 1991. The region's total elevator capacity has fallen only marginally because small primary elevators were replaced by larger elevators at the remaining grain loading points (see Figure 8.4).[32] The rate of change was slow because the $560 million transportation subsidy previously given by the federal government under the terms of Crow rate killed incentive to develop economies of scale, but deregulation has unleashed unprecedented competition and change, and accelerated replacement of the small primary elevator by inland terminals, each capable of handling 35 000 to 45 000 tonnes. United Grain Growers, for example, has erected twelve major inland terminals, and is rationalizing its elevator system by closing sixty smaller elevators, about one-third of its system. In all likelihood other grain-handling companies will soon close a further five hundred elevators, reducing the number of prairie elevators to about seven hundred by the end of the century.[33] The effects on prairie farmers have already been far-reaching. Grain producers are obliged to haul grain longer distances, sometimes as far as 60 km; in response they are rapidly turning to larger and heavier grain trucks to cut their time commitment and reduce transportation costs. Rural roads, not designed to carry heavy traffic or semitrailer grain trucks, are deteriorating rapidly. As road maintenance costs increase, rural districts pass the costs back to grain producers through increased taxes. The more efficient railway and grain-handling system will generate higher operational costs for some producers and ultimately higher costs for the region to maintain its road infrastructure. On

FIGURE 8.4 **The Changing Economies of Grain Shipments in Western Canada, 1953–1991**

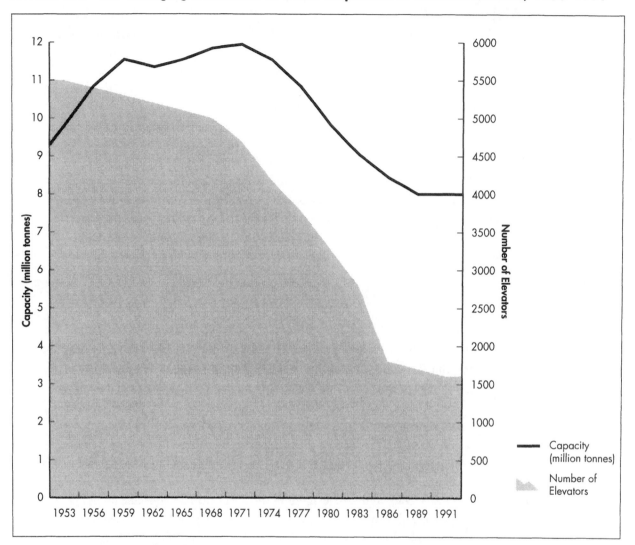

Source: Berry, "Canada's Primary Elevator System."

the positive side, a more efficient rational system of grain handling could reduce shipment charges, enabling the companies to pay the producer $5 to $10 a tonne for their grain.

Transportation policies and freight rates have some surprising ramifications in other aspects of the region's geography. From the beginning, freight rates set outside the region discouraged the processing of agricultural products within the region, and impeded the process of crop diversification. Grain production has been encouraged at the expense of livestock feeds, reducing production of livestock and forage crops, which in turn diminished the availability of animal residue needed to strengthen soil structure. Furthermore, since agricultural marketing policies have been geared toward the sale of principal crops at the expense of rotation, mono-culture, with its attendant erosional problems, has been supported by lack of markets for alternative production. Soil degradation thus has links to economic and transportation policies.

Environment

Since the dark years of the Great Depression of the 1930s, soil degradation and erosion have been a major concern in the prairies. The creation of the PFRA, withdrawal from cropping of the more marginal dryland areas, and the use of adaptive agricultural practices have done much to reduce the most obvious environmental damage (see Map 8.6). Overall agricultural productivity has increased through mechanization, application of fertilizers, herbicides and pesticides, and improvements in plant genetics. This conceals the reduction of soil potential through degradation caused by compaction, loss of organic matter, increased salinity, acidification, and water and wind erosion.

Wind erosion, like a spectre from the 1930s, erodes 5.6 tonnes of soil per hectare each year from parts of Saskatchewan. In the Peace River area, fine textured soils and hilly topography combine to create a loss of 11.5 tonnes of soil per hectare to water erosion on grain-producing fields. On a regional basis, organic matter in soil has decreased between 40 and 60 per cent since first cultivation. Over five million hectares (14 per cent of improved farmland) have lost significant amounts of topsoil through erosion, and it is predicted that an additional 1.1 million hectares will be affected by 2008.[34]

Loss of organic matter is compensated for by increased fertilizer use. Unfortunately, this presents undesirable environmental hazards: eutrification of lakes and risks of contamination of water supplies. Use of nitrogen has increased 1000 per cent since the mid 1970s, which carries grave risks of increased soil acidity.

Ironically, some traditional soil conservation measures widely used in dryland areas can themselves contribute to cropland loss. The practice of using summer fallow on large tracts of cultivated land, which is applied to 40 per cent of Saskatchewan cropland, can lead to soil salinity, resulting in a productivity loss of between 50 to 100 per cent. Salinity caused by irrigation and downward seepage of excess water costs Alberta some 370 000 hectares of cropland and $80 million a year in lost yields.

Environmental problems are not confined to soil degradation. In Manitoba's Red River Valley the practice of burning stubble from harvested grainfields prompted complaints from Winnipeg residents downwind of the burning. Eventually the province intervened, because of the perceived health risks. Working-in stubble is agriculturally a more desirable practice but also more costly. Flax growers were quick to point out that ploughing in stubble was not a viable solution for them; flax straw does not degrade quickly, and on heavy Red River

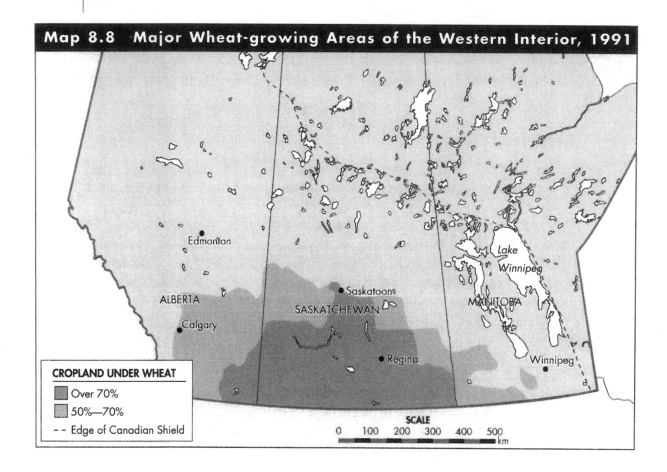

Map 8.8   Major Wheat-growing Areas of the Western Interior, 1991

CROPLAND UNDER WHEAT

■ Over 70%
□ 50%—70%
- - Edge of Canadian Shield

SCALE
0   100   200   300   400   500
km

clays it clogs agricultural machinery. As a compromise, burning is not permitted on certain days in certain districts; now Winnipeg is no longer enveloped in a pall of smoke on fine fall days.

The route to agricultural diversification is also beset with environmental hazards. Prompted in part by removal of the Crow rate, an expectation of increased production of feed grains for local feedlot operations, and opportunities for processing within the region, there has been an increase in the number of large-scale hog operations in Winnipeg's hinterland. Environmental issues are raised by communities near such operations, concerned with the odour and threat to groundwater by effluent run-off. This is particularly true of Manitoba's Interlake district, where limestone bedrock facilitates possible groundwater contamination.

Other environmental issues in the region mostly centre on primary production: rehabilitation of strip-mined lands in Saskatchewan and Alberta, the threat to the aspen ecosystem posed by strandboard plants newly located across the region's aspen belt, and the environmental effects of hydroelectric power development in Manitoba's north.

Urban Issues

To a greater or lesser extent the five major Prairie cities all confront major social issues, which see geographical expression in the decline of the retailing function of the central business district, movement of the middle class from the city centres to the suburbs, and the proliferation of suburban shopping malls. Expansion at the edge and declining density overall accompanied by a steady decline in the central city's proportion of population and employment is of particular concern to each of the cities of Winnipeg, Edmonton, and Regina. In Manitoba, for example, from 1971 to 1991, although Winnipeg grew by 15 per cent, the inner city's population declined by 13 per cent while the population of surrounding rural municipalities grew by almost 70 per cent. Increasing reliance on the automobile and a corresponding decline in the use of urban public transit have accompanied urban sprawl.

The causes of inner-city decay are interrelated and numerous: old decaying housing stock, lack of strong city and regional planning controls to check wanton suburban development, and the in-migration of people without the education and technological skills to effectively compete for employment. Poverty and all its attendant ills of prostitution, alcoholism, drugs, and crime have long given Main Street an unsavoury reputation, and merchants fear the spread of the malaise to the wider downtown area. Winnipeg's downtown, for example, has seen the injection of federal and provincial funds to redevelop the north side of Portage Avenue, the major shopping street in the city. The result was Portage Place, a downtown shopping mall that has been only partially successful as the south side of Portage Avenue is in need of upgrading.

These issues, of course, are not confined to Prairie cities; they are common throughout North America, and increasingly in Europe. Nevertheless, there is no doubt that the position of Regina and Winnipeg is unusual in that they both have exceptionally high aboriginal populations. Winnipeg, for example, now has 47 525 aboriginal people, one-third of Manitoba's First Nations' population, and the number is growing rapidly (Table 8.3). Like migrants from the Third World, Native peoples come to Winnipeg seeking jobs and better services. However, unlike migrants from other parts of the world who achieve a much higher standard of living after ten years, many Native peoples and their households become locked into a cycle of poverty and dependency.[35]

Many of the urban dilemmas facing the Prairie cities have causes beyond the control of the cities themselves. Federal and provincial social and economic policies, the nature of interregional migration flows, and even the price of energy, which has encouraged the growth of suburbia and exurbia, are all beyond the control of city politicians and planners. These problems should be considered to be of regional concern, though they have yet to be treated as such.

Table 8.3   Populations of 13 000 or more with Aboriginal Origins in the Western Interior 1991

Urban Area	Population
Winnipeg	46 000
Regina	13 000
Saskatoon	15 000
Calgary	25 000

	Province
Manitoba	107 000
Saskatchewan	94 000
Alberta	118 000

Source: Statistics Canada, 1991.

Tourism

With the exception of the Rocky Mountain resorts in western Alberta, the Western Interior is not generally regarded as an attractive tourist destination. Yet in the 1990s many communities in the region are looking toward tourism as an activity holding a good deal of economic promise. Tourism is now the third most important industry in Canada, and its importance is growing rapidly. Significantly, the sector of tourism demonstrating the most rapid growth on a world scale is cultural tourism, a sector that encompasses heritage and ecotourism.

The region is well placed to develop the ecotourist sector, especially in the northern Shield areas where, in Saskatchewan and Manitoba especially, tourists from beyond the region have for long been attracted to fly-in fishing, hunting, and wilderness adventure opportunities. Churchill, on the shores of Hudon Bay, is internationally renowned as a destination for ecotourists who prize the opportunity to observe wild life, principally polar bears, beluga whales, and a wide variety of birds, and to experience the Canadian subarctic. Ethnic and cultural festivals across the region draw from within the region, and the West Edmonton Mall has certainly enhanced Edmonton's appeal as a regional tourist destination. In Alberta, the development of Head-Smashed-In Buffalo Jump near Fort Macleod, the Kalyna Country ecomuseum in the area of Ukrainian settlement east of Edmonton, and the Tyrell Museum in Drumheller, and in Saskatchewan the Wanuskewin heritage site in Saskatoon, Batoche National Historic Park near Wakaw, and the recently designated Grasslands National Park have all contributed to the region's heritage tourism assets. Not all tourist development initiatives are initiated by governments, or dependent upon major governmental funding. Both Moose Jaw, Saskatchewan, and Boissevain in Manitoba, for example, have copied the development strategy of Chemainus, British Columbia, which successfully revived its sagging local economy by developing a tourist industry based on murals painted on building walls. Both Boissevain and Moose Jaw have had some success in developing a tourism sector in their local economy. Moose Jaw, which has also developed a mineral spa and is promoting its past links to the bootlegging trade of the prohibition era, has now become a destination for tours emanating from Winnipeg.

Tourism will continue to grow in regional importance. Predictions are that as North America ages the heritage sector will develop more rapidly. With its unique mix of cultural landscapes, its historic sites, the legacy of its polyglot pioneering past, vast wilderness areas, and surprising physical diversity, the region is well positioned to move more aggressively into competition for tourists.

CONCLUSION

In repudiating the yoke of perceived subordination to central Canada, the Western Interior established a sense of self-confidence and prowess that remains in place even though international economic factors have posed temporary setbacks to the region's development in the 1980s. This expression of regionalism is marked at the provincial level — irrespective of the political persuasion of governments in power — by on-going governmental intervention in the region's society and economy. The relocation of many corporation headquarters and investment centres to the Western Interior from central Canada suggests that the region is firmly engaged in the process of upward transition toward a state of maturity and equality with other areas of Canada, and is achieving closer integration with the advanced economies of Europe and Asia.

However, while many of the traditional adverse relationships between the region and external core areas have diminished, they have not completely disappeared. The constraining fetters of the past in the West are loosening rapidly, but they will only be removed when the distribution of decision making in Canada becomes entrenched in a viable constitution or related legislative schema, and when both international corporations and world trade are firmly managed to ensure effective regional participation, not subordination. In terms of the national and international political economy, the Western Interior is at a crossroads — vulnerable to external forces that demand wisdom and political expertise from those who represent the region's interests. Control and ownership of the region's natural resources must be wisely managed to ensure that the monetary benefits accruing from national and international markets are effectively repatriated and are equitably redistributed throughout the region.

The process of ensuring that the Western Interior benefits from the sale of its resources abroad will not be complete, however, until key elements within the region also effectively participate in its social structure and economy. Many crucial issues evident at the time of initial settlement were not resolved but were swept aside by the momentum of development; today, they have emerged as problems to test the region's political fabric and weigh on its social conscience. The exclusion of Native peoples from the mainstream of economic development and their relegation to special areas and inferior socio-economic status now pose grave problems. Native peoples face problems of identity within the structure of the urban industrial society surrounding them. Continuing rural poverty, difficulties with education and health care, and futile migration to the region's urban areas suggest that the general prosperity of the West has eluded this important group. The lack of participation by indigenous peoples in the socio-economic structure continues to tarnish the region's otherwise notable achievements, and is one of the important issues that will have to be addressed in the political arena.

The general direction of future economic development will be determined by the region's ability to effect change in several areas of discontent. As an example, despite numerous attempts to negotiate changes, the region has not been able to revise the institutionalized freight-rate structures that affect its competitive relations with other Canadian regions — especially those rates favouring the regional export of raw materials over processed or finished commodities. This structure continues to discourage the development of a diversified economic base. Grain and young (feeder) livestock, for example, are still shipped as primary commodities to related industries in central Canada where value is added through market-oriented manufacturing or finishing and processing (livestock).

The flour milling industry provides an example of this discrimination as shown by evidence produced by the Hall Royal Commission, set up in the mid 1970s to examine Canada's grain industry and grain-transportation problems. At that time, western mills were required to pay unequal interest, storage, and railway stop-off charges. The commission found that "the application of certain government programmes, of Canadian Wheat Board selling practices, and of ancillary rail charges offset the natural geographic advantage Western Mills should enjoy."[36]

The Western Grain Transportation Act passed in 1983 has alleviated some of these inequalities by abolishing the Crowsnest freight rates. The new rates and subsidies should guarantee adequate returns for railway companies so that they can expand to meet the growing needs of the region and the new opportunities in international markets. The Act also helps to protect farmers from low world grain prices and high domestic interest rates by guaranteeing that their share of the rail transport cost should not exceed 10 per cent of the 1988 price of grain.

Problems still remain, however. Subsidies paid for statutory grain transportation increase the price of grain and hence stimulate its production over other farm products. The subsidies also raise the price of feed grain in the Western Interior, thereby negating some of the natural advantages of the region's livestock producers. Further, rail transportation of grain is subsidized, unlike movement by road. Critics argue that production efficiency could be enhanced if payments were made not to the railways, but directly to farmers, who could then determine what crops to produce and what modes of transportation to use.

How then does the heartland–hinterland paradigm apply to the Western Interior in the late 1990s? Undoubtedly tensions still exist between the West and the Industrial Heartland. In the West's view, the federal government's management of subsidies and freight rates continues to hinder the movement of both agricultural and energy resources. The policies are barriers to the development of a diversified economic base. They are continuing sources of discontent and of regional alienation and seem to lock the West into its hinterland status.

The traditional relationship between central Canada and the Western Interior is changing dramatically, however. The western provinces' growing representation abroad is a telling symbol of the ability of many regions to bypass Ottawa in their dealings with the world's heartlands. The growth in trade between Western Canada and Asia demonstrates that many goods do not have to pass through central Canada on their inward or outward movements to world markets. The increasing penetration of foreign airlines and banks directly into the Western Interior further demonstrates that the region is gradually, but persistently, loosening its physical connections with central Canada. These political and economic changes signify the emergence of a new set of relationships between the Western Interior and the Industrial Heartland. Growing attention to uranium, heavy oil, deep tar sands, and northern electricity, in addition to the continuing significance of conventional oil, natural gas, and coal deposits, suggest that international interest in the region's resources will eventually expand across the energy spectrum, bringing increased regional income and associated political confidence. These in turn will put greater pressure on Ottawa for the effective redistribution of political and economic power within Canada.

Events in the 1980s, such as the patriation of the Canadian constitution, abolition of the Crowsnest freight rates, election of a national government consistent with western electoral preferences, and repudiation of the 1979 National Energy Policy may have temporarily defused some regional alienation, but they have not eliminated the discontent and the inequities. Western attitudes toward regional economic development, including full provincial autonomy in ownership and control of natural resources, are likely to find increasingly sophisticated political expression. Canadians must recognize that western frustration and the pursuit of regional goals are not simply parochial intransigence. They represent the need for a fundamental change in the region's economic status within Canada to ensure future harmonious relationships *within* the federal system. Many of the issues and problems underlying the grievances of the Western Interior, including those of its own minorities such as Native peoples and non-Native ethnic groups toward the region's provincial centres of power and authority, require resolution before attitudes become intractable.

In the late 1990s, therefore, the human and economic geography of the Western Interior as viewed through the heartland–hinterland paradigm suggests the following: (1) the region's relationships with foreign heartlands are in a state of flux, but they are likely to become more comprehensive and sectorally diverse throughout the rest of this century; (2) the region's traditional dependence on the central Canadian Heartland has lessened, but the relationship is still restrained by unresolved and deep-rooted obstacles; and (3) within the Western Interior serious spatial imbalances exist in the urban hierarchy, between urban and rural regions, and in levels of economic and social participation among important ethnic groups.

NOTES

[1] R. Keith Semple, "Quaternary Places in Canada," in John N. H. Britton, ed. *Canada and the Global Economy: The Geography of Structural and Technological Change* (Montreal and Knopton: McGill-Queen's University Press, 1996), pp. 356–7.

[2] Statistics Canada, *Canadian Agriculture at a Glance*, Catalogue 96-301, 1994.

[3] One of the most readable descriptions of the physical geography of the Western Interior is still that of W. A. Mackintosh, *Prairie Settlement: The Geographical Setting*, vol. 1 in *Canadian Frontiers of Settlement*, W. A. Mackintosh and W. L. G. Joerg, eds. (Toronto: Macmillan, 1934).

[4] Peter B. Clibbon and Louis Edmond Hamelin, "Landforms," in *Canada: A Geographical Interpretation*, John Warkentin, ed. (Toronto: Methuen, 1968), p. 72.

[5] C. F. Shaykewich and T. R. Weir, "Geography of Manitoba," in Government of Manitoba, *Manitoba Soils and their Management* (Winnipeg: Manitoba Department of Agriculture, n.d.).

[6] Laurence Ricou, *Vertical Man/Horizontal World* (Vancouver: University of British Columbia Press, 1973).

[7] Harold Innis, *The Fur Trade in Canada: An Introduction to Canadian Economic History* (New Haven: Yale University Press, 1946), pp. 386–92.

[8] See Arthur J. Ray, *Indians in the Fur Trade: Their Role as Hunters, Trappers and Middlemen in the Lands Southwest of Hudson Bay, 1660–1870* (Toronto: University of Toronto Press, 1974); and Arthur J. Ray and Donald Freeman, *Give Us Good Measure: An Economic Analysis of Relations Between the Indians and the Hudson's Bay Company before 1763* (Toronto: University of Toronto Press, 1978).

[9] Canada, *Report of the Department of the Interior 1892*, Sessional Papers XXV, No. 13 (Ottawa: 1893).

[10] *Winnipeg Times*, quoted in John W. Dafoe, *Clifford Sifton in Relation to His Times* (Freeport, N.Y.: Books for Libraries Press, 1971), pp. 103–4.

[11] Clifford Sifton, quoted in Joseph Schull, *Laurier* (Toronto: Macmillan, 1967), p. 336.

[12] Chester Martin, *Dominion Lands Policy* (Toronto: Macmillan, 1934; reprinted, Toronto: McClelland and Stewart, Carlton Library Series, 1973), pp. 46–7.

[13] Douglas Hill, *The Opening of the Canadian West* (Toronto: Longman, 1973), pp. 233–48.

[14] Kenneth H. Norrie, "The National Policy and Prairie Economic Discrimination, 1890–1930," in *Canadian Papers in Rural History*, vol. I, D. H. Akenson, ed. (Gananoque, Ont.: Langdale Press, 1978), p. 16.

15 Edward Blake, letter to the electors of West Durham, 1891. Quoted in Dafoe, *Clifford Sifton in Relation to His Times*, p. 316.

16 Clifford Sifton, "The Immigrants Canada Wants," *Maclean's*, 1 April 1922, p. 16.

17 Ibid.

18 John C. Lehr and D. Wayne Moodie, "The Polemics of Pioneer Settlement: Immigration and the Winnipeg Press," *Canadian Ethnic Studies*, 12 (1980), 87–101.

19 John C. Lehr, "The Rural Settlement Behaviour of Ukrainian Pioneers in Western Canada, 1891–1914," in *Western Canadian Research in Geography: The Lethbridge Papers*, B. M. Barr, ed. (Vancouver: Tantalus Press, 1975), pp. 51–60.

20 Canada, Department of Agriculture, *PFRA* (Ottawa: Queen's Printer, 1961).

21 Alan F. J. Artibise, *Winnipeg: A Social History of Urban Growth, 1874–1914* (Montreal: McGill-Queen's University Press, 1975).

22 Quoted in John Herd Thompson, *The Harvests of War* (Toronto: McClelland and Stewart, 1978), pp. 52–3.

23 For an excellent review of the literature on ethnic diversity in the Western Interior, see Hansgeorge Schlictmann, "Ethnic Themes in Geographical Research in Western Canada," *Canadian Ethnic Studies*, 9 (1977), 9–41.

24 B. M. Barr, "Reorganization of the Economy," in *The Prairie Provinces*, P.J. Smith, ed. (Toronto: University of Toronto Press, 1972), p. 67.

25 J. W. Simmons, *The Canadian Urban System as a Political System, Part I: The Conceptual Framework*, Research Paper No. 141 (Toronto: University of Toronto, Centre for Urban and Community Studies, 1983), pp. 9–13.

26 Semple, "Quaternary Places in Canada," pp. 356–69.

27 Ibid., p. 135.

28 R. Thompson, "Commodity Flows and Urban Structure: A Case Study in the Prairie Provinces" (unpublished doctoral dissertation, University of Calgary, 1977), p. 86.

29 Syncrude Canada Ltd., *Management Report*, 1995.

30 Statistics Canada, 1991, *Canadian Agriculture at a Glance*, p. 70–73.

31 Canadian International Grains Institute, *Grains and Oilseeds: Handling Marketing, Processing*, vol. 1, 4th ed. (Winnipeg: 1996), pp. 113–238.

32 Albert E. Berry, "Canada's Primary Elevator System," in *Grains and Oilseeds: Handling, Marketing and Processing*, Vol. I (Winnipeg: Canadian International Grains Institute, 1993), pp. 139–164.

33 *Winnipeg Free Press*, 12 April 1997.

34 Science Council of Canada: *A Growing Concern: Soil Degradation in Canada*, September 1986, pp. 7–14; see also, Paul D. Bircham and Hélène C. Bruneau, "Degradation of Canada's Prairie Agricultural Lands: A Guide to Literature and Annotated Bibliography," Working Paper No. 37, Lands Directorate Environmental Conservation Service, Environment Canada, March 1985.

35 Tom Carter, "Winnipeg: Heartbeat of the Province," in John Welsted, John Everitt, and Christoph Stadel, ed. *The Geography of Manitoba: Its Land and Its People* (Winnipeg: University of Manitoba Press, 1966), p. 144.

36 Hall Commission, *Grain and Rail in Western Canada*, Vol. 1 (Ottawa: 1977), pp. 284–311.

chapter 9

British Columbia: Canada's Pacific Province

Lewis Robinson

In the late 1970s Canada's total trade with the Asia Pacific nations began to exceed our total trade with Europe. It was a change that was to be permanent, reflecting as it did the growing importance of Asian nations in the global economy. By the early 1980s, Canadians were becoming more and more convinced that Asia in the twenty-first century, like Europe a hundred years ago, would be the world's economic heartland. The change in outlook was dramatic. For virtually all of its history, Europe and the United States dominated Canada's world trade links. Eighty per cent of Canada's external trade is still with the United States. Furthermore, we share their histories and cultural characteristics. For the first time, Canadians were facing the prospect of dependence on trade with nations whose cultures and histories were only dimly understood.

British Columbia's links with Asia Pacific nations go back a long way. Over two hundred years ago, long before Europeans knew much about western Canada, sea otter pelts were being shipped from Nootka Sound to China in exchange for tea and porcelain to be traded in Europe. A century later, when the railway reached Vancouver, the first cargo sent east on the new line was tea. The fur and tea trades were just the beginning of British Columbia's ties with Asia. In the years following the gold rushes of the mid-nineteenth century, and again during the construction of the Canadian Pacific Railway in the 1880s, many thousands of Chinese migrated to British Columbia. Today, there are many thousands more Asian migrants from all parts of the Pacific Rim. British Columbia's proximity to Asian countries —

FIGURE 9.1   ASIA–PACIFIC REGION

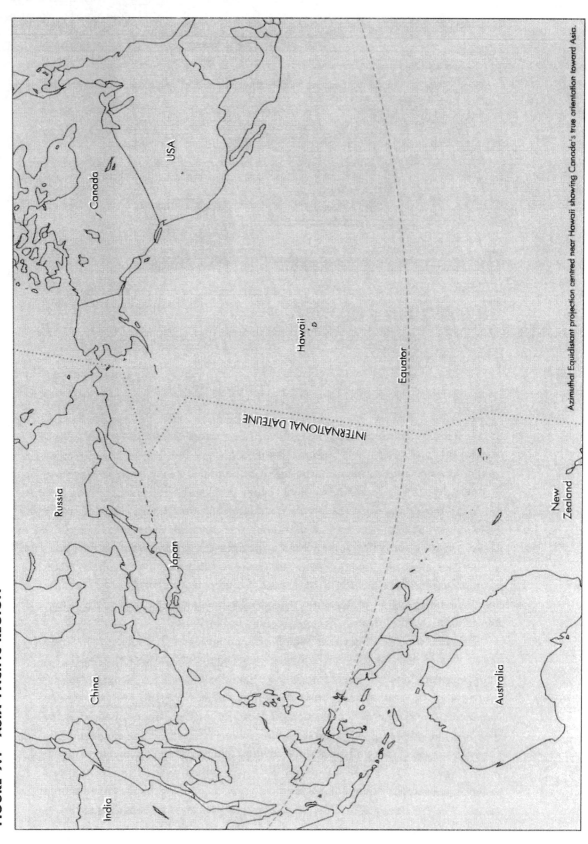

Azimuthal Equidistant projection centred near Hawaii showing Canada's true orientation toward Asia.

something not well displayed on traditional atlas maps — is the main reason for these trading relationships (see Figure 9.1).

In the middle of the nineteenth century, cut off by the mountain barriers of the Cordillera and the sparsely occupied Interior Plains from the other colonies in eastern Canada, and far from "mother England" by sea around Cape Horn, this part of Canada was isolated. The outlines of its present areal characteristics were not apparent until the end of the nineteenth century. The exploitation of nature's endowments came gradually and to different places at different times. The spotty and discontinuous settlement pattern that evolved was a contrast to the broad, westward-moving frontier which occupied the western plains.

Thus British Columbia, from an early stage of development, was a distant hinterland of Britain and far from the heartland of Canada. Now most of the province's raw materials — lumber, fish, minerals — go to external foreign markets. Manufactured goods from Canada's heartland must compete in British Columbia with imported East Asian goods. Economic, social, and travel links are strong between British Columbia and the west coast American states. British Columbia is a Pacific province; East Asian influences are strong.

Herein lies the paradox of British Columbia's place in the global economy, well articulated by the President of the Asia Pacific Foundation of Canada, a Vancouver-based institution, early in 1997: "Canada is a developed country, selling nuclear reactors to Korea, yet most of its trade with Asia, the fastest growing economic region of the world, is in fuel, wheat, forest products, and minerals." British Columbia is the funnel through which this trade flows. It is also one of the main producers of the products being shipped.

Although a major centre of technical innovation, and a new hub for air travel between east and southeast Asia and the major cities of North America, this province's economy is still based on its natural resources. Hence the focus in this chapter on forestry and other basic resources and their associated environments.

Overall, as the following pages show, British Columbia is a mountainous province. Great contrasts within small areas are characteristic of the natural environment, and there are also wide contrasts in population densities. It is a region of urban people, with agriculture entirely lacking over large areas, or confined to certain narrow valleys or floodplains. It is a region of small-scale diversity. The common physical characteristics are the rugged, mountainous landforms and the interior valleys and coastal plains. People live on these lowlands, and transportation lines go through them.

The human patterns of occupance have dispersed distribution patterns outside of the southwestern core. In the valleys of the southern interior and northward along the coast, locations of the small settlements reflect either the utilization of local resources, or are service centres along transportation routes. Economic activities developed in local regions at different times in response to expanding external markets for their resources. Most of the northern valleys are almost empty, and

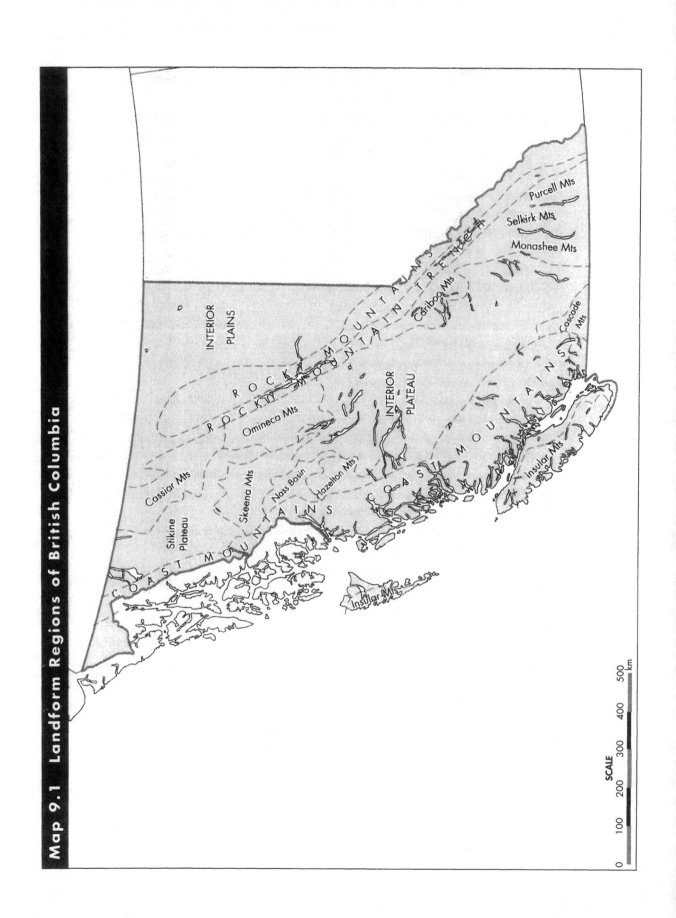

Map 9.1 Landform Regions of British Columbia

INTERIOR PLAINS

R O C K Y   M O U N T A I N S

ROCKY MOUNTAIN TRENCH

Purcell Mts

Selkirk Mts

Monashee Mts

Cariboo Mts

Cascade Mts

INTERIOR PLATEAU

Omineca Mts

Cassiar Mts

Stikine Plateau

Skeena Mts

Nass Basin

Hazelton Mts

C O A S T   M O U N T A I N S

Insular Mts

Insular Mts

C O A S T   M O U N T A I N S

SCALE

0   100   200   300   400   500
km

Spiral railway
in the Rocky
Mountains

throughout the province most of the mountain and upland areas are unoccupied. Few people live on farms, but large sections of the south-central interior are used for grazing. Throughout the southern two-thirds, extensive forests are utilized under various tree-farm management licenses where controlled cutting should maintain a continuous harvest. Similar to other Canadian regions and developed economies worldwide, the number of primary workers in the farms, forests, mines, and fisheries is declining compared with the number of people employed in a wide range of occupations in the large urban centres (see Table 9.1).

Table 9.1   British Columbia: Employment Change in Selected Industries from 1985 to 1995

Primary industries	- 10%
Transportation, storage, communications	+20%
Retail and wholesale trade	+34%
Finance, insurance, real estate	+34%
Business and personal services	+54%

Source: B.C. Statistics

THE PHYSICAL ENVIRONMENT

Landforms

The physical landscape of the Cordillera has great variety and its scale is impressive; it is one of the attractions of a valuable tourist industry. Although locally the mountains seem to be a jumbled mass of peaks and stretch endlessly to the horizon when viewed from the air, they have specific areal patterns, and are subdivided into smaller subregional units (see Map 9.1).

The Coast Mountains have striking landform variations in their numerous rugged, offshore islands, protected channels, linear fiords twisting inland into jagged alpine peaks, narrow coastal lowlands, low marine terraces, and the broad flat delta of the Fraser River. The Insular Mountains on Vancouver Island and the Queen Charlotte Islands are not as high and have wider lowlands on their east sides. The visual attractiveness of these coastal features to tourists in cruise ships is very apparent.

The Rocky Mountains are a specific line of mountain ranges extending northwest from Montana along the southern and central Alberta–British Columbia border and terminating at the broad plain of the Liard River. The western boundary of the Rocky Mountains is the Rocky Mountain Trench, one of the longest continuous valleys on the earth's surface, extending from Flathead Lake, Montana, to the Liard Plain in the Yukon. Other mountain systems to the westward, such as the Columbia and Cassiar–Omineca mountains, are separate landform regions. They have similar internal characteristics of high, sharp peaks, and narrow U-shaped valleys. The Columbia Mountain system (a name seldom used in British Columbia) consists of three parallel mountain ranges in southern British Columbia — Purcells, Selkirks, Monashees — together with the Cariboo Mountains to the north. East of the Rocky Mountains, northeastern British Columbia is part of the Interior Plains landform region — a generally level area with incised river valleys — and very different in landform characteristics and in land uses from the rest of the Cordillera.

The Interior and Stikine plateaus are undulating interior basins with generally level horizons; in some places, however, rivers have cut deep canyons into the plateau edges. Interior river valleys usually have narrow terraces of level land above the entrenched rivers. The positions, directions, and interconnections of the interior valleys are particularly important; they are the inhabited parts of the Interior region and the routes of land transport. One is always aware, within the Cordillera, how little level land there is at low elevations. This is a reminder that only five per cent of British Columbia is classed as arable land suitable for crop cultivation.

Lake sediments from last ice age, Thompson River valley

Close-up of top photograph

Climate and Vegetation

Contrasts in climate are characteristic of mountainous regions. Because most weather stations are in valley bottoms, where the settlements are, vertical contrasts have to be inferred from vegetation differences or other indirect environmental information.

The greatest amounts of precipitation recorded in Canada — more than 375 cm (150 inches) annually — fall on some of the west-facing slopes of the Insular and Coast mountains. But only 400 kms eastward, in the southern interior valleys, some of the driest stations outside of the Arctic report less than 25 cm (10 inches) of annual precipitation. Most of the precipitation falls on the coastal lowland settlements as rain during the winter months while, at the same time, snow blankets the higher elevations of the nearby Coast Mountains. These winter rains are brought by large air masses moving eastward across the Pacific Ocean. Cold air masses from the Interior Plateau can spill westward along the Fraser Valley, raising and chilling the usual mild, moist Pacific air masses, and resulting in great amounts of snow being dumped on the coastal cities for a few days.

Coastal British Columbia has the mildest winter temperatures in Canada, with January averages of about 0°C. However, the linear valleys of the Interior are open to the southward penetration of cold air masses, and settlements there record many winter days of below-zero temperatures. Indeed, winter temperatures in the interior of British Columbia can be similar to those felt across the Prairie provinces.

In summer, the southwest coast is normally cool, about 15°C average in July, and sunny for many weeks when high-pressure ridges form over it and deflect Pacific storms northward. A few hundred kilometres eastward, however, residents in the southern valleys can experience some of the hottest temperatures recorded in Canada.

Vertical contrasts in climate are shown by the horizontal zonation of vegetation on mountain slopes. The mild and wet coastal climate has nourished the largest trees in Canada. Because temperatures decrease with altitude, and steep, rocky slopes have little soil, tree size and density decrease at about 1500 m; alpine slopes above 2000 m are treeless. In northern British Columbia, the upper tree line can be about 1000 m above sea level, leaving only narrow strips of forest along the river valleys. Although grassland is seen through the dry valleys of the southern Cordillera, it occupies only narrow strips or small basins, and the total grassland area for ranching is not large.

Drainage Basins

Water is one of the valuable natural resources of British Columbia. Its distributional patterns of abundance and scarcity can be studied as functions of precipitation upon the surface and as run-off in the major river basins. Climate and landforms are functionally related. The Cordillera has six interior river basins; several small, separated river basins along the coasts; and the Peace–Mackenzie basin in the northeast. The positions of these river basins, and the variations in their amounts of annual and seasonal run-off, are important in understanding both the distribution of hydroelectric power and the problems and possible conflicts with salmon fishing.

What are our perceptions of river-valley flood plains as desirable, or dangerous, places on which to live? Is flooding in the Cordillera an accepted natural hazard because of the wide regional and seasonal variations in river run-off? Flood plains, such as along the lower Fraser River, have been recognized as attractive for agriculture because of their flat land and fertile soils, but the natural flooding that deposited the good soils is a continuing process. Can this aspect of nature be controlled by dikes, levees, and upstream dams?

EARLY RESOURCE UTILIZATION

British Columbia has little uniformity over large areas in either the physical landscape or in settlement patterns and the use of the land and its resources. The variety in the natural environment offered wide choices of natural resources for the Native peoples, and for Europeans who entered the region near the close of the nineteenth century. Except for the cities in the southwest, other settlements were based mainly on the exploitation of a particular natural resource. The use of nature's endowment came gradually and to different places at different times. The spotty and discontinuous settlement pattern that evolved contrasted with the broad westward-moving frontier that characterized occupation of the Interior Plains.

A dual pattern evolved: coastal settlement, which slowly penetrated inland; and interior settlement from the east which connected to the coast. After the middle of the nineteenth century, small sea-oriented settlements were established in the southwestern corner, but at the same time, older dispersed settlements of the Interior persisted as remnants of the land-based fur trade or mines. Present geographical patterns evolved through increasing functional connections which tied the urban areas of the coast with the small centres of the Interior, and the province as a whole with the rest of North America and the Pacific countries.

Fur-bearing animals of the Interior forests were the first natural resource utilized by Europeans. Fur traders crossed the Interior Plains and made contact with Native Indian groups in the interior early in the nineteenth century. A few decades later, traders arrived by sea and the appeal of their trade goods increased the hunting and trapping activities of the Native peoples of the coast. Coastal trading centralized in such posts as Fort Langley, Victoria, and Nanaimo.

Gold was the next natural resource to attract settlers to the Cordillera. The gold rush of 1858 brought prospectors, miners, and transport facilities to the central Fraser River, and to the western slopes of the Cariboo Mountains. At the same time, commercial activity and administrative and governmental control were located on the coast, in Victoria and New Westminster. New transport lines, such as the Cariboo Road, linked the separated exploitation and management regions.

The heartland–hinterland relationships in the Cordillera had their beginnings in the 1860s. When British Columbia became part of Canada in 1871, the largest clusters of European population were concentrated on the southwest coast; this geographical pattern remained the same for the next hundred years, although, of course, numbers greatly increased.

Although the Cordillera is not a major agricultural region, some parts were originally settled solely for agriculture. Some early settlers came to British Columbia, as to other parts of Canada, for the main purpose of owning a piece of land and rearing a family in a rural setting. Most farm settlers, however, intended to develop commercial agriculture as quickly as possible, selling surplus food to mining and forestry workers, or to the increasing urban population in the southwest.

The most successful of the Interior farmers were in the Okanagan Valley, where irrigation transformed the dry environment to produce fruit for markets throughout western Canada. Small fruit-farms were established at the northern end of the Valley in the early decades of the century, sometimes by wealthy British immigrants. Orchards producing apples, cherries, peaches, pears, and other fruits occupied narrow former glacial-lake terraces as well as small deltas and alluvial fans deposited by tributary rivers. Amid each agricultural area a small service-centre town arose. These towns were evenly spaced along the Okanagan Valley, each supplying and serving a surrounding agricultural hinterland similar to the service-centre relationships on the Interior Plains.

Agricultural settlement in the Lower Fraser Valley, and on southeastern Vancouver Island, was directly related to the establishment and growth of the two metropolitan cities, Vancouver and Victoria. As these cities increased in population, the area of farmland expanded on the nearby lowlands and now occupies about 10 per cent of the cultivated farmland in British Columbia. These relatively small areas of level land with suitable soils not only had the advantage of nearby large markets, but they had the longest frost-free season in Canada.

Some of the crops of the Lower Fraser Valley fit local environmental conditions. For example, dairying and vegetable production are concentrated on the fertile flood-plain lowland along the lower Fraser River, whereas poultry and small fruit are located about 100 m above the river, on the poorer soils laid down on an older, raised glacial delta. Cranberries and blueberries are grown in the acidic soils of the bogs after peat is removed.

Agriculture

Although the agricultural crops on the eastern side of Vancouver Island are similar to those in the Lower Fraser Valley, they have different distribution patterns and smaller local markets. Most island soils are derived from glacial and alluvial deposits carried eastward by short rivers flowing from the mountains. Soils vary greatly in quality within short distances; good soils are dispersed in small pockets. Because urban markets are also dispersed, small agricultural clearings are mainly near cities such as Victoria, Nanaimo, and Courtenay.

The wide range of crops that can be grown in the southwestern corner contrasts with the limited choices on the Interior Plateau near Prince George, where the frost-free season is usually less than 100 days. The Interior Plateau and southern valleys are open to cold air masses that flow southward in early and late summer along the north–south grain of the topography.

Almost half of the improved farmland in British Columbia is in the Peace River area of the Interior Plains landform region. The main crops grown there — wheat and other grains — are similar to those in adjoining Alberta. Most of the grain is transported by rail to the ports of Vancouver or Prince Rupert, but some becomes feed for local livestock.

Although the total amount of agricultural land is relatively small in British Columbia, additional farmland is still being cleared and cultivated. Most of the new farmland has been occupied on the Interior Plains of the northeast, but new areas have also been cultivated on the Nechako Plateau, west of Prince George, as increased forestry activity promoted the growth of central interior cities. Interior expansion has been partly balanced, however, by decreased farmland in the Lower Fraser Valley as a result of the urban spread of metropolitan Vancouver.

NATURAL RESOURCE UTILIZATION TODAY

Forestry

Variety in British Columbia's landscapes is intensified by contrasts in vegetation. Extensive stands of tall conifers clothe the lower slopes of the Coast and Insular mountains and the west-facing slopes of Interior mountain ranges; smaller trees spread endlessly across the Interior Plateau. In the lee of the mountains, however, and in the dry southern Interior valleys, tawny bunch grass and even cactus indicate the sparseness of the precipitation and vegetation cover. In winter, the green of trees, grass, and shrubs on the mild and wet coast contrasts with blue skies, cool temperatures, and snow-covered landscapes of the Interior. Human impact upon the natural vegetation in terms of agricultural clearing and forest logging has modified only small parts of the total natural environment, but in some local areas these patterns of utilization constitute significant landscape changes.

Forestry has been the main income-generating segment of British Columbia's economy. Cutting of large trees for lumber started in the southwest, and the wood-processing industry is still concentrated there. In this Georgia Strait region, transportation facilities were available for assembling logs in preparation for export to world markets. The urban region in the lower Fraser was also a small market in itself.

The easily accessible forests on the southwest coastal lowlands were first processed in small local mills for export. As more distant forests along the coast were cut, a log-transporting technology was developed to carry logs to large mills

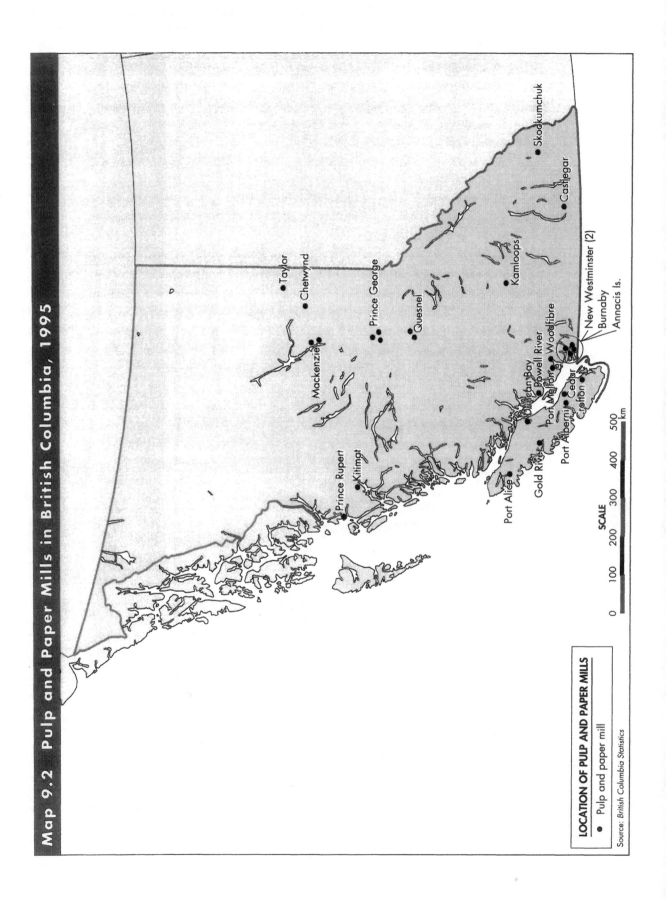

Map 9.2 Pulp and Paper Mills in British Columbia, 1995

Taylor
Chetwynd
Prince George
Quesnel
Mackenzie
Kamloops
Skookumchuk
Castlegar
New Westminster (2)
Burnaby
Annacis Is.
Woodfibre
Powell River
Duncan Bay
Port Mellon
Port Alberni
Crofton
Cedar
Gold River
Port Alice
Kitimat
Prince Rupert

SCALE

0   100   200   300   400   500
                                  km

LOCATION OF PULP AND PAPER MILLS

● Pulp and paper mill

Source: *British Columbia Statistics*

in or near the southwest cities. The coastal hinterland supplied the raw materials and the southwest heartland did the processing. The mills and settlements around the Strait of Georgia, and lumber camps northward along the Inside Passage, became part of an integrated, functional region linked together by movements and services related to the forest industry. The number of mills decreased after 1950 and the remaining ones became larger. The enormous sawmills in the southwest now produce much of the lumber and plywood in Canada.

Pulp and paper mills were established on the west coast about 1910, more than a decade later than the first mills in the Canadian Shield. They used inexpensive coastal water transportation to assemble raw materials and to move their finished products. The market for the pulp and paper was mainly in California — a much smaller market than that of eastern United States for eastern Canadian mills. Many of the early pulp and paper mills were in single-industry, resource-based towns similar in form and function to those in the Canadian Shield. These older mill towns now contrast in visual appearance with the attractive new pulp and paper mill towns, such as Gold River and Mackenzie. (See Map 9.2.)

Corporate and functional integration of the wood product industry was one of the major developments in the coastal forestry industry after 1950, and this affected distributional patterns of processing plants. Pulp and paper mills, sawmills, plywood operations, and the manufacture of other wood products could be integrated by one company into one large plant or into adjoining mills. Much of this forestry activity was concentrated in the Georgia Strait region: logs were brought from as far north as the Queen Charlotte Islands by rafts and barges; lumber was moved from sawmills to the major coastal ports for export; sawdust and wood chips from sawmills were carried by barges to pulp and paper mills within the region; newsprint rolls were picked up at the mills by coastal or ocean ships. The many operations of the coastal forestry industry were linked together over the sheltered, calm waters of Georgia Strait.

As a result of more than a century of forest-cutting on the lowlands bordering Georgia Strait, much of the original tall old-growth forest is gone. Controversy is common as to whether and where some of the remaining large trees can be preserved as heritage sites, to show future generations what the forest looked like. Since reforestation has accelerated in only the past few decades, replacement of this forest will take another fifty to seventy years. Forest-cutting will undoubtedly decline in future years in the Georgia Strait region, affecting the economy of the British Columbia heartland.

After 1950, increased world demand, plus improved rail and road transport into the untapped forest reserves of the Interior, permitted areal expansion of forestry. In the first half of this century most of the sawmills were located in valley settlements on or near the few railways. The Interior forest industry seldom used rivers for transportation as was done in eastern Canada. Because population is sparse, local demand has been minor except near the mining settlements, such as in the southeast.

Trucking logs
to mills

The forest industry, along with mining in some places and agriculture in a few other places, became the economic base of much settlement throughout the Interior. Some former mining towns, such as Grand Forks and Nelson, survived by finding a new resource base in forestry after the old mines closed. The major expansion of sawmilling in the central Interior came after 1950, as a result of the extension of the B.C. Railroad (then called the Pacific Great Eastern Railway) north of Quesnel to Prince George, and beyond to the Peace River area. The small, often portable, sawmills in the woods gradually closed, and bigger sawmills were built in towns at transport junctions, such as Prince George and Kamloops, and in cities in the southeast such as Cranbrook. The latter were close to direct-shipping markets in the Interior Plains and west-central United States.

The Interior forestry industry became more diversified and centralized by the addition of pulp and paper mills after 1961. It went through a corporate and functional integration similar to that which occurred earlier in the coastal industry. Compared with the concentration of mills on the southwest coast, the distribution of Interior pulp and paper mills is still dispersed, partly because of the spread of the areas in which companies have cutting rights.

Fisheries

Five main species of Pacific salmon constitute most of the west coast fish catch, and the industry has adapted to the natural habits and migration of these fish. Halibut, herring, and other fish are also caught in fewer numbers with less value. Salmon enter river mouths in late summer, heading for spawning waters in Interior lakes and shallow headwater tributaries; fishers assemble in the waters close to these river mouths to catch the salmon en route. Salmon canneries were established at or near the mouths of many rivers along the coast late in the nineteenth century, but the greatest concentration was near the Fraser and Skeena rivers, which had the largest drainage basins and therefore the most fish production.

As fishing technology gradually improved, larger and faster fishing vessels, with better gear and greater capacities, could harvest a wider area away from the river mouths. As the catching area which supported a processing cannery increased, fewer canneries were needed. The linear dispersal of about one hundred canneries early in this century decreased to about ten in the 1970s. The central coast now has no salmon canneries. As with the forest industry, the changing distribution of fish processing has been the result of both corporate integration and new fishing technology. The lifestyles of fishers have changed accordingly; most now live in large coastal cities. The few processing plants at the mouth of the Fraser River are near the port of Vancouver, from which processed fish can be shipped to world markets. Prince Rupert, at the mouth of the Skeena River, is the second largest salmon-producing river of the province, but the port does not attract the volume or variety of transportation services that metropolitan Vancouver provides.

Fish farms to meet growing demand, and located near the markets and transportation of the southwest heartland, became significant producers of salmon in the 1980s. Most of the fish farms (and also oyster beds) are located along the sparsely populated coast northwest of Vancouver. As with other "farming" of "wild" species of animals the industry has much to learn about the containment and feeding of large numbers of fish in a small area, and the control of diseases and "natural" calamities among such numbers.

A number of conflicts are of concern to the fishing industry. For example, some of the fishery is shared with American fishers, and there is disagreement as to how to share it. Should Canada have priority in trying to catch salmon spawned in the interior lakes of British Columbia? Commercial fishers also compete with an important sport-fishery, which is part of British Columbia's valuable recreation industry. Native peoples claim rights to catch fish along interior rivers after the fish escape coastal nets, and want priority in certain coastal areas. Because salmon live part of their lives in the ocean, and part in the inland rivers, the management of the fishery is complicated by the regulations of both federal and provincial authorities and a lack of international control of the Asiatic fishing industry, which has taken large catches from the waters just west of British Columbia.

One of the dormant, but potential, conflicts is with the hydroelectric power generating industry for use of certain rivers. The Fraser River, for instance, yields the largest number of salmon in its many tributaries, but also has the largest potential for generating hydroelectric power, if dammed in its lower canyon or upstream north of the junction of the Thompson River. Provincial government policy has given priority to fish production in the river, since hydroelectric power may be obtained from alternative sources and sites. Similarly, the Skeena and Stikine rivers have been maintained as salmon-producers, although both could produce large quantities of electric power for future industries along the northwest coast. In addition, future use of the Stikine River is complicated by the need for political agreements with Alaska through which the river empties. These political problems may be greater than the engineering ones.

Mining

Mining has been significant in the economy of British Columbia since the Cariboo Gold Rush after the middle of the nineteenth century. Most of the geological ages are represented somewhere in the province. For example, young, tilted sedimentary rocks contain coal in the Rocky Mountains, whereas nearby in the Purcell and Selkirk mountains, old Precambrian rocks produce lead, zinc, and other metals (see Map 9.3).

From 1890 to about 1905, the southeastern region of the Kootenays was one of the first and most important mining areas of Canada. The mines and their supporting communities were linked together by rail transport; local agriculture on narrow river terraces produced food for the mining towns; local forestry supplied building material; rivers were dammed early in the century to supply electric power. As a mining region, the Kootenays still function around the large smelter-refinery at Trail, where a variety of metal ores are processed, both from nearby and from as far away as the Yukon.

Former copper mine near Kamloops, BC

Map 9.3 Operating Mines in British Columbia, 1995

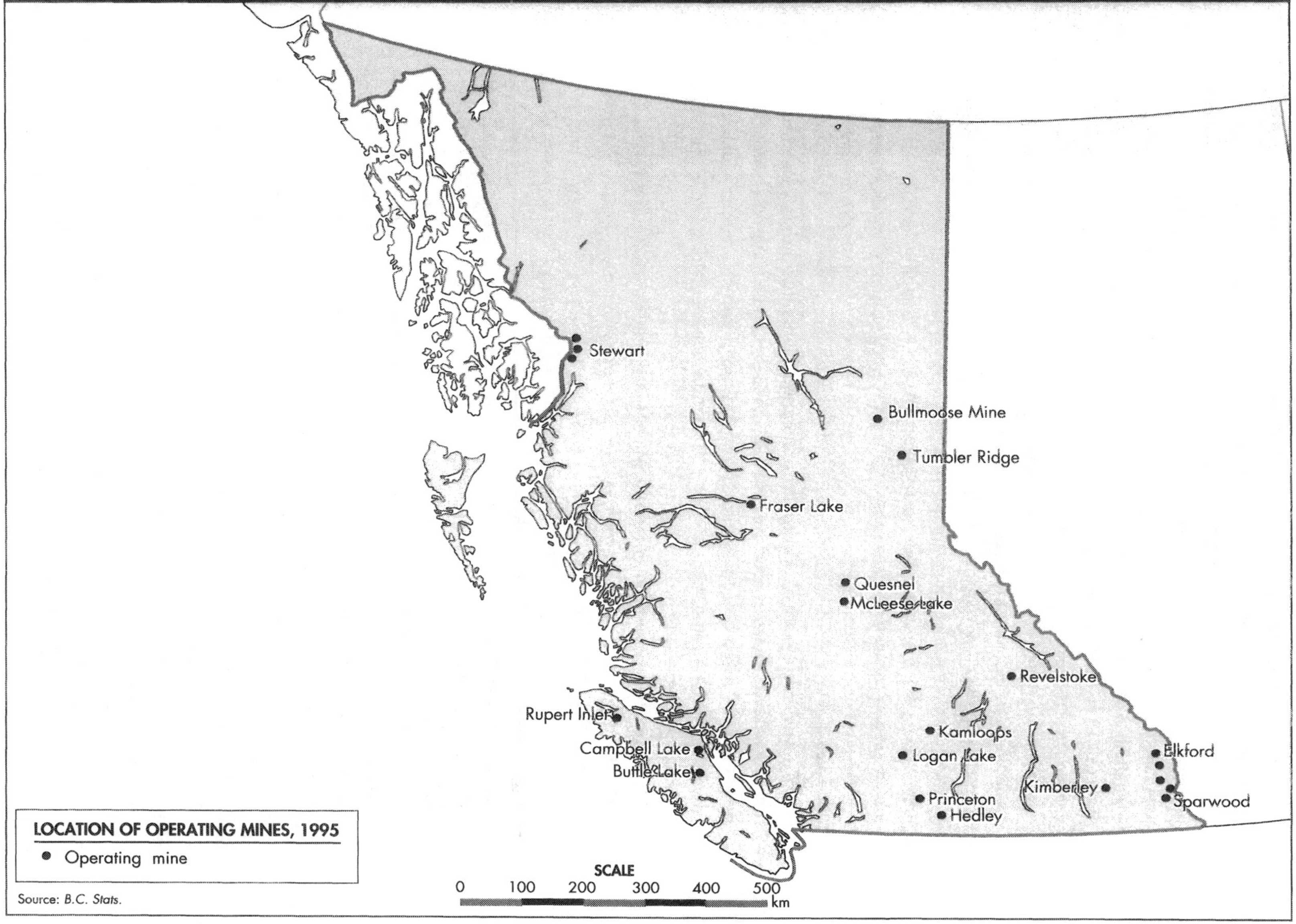

Sparwood
Elkford
Kimberley
Revelstoke
Kamloops
Logan Lake
Princeton
Hedley
Bullmoose Mine
Tumbler Ridge
Quesnel
Mcleese Lake
Fraser Lake
Stewart
Rupert Inlet
Campbell Lake
Buttle Lake

SCALE

0   100   200   300   400   500 km

LOCATION OF OPERATING MINES, 1995
● Operating mine

Source: B.C. Stats.

This early Kootenay economy was not connected by rail to the coast until 1915. The Kootenay region then had closer economic links with the adjoining mining region in the American Cordillera. Since then, southeastern British Columbia has gradually been brought into closer economic relations with the southwest core of the province, particularly after highways were improved after 1950. These links were strengthened in the 1970s when coal from the southern Rocky Mountains, destined for export to Japan, was transported by unit trains to Roberts Banks, a new artificial port built off the mouth of the Fraser River. By the late 1980s, however, the south-eastern coal mines were facing increased competition from newly opened strip mines around the new, planned town of Tumbler Ridge in the north-central Rockies. These latter mines were also exporting to Japan via the port of Prince Rupert. For more than a century, mining was a major impetus in the economy of the Kootenays; abandoned mines, slag piles, derelict houses, and "ghost towns" — as well as modern efficient mines and new planned towns — are significant elements in the regional landscape.

Because nearly all minerals are produced for export, the mineralized areas of the Coast and Insular mountains, accessible to ocean transport, continue to support producing mines. As in other mining regions, individual mines close and new ones open, but the coastal area as a whole remains a producer. In the early 1970s, the capital and markets of Japan provided the stimuli for the opening of small iron mines on the coast, and copper and molybdenum mines along the edges of the southwestern Interior Plateau.

Unlike other non-ferrous mines elsewhere in Canada, recent Cordilleran mines are enormous open pits where low-grade ore, such as copper, is extracted by the use of improved technology in equipment and facilities. Such pits leave small scars on the landscape, but are generally in areas with few people or with significant competing forestry or agricultural land uses.

The differences in physical appearance and in amenities between the old and new mining communities are remarkable. In the southwestern Interior Plateau, mining towns that flourished and died near Princeton early in this century have their modern counterpart in the planned town of Logan Lake. On the northwest coast, mines have opened and closed in the Portland Canal area for more than seventy-five years. Similarly, mining of different metals at different places has been continuous on Vancouver Island for more than a century.

None of Cordilleran mineral resources go to eastern Canadian markets. Mineral production may partially depend upon eastern heartland companies for capital, but not for consuming markets. Economic links with the United States and with Pacific Rim countries, particularly Japan, are strong.

The development of petroleum and natural gas in the northeast is part of the economy of fuel production on the Interior Plains. Oil and natural gas moves by pipelines from the Peace River region to markets in the southwest and adjoining American states. The concept of a natural resources hinterland supplying a consumer heartland is again illustrated.

Hydroelectric Power

The Cordillera has the natural endowments required for hydroelectric power generators — heavy precipitation, sloping landforms, lakes for water storage, and numerous rivers. Private companies and provincial power authorities have built dams and installed generating equipment to develop hydroelectric power, which is transmitted to urban and industrial markets. Despite the vast amount of precipitation that falls on the Cordillera, only two of the Interior river basins are substantially developed — the Columbia and Peace.

Water power development has shown one pattern of clustered concentration and another of dispersal. Throughout the history of settlement in British Columbia, the largest cities, major rural population, and most industries have been concentrated in the southwest corner. Hydroelectric power for these markets was supplied from small sites in the Coast Mountains near Vancouver, or on short rivers on southern Vancouver Island. As transmission technology improved, power could be produced at more distant sites, such as Bridge River east of the Coast Mountains. Although the dispersed, separated, small river basins of the Coast region were the first to be harnessed, most of those suitable for hydroelectric power development are almost completely utilized.

Elsewhere, power developments were dispersed in three other corners of the province, and at first supplied only local needs. The availability of relatively inexpensive hydroelectric power throughout the province encouraged other resource developments. For example, power for the large smelter-refinery at Trail and for cities in the southeast was being produced early in the century from several plants on the Kootenay River. Later, in the 1960s to 1980s, power and water-control dams were

W.A.C. Bennett
Dam on Peace
River, BC

built at Castlegar, Revelstoke, and Mica Creek on the Columbia River to supplement power production in Washington State and to provide for the needs of Vancouver. The Columbia River basin produces more electric power than any B.C. river.

In the northwest, the Nechako River was dammed in the early 1950s and diverted through a tunnel beneath the Coast Mountains to produce power at Kemano for the large aluminum smelter nearby at Kitimat. In the late 1960s, power was developed in the fourth corner of the province, on the Peace River of the northeast, partially as a result of improvements in long-distance transmission technology. Power developments along the Peace River did not conflict with a fishing industry, but downstream changes in water levels are a source of complaints from valley residents in Alberta. By the 1970s, the two distribution patterns of concentration in the southwest, and dispersal in three other parts of the province, became part of one large provincial distribution network.

Recreation and Tourism

The varied and spectacular natural environments of the province may prove to be one of its most valuable "natural resources." The greatest intensity of recreational use — and of abuse — is in the southwest near the large centres of population, an area also accessible to still greater numbers from the American states on the west coast and visitors from East Asia.

Aerial view
of Kitimat

The natural environments of the popular Rocky Mountains are duplicated in other parts of the Cordillera, such as in the Coast Mountains north of Vancouver. Governments and the public face the continual conflict of preserving wilderness areas from the population pressures which decrease their "wilderness" character and appeal.

As road transport improved in the Interior in the 1950s and 1960s (there is still no continuous road along the coast), other interesting environmental areas became available to visitors, such as the Okanagan Valley and the Cariboo. However, many parts of west-central and northern British Columbia are still without roads, and can be seen only by a few hardy visitors seeking wilderness environments by using chartered air services.

GEORGIA STRAIT REGION: Heartland British Columbia

Much of the resource-based activity throughout British Columbia is linked directly or indirectly to Metropolitan Vancouver. This city, and others in the Georgia Strait urban region, grew as the Cordillera's natural resources were developed. The raw materials and power of the Interior and Coastal hinterlands either supplied the industries of the southwestern heartland, or they were funneled through the southwestern ports to markets elsewhere. Although resource-based extractive activities are dispersed throughout the province, head offices, management, and financial activities are concentrated in Metropolitan Vancouver.

Vancouver harbour

Almost half of the population of British Columbia lives in Greater Vancouver, and another 25 per cent lives nearby in the Lower Fraser Valley and on southern Vancouver Island, including the capital, Victoria. Vancouver is the focus of an urban system which has evolved around the shores of Georgia Strait — a functioning, multicentred urban region. In this concept, Georgia Strait links the urban centres rather than separating them.

When British Columbia became a British colony in 1858, the strategic site of New Westminster on the Fraser River was chosen in 1859 as the capital (at that time there were two separate colonies of British Columbia and Vancouver Island); this fort and administrative centre could control the water entrance to the Interior gold fields. The expected future growth of New Westminster, based on the apparent advantages of its site and geographical position, ceased after 1886 when the nearby tiny, sawmill-based village of Granville, on Burrard Inlet, was chosen as the western terminal of the Canadian Pacific Railway. The large land grants given to the CPR by the new city of Vancouver were to be influential in the areal growth of the city for the next century.

Vancouver and Victoria were competing cities during the last decade of the nineteenth century. Victoria was the ocean gateway to the Cordillera, a cultural centre, and the political capital of the province. Vancouver became the industrial and commercial city closer to the interior markets. Vancouver also became the transshipping point and main port for western Canada.

By 1900, the future urban land-use patterns of Vancouver had been established. An industrial zone expanded eastward along the railway and along the harbour, more concerned with supplying and assembling goods for regional resource developments and for local consumers than with a trans-Pacific trade, which did not materialize as expected. Directly south of the city's commercial centre, a second industrial zone spread around False Creek. Because it was too shallow to be used by ocean vessels, the shores of this protected water body became the focus of Vancouver's sawmill industry.

Throughout the next decades, Vancouver's economic base, except for the sawmills, was not dependent on primary processing of the natural resources of its hinterland. Much of the salmon canning was located along channels of the Fraser delta rather than in Vancouver itself; none of the mineral resources of the southeast or those from along the coast were processed in Vancouver. As thousands of people flooded into the city early in this century, some manufacturing developed to supply local consumer goods. These industries were protected, in one sense, by the transportation costs of goods imported from eastern Canada by rail or from Britain by sea.

The commercial core developed two nodes — the older, eastern section catered to the lower-income population in the East End, and new specialty stores served higher-income people in the West End. In these early days, working-class people lived in small houses on narrow lots in the eastern part of the city, near the industrial areas, whereas commercial and management people built their larger houses on larger lots west of the city centre. This income and social areal differentiation between east and west remained in Vancouver and is still apparent in local and provincial politics. Vancouver's outward residential spread was different from that of cities in eastern Canada in at least two ways: expansion was not onto agricultural land but into low-value forest land that had already been cut-over in the 1880s; and people were able to build or buy their own homes, tenements were few, and the city had a high percentage of single-family detached homes.

Along Burrard Inlet an industrial zone which had developed in the pre-World War I period spread eastward during the 1920s, when grain elevators and oil refineries were built on the waterfront. Transport, storage, transshipping facilities, and other industries spread to the North Vancouver waterfront in the 1950s. Many of the raw materials of western Canada, such as wheat, oilseeds, lumber, mineral concentrates, coal, potash, and sulphur now flow through the port facilities of Metropolitan Vancouver en route to world markets.

The second industrial zone around False Creek also expanded eastward, occupied by land-transportation oriented industries and facilities. Much of this industrial area changed in land use during the 1980s; most of the CPR maintenance

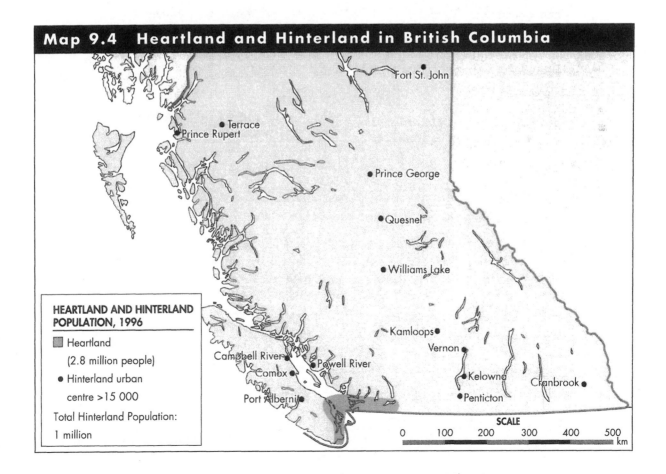

Map 9.4   Heartland and Hinterland in British Columbia

HEARTLAND AND HINTERLAND
POPULATION, 1996

▨ Heartland
   (2.8 million people)
● Hinterland urban
   centre >15 000
Total Hinterland Population:
1 million

Fort St. John
Terrace
Prince Rupert
Prince George
Quesnel
Williams Lake
Kamloops
Vernon
Campbell River
Comox
Powell River
Kelowna
Cranbrook
Penticton
Port Alberni

SCALE
0   100   200   300   400   500
km

yards on the north side of False Creek were removed to become the site of Expo '86, and later of high-density housing. Old homes of former industrial workers on the south side of the shallow inlet were replaced by modern condos, markets, parks, walks, and marinas.

The third industrial zone, along the Fraser River, was established in the last century, consisting mainly of sawmills and fish canneries. This industrial strip was partly filled in during the 1930s and the process accelerated after the 1950s; it became the main wood-producing area in Metropolitan Vancouver. Industry spread to the suburbs of Vancouver after the 1960s, similar to the outward areal spread of industry in other Canadian cities.

As the economic character and functions of Vancouver changed, the city came to qualify as a commercial, service, and management city. Policy decisions concerning resource development throughout the hinterland of the province were made in high-rise office towers of the commercial core.

The shift to tertiary forms of economic activity throughout the province is evident in two ways: first in the increased concentration in the provincal heartland (see Map 9.4); and second, in the labour force. Eighty per cent of the population increase in the province between 1985 and 1995 was added to the heartland, a disproportionate number in terms of earlier patterns. Table 9.1 shows the percentage changes in employment in several sectors from 1985 to 1995. The emphasis in numbers of people employed has clearly shifted from primary industries — agriculture, fishing and trapping, mining, and forestry — to service activities. British Columbia's gross domestic product (GDP) for 1995 reflects these changes (see Table 9.2).

Table 9.2   British Columbia: Gross Domestic Product by Industry, 1995 ($ millions)

Agriculture	850
Fishing and trapping	260
Forestry and logging	3 080
Mining	1 960
Manufacturing	11 050
Construction	6 700
Utilities	2 040
Transportation, storage, communication	8 100
Retail and wholesale trade	11 580
Finance, insurance, real estate	19 480
Business and personal services	24 470
Public administration and defence	4 550
Total	94 120

Source B.C. Statistics

Map 9.5 Net Population Movements to British Columbia, 1995

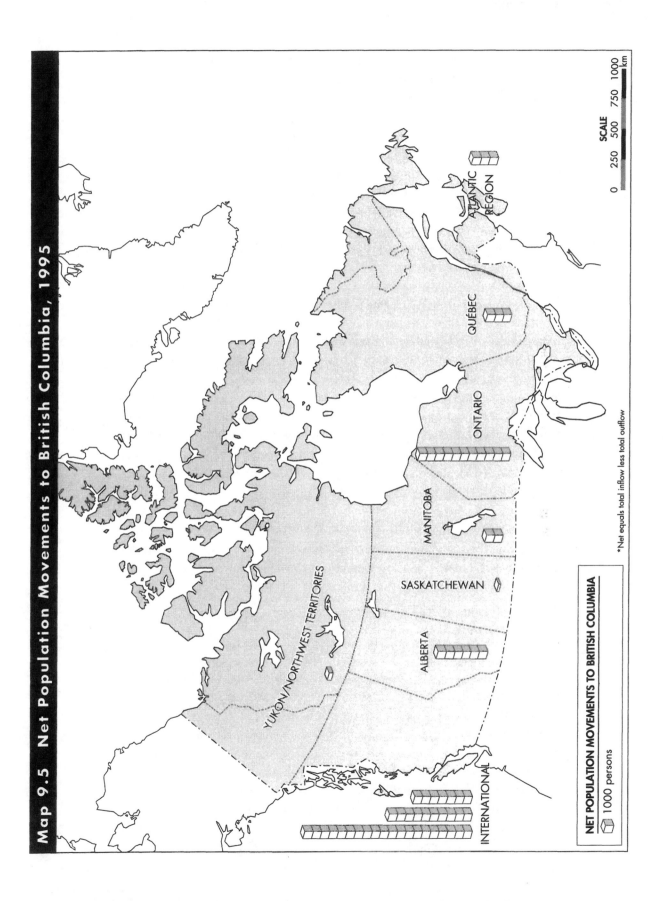

ATLANTIC REGION

QUÉBEC

ONTARIO

MANITOBA

SASKATCHEWAN

ALBERTA

YUKON/NORTHWEST TERRITORIES

INTERNATIONAL

NET POPULATION MOVEMENTS TO BRITISH COLUMBIA

⬡ 1000 persons

*Net equals total inflow less total outflow

SCALE

0   250   500   750   1000
                              km

Technological developments of the kind that have changed the economic face of Canada everywhere are one reason for the drop in employment in primary industries. At the same time, new industries, the needs of the trading sector, and demographic shifts all have created their demands for services. Demographic changes are particularly significant in British Columbia. This province has a history of attracting retired persons, and between 1985 and 1995 the percentage of people of retirement age in the province grew by 40 per cent while the population as a whole increased by 26 per cent (Map 9.5). This age-group requires a substantial volume of services, particularily in the health-care sector.

The "British" character of Vancouver also changed after the 1960s. Eighty per cent of the residents of Metropolitan Vancouver were of British racial origin in 1941, but Mediterranean immigrants, particularly Italians and Greeks, reduced this to 53 per cent by 1981. Thousands of Asians came to Vancouver after immigration laws were changed in 1967. South Asians, especially from India and Pakistan (mainly of Sikh religion), clustered into the southeastern part of the city and the adjoining Fraser Valley. Filipinos came to the central east side. Chinese have always been a "visible minority" in Vancouver, having their own "Chinatown" east of the city centre, but additional thousands have entered Vancouver and adjoining Richmond in the past two decades. They came from East and Southeast Asia as well as China. Most Chinese-origin people live in the multicultural east side of the city, but wealthy Chinese became noticeable in the "British" west side of Vancouver in the 1990s.

The extent of the ethnic change is best illustrated in the lives of school-age young people in Vancouver; more than half of them use English as their second language. Table 9.3 shows the nature of the shift in the course of one generation.

The urban character of the adjoining Lower Fraser Valley has also changed. Service centres, such as Chilliwack and Langley, arose amid the farmland of the valley, having the same functions as agricultural service-centre towns all across Canada. Western Fraser Valley towns have taken on commuter and industrial functions as part of the outward spread of Metropolitan Vancouver.

Native Peoples

Of even greater significance in the economic life of both heartland and hinterland is the renewed relationship with Native peoples. British Columbia's relations with its aboriginal peoples are unique. Nowhere else in Canada are there so few treaty arrangements. Hundreds of Indian bands are discussing land claims that date back to the early days of the province's history. In 1996, tentative agreements were proposed on one of the biggest of these claims, that of the Nisga'a. If approved approximately 1900 square kilometres of land in the Nass River area is to be transferred on an ownership and self-government basis. Additionally, there will be a cash payment of $190 million. The plan includes surface and subsurface control of land together with fishing and hunting rights. Canada can expect to see many more agreements of this kind. Table 9.4 shows Native population in British Columbia's hinterland.

Table 9.3   Ethnic Origins of Residents in the Greater Vancouver Area, 1961, 1991

	1961	1991
United Kingdom	60%	36%
Other European	33%	28%
Asian	4%	31%
Other	3%	5%

Table 9.4   Major British Columbia Urban Centres with Populations Including 1300 or More Persons of Aboriginal Origin, 1991

Area	Population
Vancouver	43 000
Total for other cities of British Columbia	121 000

Source: Statistics Canada, 1991

Victoria and Vancouver Island

Victoria was the first major European settlement along the west coast of what later became Canada; it was selected by the Hudson's Bay Company in 1843 as a trading post and chief administrative centre. Its commercial and governmental population grew when it became the sea entrance to the gold rush to the interior in 1858–62. Victoria remained the largest city in southwestern British Columbia, with its government and business residents, until surpassed by Vancouver in the 1890s.

Victoria is still classified as a government and institutional city, but it is also a service centre for southern Vancouver Island. The city has attempted to maintain an "olde English" charm in its business centre and in some residential sections to attract tourists. Its mild winter climate, with less rainfall than the coastal areas on the east side of Georgia Strait, has attracted many retired people from the Prairie provinces and eastern Canada.

North of Victoria, Nanaimo is the chief wholesale and distribution centre for central Vancouver Island. Forestry is its industrial base, with both sawmills and a pulp and paper mill. Both Nanaimo and Victoria are linked to Vancouver with frequent daily ferry and air service. The complementary urban functions of these cities, and the high degree of interaction between them, strengthens the concept that the communities around southern Georgia Strait are one large urban region.

HINTERLAND BRITISH COLUMBIA

The other cities of British Columbia are subregional centres (see Map 9.4). The linkages between these cities and Vancouver were strengthened after 1950, but the interaction between the hinterland cities is still minor. In British Columbia, the heartland–hinterland concept is known by the terms of "the coast" and "the interior;" to people who live around Georgia Strait, the term "coast" usually means only the southwestern coast.

Prince Rupert is the only major city on the north coast. The city and its interior hinterland has had many decades of hopes but little development. It was laid out early in this century as the northern terminal of the (now) Canadian National Railway — the closest port to the expected trans-Pacific trade with East Asia. But the port and terminal could not compete with the volume of products that moved through Vancouver; traffic and trade remained minor. Prior to 1950, the hinterland east of Prince Rupert had little forestry and agricultural settlement along the railroad. Local processing of fish and forest products, plus aluminum smelting and chemicals at nearby Kitimat, have maintained the subregional economy, but this northern gateway never became a duplicate of Vancouver. Prince Rupert may yet become the northern outlet for which it was created, if minerals and forest products of the northwest are joined by coal and grain flowing to the port from northeastern British Columbia. The amount of incoming traffic, however, will probably remain small.

The two largest regional centres in the Interior Plateau are Kamloops in the south and Prince George in the north. Kamloops occupies a strategic transportation position at the junctions of the North and South Thompson rivers; as well as the Canadian Pacific and Canadian National railways and the Trans-Canada and Yellowhead highways. It is a service and supply hub, and a processing centre for a nearby hinterland which includes mining, ranching, and forestry. Many branch offices of the provincial government are located in this central place. A regional name, Cariboo, is applied loosely to the adjoining area where the natural environments attract tourists to its dry grasslands, wet forested uplands, gently rolling plateaus, and deeply cut river canyons.

Prince George also occupies a crossroads position. River, road, rail, and air routes all meet there. Its economy is based mainly on the forest-processing industry, consisting of large sawmills, woodworking, and pulp and paper mills, but it is also the service centre for the forestry and agricultural activities of the central Interior Plateau. Prince George occupies a gateway position for northern British Columbia; products and people from the Peace River area funnel through Prince George to and from heartland Vancouver. Northwest of Prince George, the sparsely occupied northern Cordillera has yet to be well integrated into the economy of southern British Columbia.

The Okanagan Valley is the most populated of the southern interior valleys. Its three evenly spaced cities, Penticton, Kelowna, and Vernon, vie for dominance in central place and agricultural-processing functions. The economy of the Okanagan Valley diversified after the 1950s, when highway connections to the southwest coast were greatly improved. The warm summer temperatures and attractive sandy beaches of the Okanagan became a popular hinterland of Metropolitan Vancouver. Tourist seasonal activity, which unfortunately often conflicted with the summer labour needs of the fruit industry, was expanded to winter occupations with the opening of ski lodges on the snow-covered slopes above the valley. In addition, the orchard-based agricultural economy was diversified by planting increased acreages of grapes for the wine industry, and by increased production of vegetables. As manufacturing plants came to the Okanagan cities, urban expansion caused urban–rural land-use conflicts. With limited amounts of water, and a limited area of level land available for both city and farm use, future urban or rural growth requires careful compromises and planning decisions by local governments and the public.

The southeastern valleys are collectively known as "the Kootenays." Clustered and linear settlement in each valley, based on mining and forestry, are supported by local agriculture and water power. The amount of level land for urban use has been small at the sites of each settlement. The group of centres from Trail to Castlegar to Nelson, near the junction of the Columbia and Kootenay rivers, is beginning to achieve some subregional dominance. A secondary node of Cranbrook–Kimberley is the focus in East Kootenay. Southeastern British Columbia has not been well integrated into the provincial economy in the past, and its rate of economic growth has been slower than other parts of the Interior hinterland.

Northeastern British Columbia, known as the Peace River region, is probably the best defined region in the province. It is part of the Interior Plains, with the visual landform boundary of the Rocky Mountains to the west; internally, it has a grain and livestock, petroleum and natural gas economy that is entirely different from the rest of the province. The Peace River region was tied to the economy of Alberta and was part of the hinterland of Edmonton until the 1950s, when new road and rail transport, together with pipelines and transmission wires, linked it to the hinterland of Prince George and Vancouver. The functions of a regional centre are shared by Dawson Creek and Fort St. John. Dawson Creek was the original "central-place prairie town" with the rail terminal and most of the grain elevators. However, the bridging of the Peace River during construction of the Alaska highway in the 1940s opened up new agricultural, forestry, and natural-gas frontiers north of the river, and increased central-place activities in Fort St. John.

There was little coordinated, integrated development in the British Columbia hinterland prior to 1940. Resource development came at different times, at different places, for different purposes. The Kootenays was a mining region; the Cariboo was a ranching region; the Okanagan was a fruit region; the northwest coast was a fishing region; the central interior was a forestry region. The growth or decline of towns depended on the success or failure of the local mine, forest, or fish industries. Interior settlements had much more interaction with Vancouver than with each other. However, regional isolation slowly changed to functional integration after about 1970, as transportation in the Interior improved, and the hinterlands of the southwestern heartland were brought closer together into an economic system.

TRADE

Asia-Pacific Region

Newly industrialized nations such as South Korea, Singapore, and Thailand are creating new demands for Canadian resources and products, reinforcing the long-standing role of Japan. The continuing volume of trade with Asia-Pacific nations represents dramatic growth. A look at the changes over a five-year period (see Table 9.5) reveals two developments: first, a shift of the second-level export destinations from European to Asian countries; and second, a doubling in the cash value of shipments to these Asian countries compared with little change in the total for European nations.

World Trade in Perspective

Throughout its history, British Columbia was always oriented to staple production and dependent on the export of these staples in world trade. As we take a reflective look at the evolution of that trade, we can see first a succession of natural resources — fur, gold, coal, lumber, base metals, pulp and paper, hydroelectricity, and natural gas. The exploitation of these resources and the development of hydroelectricity, and the accompanying shaping of society and economy in British Columbia, reinforced lines of communication and strengthened business ties with Europe. It also fostered commercial connections with California. The flow of financial capital, labour, and merchants from both places was substantial, but these early ties were limited by difficulties of isolation and the high costs of exporting products.

It is easy to see why there was an initial emphasis on commodities such as furs and gold, or the interest in canned salmon. The high value and limited bulk of these commodities could sustain the considerable costs of transportation. The concentration of settlement in the southwestern corner of the province, still important today, also took shape in this early period.

The arrival of the Canadian Pacific Railway (CPR) in Vancouver in 1886 opened lines of communication and trade with the North American economic community. American capital and mining technology were initially prominent in the Kootenay mining boom of the 1890s, but after 1897 British and eastern Canadian financial capital played the leading role. Canadian funds were similarly important in the reorganization of the coast salmon canning industry in 1902, by which time the banks and manufactured products of the Industrial Heartland had made significant inroads, replacing once important British institutions. Eastern Canada also supplied considerable entrepreneurial talent and labour.

Table 9.5   British Columbia: Value of Exports by Destination, 1991 – 1995 (Ranked by $ Values for 1995 in Millions)

	1991	1992	1993	1994	1995
United States	6 618	7 993	10 061	12 355	13 292
Japan	4 046	4 133	4 946	5 658	6 768
South Korea	632	509	581	738	982
Germany	449	428	350	466	694
Italy	309	304	267	409	646
China	229	244	284	305	499
Taiwan	263	265	347	361	482
Belgium	277	262	234	305	417
United Kingdom	571	543	311	321	386
France	230	168	139	169	318
Australia	224	222	224	262	298
Hong Kong	94	125	149	192	271
Netherlands	189	152	135	99	185
Brazil	87	54	52	103	181
Indonesia	73	75	75	92	173
Thailand	62	91	76	91	126
Philippines	77	81	74	95	105
Spain	112	95	70	74	97
Chile	47	18	41	69	86
India	47	52	38	58	77
Malaysia	31	24	28	33	65
Mexico	28	45	47	77	54
Venezuela	33	40	37	29	53
South Africa	8	11	24	26	53
Singapore	38	46	69	40	51

Source: BC Statistics

Patterns of integration within the world-economy changed again following World War I. The volume and diversity of staple exports increased, but trade patterns often shifted dramatically. The former lumber market of the Western Interior had declined appreciably during and following the war as settlement there subsided, but the slack in this trade was more than compensated for by expanding overseas markets and increased demand in the United States. In addition, pulp and paper — representing a new generation of staples — was becoming an important export commodity and the basis for resource-town development along the coast.

Coal mining community, Vancouver Island, c. 1889

For the first time, too, grain from the Western Interior flowed through British Columbia's ports in significant quantities as a direct result of the opening of the Panama Canal in 1914. The salmon canning industry also benefitted from this development, although it never regained its late nineteenth-century role as the leading provincial export. The Kootenay boom subsided early in the twentieth century, and although coal from the Fernie district and lead and zinc from Kimberley were still important export commodities, further advances in this sector occurred only after World War II.

All of these export trades suffered a brief decline at the beginning of the Depression, then quickly recovered, only to decline again during World War II. These swings illustrate both the resiliency and the weakness of British Columbia's position in the world-economy. Throughout the interwar period, while traditional sources of financial capital remained constant, new and significant sources were developed within the region, notably in the forest industry, but there were no regional banks, and other local financial institutions were few in number and small in scale.

By the early 1960s, the United States was still the major destination of the province's staple exports; the United Kingdom remained second; Japan came next; and Australia, which took less than 5 per cent of the province's exports, was a distant fourth. To the United States went chiefly forest products, some minerals, and natural gas. The United Kingdom bought lumber and plywood. Japan was a new purchaser of copper, coal, and forest products. Australia was interested in small quantities of lumber, aluminum, wood pulp, newsprint, and asbestos fibres. By the early 1980s, however, several new trends in staple exporting were evident: the United Kingdom market was declining relatively, Japan was importing more forest products, coal, and other mineral products than ever before, and Western Europe, India, and Australia were increasing their imports of minerals.

The case of Japan shows that countries desperately in need of raw materials will seek out stable political environments. As political tensions in the world-system mounted in the postwar period, Japan increasingly looked eastward to the resources of British Columbia to fuel its industrial expansion. Exports of coal to Japan rose rapidly from 640 000 to 8 300 000 tons between 1962 and 1976, but Japan also imported considerable quantities of copper from the province, despite the fact that it held copper interests elsewhere — for example, in the Philippines, Uganda, and Indonesia.

In 1980, an agreement was signed to open the Sukunka coal fields north of Prince George, proposed jointly by the provincial government and several Japanese multinationals. These rich deposits supplement coal already drawn from the East Kootenays, where Japanese demand promoted the first direct foreign investment by Japanese steelmakers in an overseas source of coking coal.

CONCLUSION

Peripheral west coast location has always affected British Columbia's development. When eastern Canada was being settled by people from across the Atlantic Ocean looking for farmland, British Columbia was unknown to Europeans apart from a few traders. The heritage and influence of large numbers of farmers, who were part of eastern Canada's population, were never significant in British Columbia. Its population has always been mainly urban — whether "urban" meant a fur-trading post, a mining town, logging camp, or a fish cannery early in the century, or the present urban complex around Georgia Strait.

Isolation from Britain was a social and economic problem in the nineteenth century, and separation from the heartland of eastern Canada was equally real for the first half of the twentieth century. Without a large consumer market, the development and use of regional raw materials from the forest, sea, and rocks were usually dependent on markets, transportation, and capital investment from external places. Secondary manufacturing came very slowly and was often by branch plants opened by eastern Canadian manufacturers. Only by the 1970s did British Columbia reach a threshold of population density and strategic resource potential to become more economically independent.

British Columbia has distinctive variety within small areas in climate, vegetation, soils, land use, and settlement. Its heartland in the southwest is functionally connected with coastal and interior hinterlands. The provincial economy stresses the use of natural resources and has many characteristics similar to those of the Canadian Shield and the Atlantic provinces. The urban development of the southwestern Cordillera is like that of metropolitan Montréal and Toronto, but it lacks the wider network of interconnected large cities that is one of the characteristics of the Great Lakes–St. Lawrence Lowlands. Within the province, the two main subregions are the southwest coast with its distinctive climate and urban concentration, and the growing resource-oriented settlements of the Interior and north coast. These regional subdivisions were fundamental in the past history of occupation, but improved transport after 1950 decreased the effects of separation. New areal patterns are now evolving.

R E F E R E N C E S

Barker, Mary. *Natural Resources of British Columbia and Yukon* (Douglas and McIntyre, Vancouver, 1977).

Harris, Cole. *The Resettlement of British Columbia: Colonialism and Geographical Change* (University of British Columbia Press, Vancouver, 1997).

Hay, John, and Oke, Tim. *The Climate of Vancouver* (Department of Geography, University of British Columbia, 1994).

Koroscil, Paul M. "British Columbia: Geographical Essays." (Simon Fraser University Press, Burnaby, 1991).

Robinson, J. Lewis. *British Columbia: One Hundred Years of Geographical Change* (Talonbooks, Vancouver, 1973).

Wynn, Graeme, and Oke, Tim. *Vancouver and Its Region* (University of British Columbia Press, Vancouver, 1992).

Environment Canada, Lands Directorate, Ottawa.

    1974.   *Water Resources and Related Land Uses in the Strait of Georgia-Puget Sound Basin.*

    1976.   *The Urbanization of the Strait of Georgia Region.*

    1985.   *Okanagan Fruitlands: Land Use Change Dynamics.*

Western Geographical Series, Dept. of Geography, Univ. of Victoria.

    No. 12, 1976. *Victoria: Physical Environment and Development*

    No. 16, 1978. *Vancouver: Western Metropolis.*

    No. 17, 1979. *Vancouver Island: Land of Contrasts.*

    No. 22, 1987. *British Columbia: Its Resources and People.*

B.C. Studies, University of British Columbia Press, Vancouver.

No. 69-70, 1986. "Special Issues on Vancouver, 1886-1986," pp. 11-325.

No. 73, 1987. "Management of Natural Resources in B.C." pp. 14-42.

No. 76, 1987. "Forest Conservation in B.C., 1935-85," pp. 3-32.

No. 78, 1988. "The Impact of Changes in Transportation Technology on the Development of B.C.'s Fishing Industry," pp. 28-52.

No. 79, 1988. "Vancouver: Changing Geographical Characteristics of a Multicultural City," pp. 59-80.

No. 108, 1995. "Visions of Agriculture in B.C." pp. 29-59.

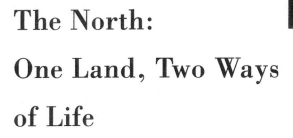

chapter 10

The North:
One Land, Two Ways
of Life

Peter Usher

The North is a region that has been of enduring fascination to many Canadians, but few have any experience of it and even fewer live there. The North comprises a large part of our country in area, and differentiates Canada from most other nations. Not only do the North's physical, biological, and human attributes differ from those of other parts of the country, but they have also inspired a special view of the future. It is part of Canadian mythology that our identity and purpose lie in the North; it's our last frontier and our last wilderness.

THE NORTH AS A REGION

Our focus in this chapter is on the territorial North — the Yukon and the Northwest Territories (NWT). This is at once a political and an economic definition, and one which emphasizes the North's special status within Canada. It is a hinterland unlike any of the regions south of the sixtieth parallel. Politically, these territories have not had the full range of self-governing institutions that the provinces enjoy, and the central government in Ottawa has guided their development to a degree unparalleled elsewhere in Canada. Unlike the provinces, the territories do not have jurisdiction over their mineral and energy resources. For a long time, official policy was that these resources are owned and managed by the federal government in trust for all Canadians. In recent years there has been a shift toward devolution and regional autonomy, although control over economic resources and political institutions continues to be a major issue in the North.

The territorial North is indeed distinct from the rest of Canada. The Yukon Territory, with an area of 536 000 square kilometres, and the Northwest Territories, with an area of 3 367 000 square kilometres, contain 40 per cent of Canada's land surface. The Arctic coastline is longer than the Atlantic and Pacific coastlines combined. The same applies to the area of Canada's Arctic territorial waters. Although estimates vary, some geologists believe that the territorial North holds as much as 50 per cent of Canada's potentially recoverable oil and gas and perhaps 40 per cent of its other potential mineral wealth. However, the territorial North cannot support significant agricultural, forest, or fishery production. The region receives relatively little solar radiation. It is characterized by low temperatures, low precipitation in most areas, poorly developed soils, which are often permanently frozen (permafrost), low species diversity, and for most of the region cold and relatively stable marine waters and oligotrophic lakes that are frozen over for much of the year.

Accordingly, the North is a region of low biological productivity. At the same time, with populations of 27 655 and 57 435 respectively, the Yukon and NWT accounted for only 0.3 per cent of Canada's population in 1991. Of the total population in both territories, 23 and 62 per cent were, respectively, persons reporting aboriginal origins (Table 10.1).[1] The Northwest Territories are unique in Canada in that the aboriginal population outnumbers non-aboriginal inhabitants. There is a high turnover of the non-aboriginal population, for many are transient rather than permanent residents.

Table 10.1 Ethnic Composition of the Population of the Territorial North, 1991

	Ethnic group	NWT	Yukon	Total
1.*	Inuit	18 430	60	18 490
2.*	Indian	8 665	3 550	12 215
3.*	Métis	2 315	165	2 480
4.*	Total of rows 1-3	29 410	3 775	33 185
5.**	Total aboriginal	35 390	6 385	41 775
6.***	Other	22 045	21 270	43 315
7.	Total	57 435	27 655	85 090

Source: Census of Canada, 1991

*Single response only

**Total with aboriginal origins, single or multiple response

***Calculated as total population (row 7) minus total aboriginal (row 5).

Map 10.1 Nordicity Zones Defined by Hamelin

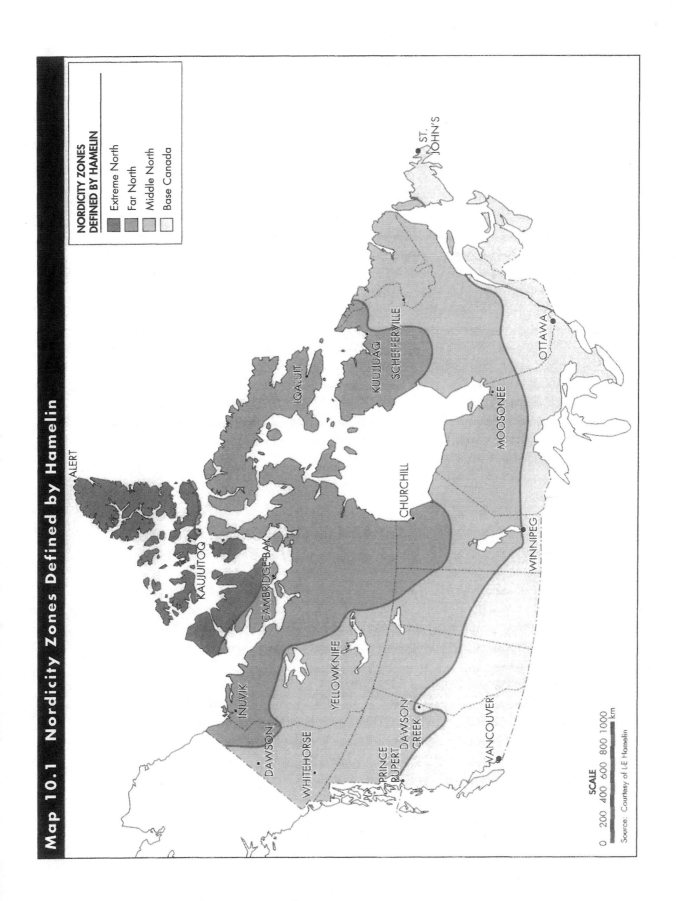

NORDICITY ZONES
DEFINED BY HAMELIN

- Extreme North
- Far North
- Middle North
- Base Canada

ALERT

KAUJUITOQ

CAMBRIDGE BAY

IQALUIT

INUVIK

DAWSON

WHITEHORSE

YELLOWKNIFE

PRINCE RUPERT

DAWSON CREEK

VANCOUVER

CHURCHILL

KUUJJUAQ

SCHEFFERVILLE

MOOSONEE

WINNIPEG

OTTAWA

ST. JOHN'S

SCALE

0   200 400 600 800 1000
                        km

Source: Courtesy of l-E Hamelin

Yet the territorial North has similarities with the northern parts of the provinces. The northern areas of all but the Maritime provinces also constitute regional resource hinterlands and, as felt by many residents, political colonies of the more populous southern regions. The same sense of remoteness from the centres of power that is felt in the territorial North is also felt by residents of, for example, the northern areas of British Columbia, Manitoba, Ontario, and of Labrador. To many southern Canadians, both the territorial and provincial parts of the North seem isolated, cold, dark, and uninviting (Map 10.1).

The ways in which contrasting perceptions of the North have developed are considered in this chapter. The view of the North as a distant and distinctive hinterland — but one which somehow illuminates the future of the nation — is the metropolitan view of the North. It is characteristic of the North's hinterland status that we know more about how southerners view the North than of the regional consciousness of those who live there. There are, however, essentially two other views of the North held by its inhabitants. One is the frontier view, commonly held by non-aboriginal residents. This view tends to unite these people, often of diverse origins and backgrounds, in terms of their reasons for being there and the nature of their lives there. The other is the homeland view,[2] a perspective held chiefly by aboriginal northerners, who view the North neither as a hinterland nor as a frontier, but as their ancestral home which has undergone a continuing transformation due to the progressive encroachment of non-aboriginal peoples and their social and economic institutions.

Physical Definition

How then are we to define the North? What are its boundaries? The territorial boundary at the sixtieth parallel is a political one that does not differentiate the character of the physical environment or the nature of human settlement north of that line. Similarly, the Arctic Circle is merely the latitude north of which the sun does not rise on the shortest day of the year and does not set on the longest. Lines of latitude tell us relatively little. Places such as Great Whale River on the eastern shore of Hudson Bay and Hopedale on the Labrador coast, both just north of 55°N, are more barren and cold than Dawson City in Yukon, or Fort Good Hope in the Western Interior, which are at 64°N and 66°N respectively.

Geographers and other natural scientists have therefore used certain bio-climatic criteria to define the North. They have identified two major natural environments: the Arctic and the Subarctic. The boundary between the two is usually defined as either the 10°C isotherm for July mean daily temperature on land (surface waters consistently at or near their freezing point for the marine environment), or the tree line (in fact not a line at all, but a zone in which trees become fewer and smaller, until they are finally found only in the most favourable sites). The 10°C isotherm for July is nearly coincident with the tree line (Map 10.2).

Map 10.2 The North: Some Physical Limits

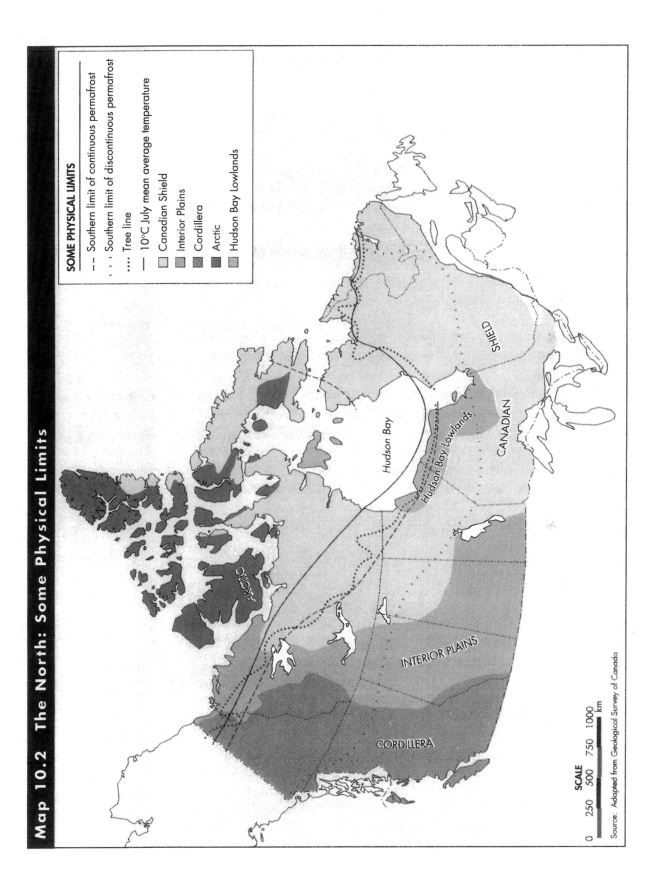

SOME PHYSICAL LIMITS

-- Southern limit of continuous permafrost
... Southern limit of discontinuous permafrost
.... Tree line
— 10°C July mean average temperature

- Canadian Shield
- Interior Plains
- Cordillera
- Arctic
- Hudson Bay Lowlands

Hudson Bay

CANADIAN SHIELD

Hudson Bay Lowlands

INTERIOR PLAINS

ARCTIC

CORDILLERA

SCALE

0   250   500   750   1000
km

Source: Adapted from Geological Survey of Canada

The Arctic is treeless, whereas the Subarctic is commonly considered to consist of the northern forest zone, ranging from closed-crown boreal forest in the south, to open woodland in which tundra vegetation predominates in the north. Coniferous species, especially spruce, dominate the Subarctic forest. Subarctic summers can be occasionally hot, but are always brief (not more than four months with a mean temperature exceeding 10°C). The winters are long and cold, although not as extreme as in the Arctic. The Arctic therefore includes only the north coast of the Yukon and the northern and eastern parts of the NWT, but also parts of northern Manitoba, Ontario, Québec, and Labrador. The Subarctic includes all of the remainder of the territorial North, plus the northern parts of all of the provinces except the Maritimes. A major physical attribute of the northern environment is permafrost, soils or rock frozen at all times of the year. In the Arctic, permafrost tends to be universal, whereas in the Subarctic it is generally discontinuous or sporadic.[3]

One of the distinctive features of the northern environment and its physical processes, therefore, is ice. On land, ice is not simply the result of a cold and long winter. Ground ice is a significant geomorphic agent. The occurrence of permafrost results in a distinctive hydrologic and groundwater regime, in which water cannot flow at depth at any time of year. Consequently, surface water must either run off entirely or saturate the active layer above the permafrost. Neither wells nor buried sewage lines are possible in such conditions. Permafrost with high moisture content, if melted, subsides unevenly, with the result that the construction of buildings, airfields, roads, and pipelines must be done so as to maintain the thermal regime at depth. Clearly, permafrost poses special problems for construction.[4]

Utilidors,
Churchill Falls

At sea, the presence of heavy ice for much or all of the year is a distinctive feature of the Arctic environment. The southern Beaufort Sea in the Western Arctic, and Hudson Bay, Hudson Strait, and Davis Strait in the Eastern Arctic, are ice-free for approximately 100 days. From October, however, they are generally covered by ice reaching a maximum depth of two metres by late winter. This cover is punctuated by pressure ridges of jumbled, tilted, and rafted ice which pose formidable obstacles to travel, whether by dog team, snowmobile, or icebreaker. In less favourable waters, for example in the Central and High Arctic, the ice pack often fails to dissipate completely during the summer, with the result that heavier, multi-year ice is frequently encountered. Northwest of the Arctic Islands, there is a permanent cover of ice, in constant counter-clockwise motion in the western part of the Arctic Ocean. Icebergs, which consist of freshwater ice breaking off the glaciers of Greenland and Ellesmere Island, are encountered off the Eastern Arctic coast, and drift as far south as Newfoundland before melting completely. The existence of these various forms of ice in the ocean pose unique problems and possibilities for the human use of the Arctic seas.

One effect of the long duration of ice and snow cover is that the primary production (vegetation on land, plankton in the marine environment) is limited to a relatively short but intense season. Biological productivity in the North is very low compared to that of temperate latitudes. Arctic and Subarctic ecosystems are characterized by low species diversity and relatively simple food chains. Fish and ani-

Asbestos mine, Clinton Creek, Yukon

mal populations consist of slow-growing, late maturing individuals and are thus vulnerable to over-exploitation. Some animal species also exhibit population cycles of dramatic amplitude. Outstanding examples include lemmings and arctic foxes in the Arctic, and hares and lynx in the Subarctic. The larger arctic mammals, whether marine or terrestrial, tend to travel great distances, either migrating seasonally in herds, or wandering as individuals over large territories. In the Subarctic, by contrast, and especially in the more densely wooded areas, the larger terrestrial mammals do not range as widely. Fire, however, is a continuing ecological determinant there, and there is, over time, a shifting distribution of ecotypes, and associated animal populations, from recent burns to mature forest. These characteristics, along with the instability of permafrost soils and slow vegetative recovery, make the North's environment and ecosystems particularly vulnerable to disturbance and contamination.

Human Occupation

The environmental differences between Arctic and Subarctic are reflected in the evolution of the two distinct aboriginal peoples who have occupied the North for many centuries. The Arctic is inhabited by the Inuit, a people traditionally oriented to the harvesting of marine resources and large migratory herds of terrestrial mammals. They developed a specialized technology, capable of converting such locally available materials as stone, bone, hides, animal oils, and snow into shelter, clothing, heat, light, tools, and transport. United by a common language (but with widely varying regional dialects), Eskimo peoples extend from the eastern tip of Siberia to eastern Greenland. Only about 20 per cent live in Canada, where they are known as Inuit.

The Subarctic is occupied by various Indian peoples who belong to two major linguistic groups, the Athapaskan in the west and the Algonkian in the east. Within these groups there are a number of distinct peoples and languages. All, however, are peoples of the forest, and all had similar technologies based on the snowshoe, the birch-bark canoe, and the toboggan, relying chiefly on the harvesting of a combination of large and small game, and fish.

Both Inuit and Indians originally derived virtually all of their food and clothing from fur, fish, and game. They were primarily hunters, relying on meat and fish for sustenance. Both peoples ranged widely over the landscape in pursuit of game rarely found in sufficient concentration to allow permanent settlements of any size. Residential groups consisted of a few families, all engaged in primary production. There was no concept of private land or resource ownership in the Western sense, although there were systems by which access to land and resources was recognized among groups and regulated and allocated within them.[5]

To non-aboriginal peoples who arrived in recent centuries, however, the distinctions between Arctic and Subarctic have been of less consequence. Neither environment has offered the possibility of significant settlement based on agriculture. Moreover, as one progresses northward in the Subarctic, forest potential becomes extremely limited. Both Arctic and Subarctic have been perceived as sources first of animal wealth, in the form of furs, oil, baleen, and ivory, and more recently of mineral and energy wealth. Despite significant differences between the Arctic and Subarctic in terms of bio-climatic characteristics, animal populations, and aboriginal occupancy, the problems of colonizing and settling each are similar in terms of the difficulty of the environment, the distance from settled and familiar parts of the world, and the inhabitation by a native population.[6]

The absence of agriculture in the North has had enduring consequences for the patterns of settlement and land use there. The land was never widely or densely settled by whites, and never subdivided or fenced off in private hands. Until recently, the landscape was not significantly altered by human exploitation. Aboriginal northerners have thus continued to be able to hunt, trap, fish, and travel widely and with minimal obstruction over their customary territories, even as they moved from their seasonal camps to live in permanent settlements. Those conditions also served to maintain both a geographic and a social isolation from the settled parts of Canada, again until very recently.

The circumstances of non-aboriginal migration to the North have continued to change. The early explorers were almost all sponsored by national governments. During the commercial era, whites came as company traders, as missionaries for the major churches, and later, as members of the national police force. On the southern fringes of the Subarctic, small farmers, loggers, and prospectors were able to settle, and for a time, white trappers as well, although for many years now northern fur resources have been allocated chiefly to Native peoples. Independent traders and small businessmen are relative latecomers to the North, especially to the Arctic. Whites have come to the North in significant numbers only since World War II, chiefly in the employ of governments (especially in the Arctic), or to work in single-purpose resource towns (chiefly in the Subarctic). Most whites, however, migrate to the North as temporary residents, either to fill a particular job or to achieve a certain financial objective. Most eventually move south again. Few see their descendants establish themselves in the North; fewer still die there of old age.

Out of this northern experience there have emerged two viewpoints of what it is to be a northerner. Aboriginal people see the North as their ancestral homeland; a heritage they will never lose, wherever they might be. Their sense of community and identity is rooted in common descent and a shared history. They have a growing political consciousness, particularly concerning the question of control over land, resources, and political institutions, and the impact of development initiatives by government and large corporations. Theirs is the consciousness of the North as a homeland.

For non-aboriginal northerners, community and identity are generally matters of choice and intent. People can and do define themselves as northerners on the basis of even a few months' residency. There is a growing regional consciousness among non-aboriginals in the North, fed in part by their small numbers, by their mobility, and by their restricted occupational structure. The common experience of a harsh climate, high prices, limited amenities, and isolation from family and friends in the south unites non-aboriginal residents. They are also united by a sense of common purpose, a feeling that they are indeed modern-day pioneers, the advance guards of civilization and development in a difficult land. Theirs is the experience and the consciousness of the North as a frontier.

Yet the regional consciousness of both aboriginal and non-aboriginal residents is fragmented because the North is not an autonomous region with strong internal linkages. Its various parts are tributary to southern centres. Yukon is part of Vancouver's hinterland, the Mackenzie Valley part of Edmonton's, the Keewatin part of Winnipeg's, and Baffin Island part of Montréal's. Contemporary air transport routes reflect this situation, building on patterns established earlier by surface transport. Even satellite communications links emphasize this regional dependence. Regional administrative centres, established at the northern termini of these metropolitan linkages, have tended to reinforce north–south orientation at the expense of east–west ties. To overcome this fragmentation, both aboriginals and non-aboriginals have attempted to develop pan-northern institutions and linkages. Native political and cultural organizations such as Inuit Tapirisat of Canada have arisen since the early 1970s, and have attempted to unite the interests of their members over broad areas of the North. The NWT government long attempted to focus the consciousness of all northern residents, from Aklavik to Iqaluit (2900 kilometres west to east) and from Grise Fiord to Sanikiluak (2250 kilometres north to south) on a common administration based in Yellowknife. Despite these efforts, links remain tenuous, and the quest for a greater degree of shared consciousness remains largely unfulfilled; in 1999 the NWT will be divided into two separate territories.

THE PROCESS OF NORTHERN DEVELOPMENT

Canada's North, since the time of European exploration, has undergone three major phases of incorporation under metropolitan control, and is now entering a fourth. The timing and duration of these phases have varied considerably from one region to another, and the transition from one phase to the next has often been gradual and even indistinct. Nonetheless, the following characterization applies broadly throughout northern Canada, and will be described with particular reference to the territorial North.

Each of these phases of metropolitan incorporation has had distinctive consequences for the pattern of settlement and economic activity in the North. The initial phase consisted of discovery and commercial penetration first by Europeans and later by southern Canadians. The second was marked by the establishment of administrative colonialism by the Dominion government in Ottawa. The third phase was led by a great increase in government activity in the North itself, for the purposes of administration of lands and resources and of aboriginal affairs, and the provision of health, education, welfare, and economic infrastructure. The objective was the promotion of large-scale mineral and energy resource development by private capital, which would become the leading force of regional economic growth.

Phase One: Discovery and Commercial Penetration

The first phase, beginning with the voyages of discovery, led to an awareness of the North in Europe, to the establishment of direct trade relations with the indigenous peoples, and to the exploitation of the North's marine resources (chiefly the whale fishery). Aside from probable sporadic visits by Norsemen in the eleventh century, these voyages (mostly British) began as early as the sixteenth century in the Davis Strait area. A period of intensive British naval exploration for the Northwest Passage occurred from the 1820s to the 1850s, although complete passage by a single vessel did not occur until 1906. Explorations continued into the early twentieth century in the more inaccessible parts of the central Arctic and the interior of Keewatin.

Throughout most of this phase, the North was a European hinterland, but it differed from southerly parts of the Americas chiefly because its access and habitation were more difficult. It therefore became a frontier of exploitation rather than of settlement, requiring little or none of the apparatus of colonial administration. It was also, with certain exceptions — such as the early French fur trade in the eastern Subarctic and the nineteenth-century American whale fishery in the Beaufort Sea and Hudson Bay — a British hinterland.

The fur trade, the first great staple to be exploited, was organized by large trading companies, the chief of which was the Hudson's Bay Company of England. This enterprise required only isolated trading forts that, except for local meat provisioning, were supplied entirely from overseas. Aboriginal labour produced the furs. Indeed, the company traders could afford, by and large, to wait for the Indians to bring their furs to the forts. Not until the Montreal-based North West Company was established in the 1770s was the fur trade aggressively brought to the Indians, and the northern interior explored by whites. The early fur trade was based on the beaver. The Subarctic was therefore drawn into this trade much earlier than the Arctic, where beaver are absent. The fur trade reached Great Slave Lake in the late 1700s and was then extended down the Mackenzie River and into the Yukon in the nineteenth century. The Inuit did not become involved in fur trapping until the early twentieth century. For them, the Arctic fox became the staple commodity.

Until 1870, the North was not a part of Canada. Rupert's Land, comprising the entire area draining into Hudson Bay, as well as the North Western Territory (since the absorption of the North West Company in 1821), was administered by the Hudson's Bay Company under monopoly charter from the Crown. The Arctic Islands had been claimed by Great Britain during the previous century. With the sale of Rupert's Land to Canada in 1870, and the transfer of the Arctic Islands to Canadian jurisdiction in 1882, control of the entire North passed to the new Canadian nation. A new era of commercial development soon began.

Direct overseas control of the North ended, but some features of the era remained. The major coastlines and rivers of the North had been charted, due as much to the quest for the Northwest Passage as to the pursuit of the fur trade. Initial metropolitan penetration had encouraged Native peoples to produce goods for exchange as well as subsistence, which previously had been the primary objective of production. European trade goods were in most instances sufficient to induce this change, because aboriginal people regarded metal knives, traps, firearms, pots, and similar items to be of great advantage in harvesting and processing fur, fish, and game. Aside from this reorientation, however, European economic interests sought neither to displace the North's aboriginal inhabitants, nor to disrupt their way of life. Nonetheless, both sometimes happened, chiefly through the spread of epidemic diseases and the establishment of intermediary trade relations among groups, both of which caused major demographic shifts. The colonial powers did not need an elaborate administrative structure either to enforce a new system of law or to protect a large settler or resident population. Religious interests did seek to convert the aboriginal inhabitants to Christianity, but as long as the fur trade and whaling interests prevailed, they were given little opportunity or encouragement to do so. (Northern Labrador, where the Moravian Mission controlled both religious and commercial activities from the late eighteenth century onward, was an exception.)

The Canadian government's immediate objective in purchasing the territory of the Hudson's Bay Company was to gain access to the prairies for agricultural settlement and development. Obtaining the Subarctic in the bargain was almost incidental. The subsequent acquisition of the Arctic Islands was spurred largely by the purchase of Alaska by the United States from Russia in 1867, and hence Canada's fear of further American encirclement. As Edward Blake, a former premier of Ontario and a leading nationalist and expansionist of the day, observed in a report to the Colonial Office in London: ". . . the object in annexing these unexplored territories to Canada is, I apprehend, to prevent the United States from claiming them, and not from the likelihood of their proving of any value to Canada. . ."[7] Thus, the new Dominion had, in a short time, acquired half a continent. Most of this acquisition consisted of land that, in the popular view, must have ranked among the most fearsome and inhospitable environments on the face of the globe. For the new country, most of whose people lived between the Atlantic Ocean and the Great Lakes, the first task was the construction of the transcontinental railway and the settling of the prairies. It could afford to do little else. With no external threat, it did not have to. The North could remain as a vast reserve for the future, its land and inhabitants left undisturbed until national policy deemed otherwise.

The transfer of Rupert's Land terminated the fur monopoly, thereby opening the region to settlement and other forms of economic exploitation. In areas accessible to the transcontinental railways or to major rivers, numerous but small mining, logging, and commercial fishing enterprises were established over the next several decades. In the less accessible areas, however, commercial resource exploitation remained impractical unless the commodity had an extraordinary value relative to its mass. This had been the basis of the fur trade, and thus the first mineral development north of the sixtieth parallel began with the Klondike Gold Rush of 1898.

In anticipation of further settlement and development, the provincial boundaries were gradually extended northwards, to Hudson Bay in the east, and to the sixtieth parallel in the west. The present political division of the North dates from 1912 (except for the Québec–Labrador boundary, which was defined in 1927). The Yukon was created as a separate territory in 1898. Railways pushed north in the new provincial extensions: to the Athabasca River in the early 1920s, to Churchill in 1929, and to Moosonee in 1932. The short but difficult railway from Whitehorse to tidewater at Skagway, Alaska, was completed in 1900.

During the first phase, the state played a modest role in the direction of northern development. Prior to 1870, Britain promoted exploration and asserted sovereignty, leaving trade, resource extraction, and settlement to private initiative. For the next seventy-five years, Canada exercised its authority chiefly in two ways: by asserting national sovereignty, and enacting Indian treaties where it thought this necessary.

The first permanent federal presence in the territorial North, the North West Mounted Police (NWMP), came as a response to the gold rush in the Klondike.

White Pass railway, Skagway

During the early twentieth century, several more NWMP detachments were established across the North and in the High Arctic. The federal government also dispatched expeditions to conduct geological surveys, scientific research, and exploration. The chief purpose was to demonstrate sovereignty. The Americans and the Danes had also sent expeditions to the Far North, creating the fear that Canadian sovereignty there could be challenged. The Dominion government maintained no other presence in the North, however. Even fifty years after Confederation, there were no publicly employed doctors, teachers, or administrators in the Northwest Territories.

The discovery of oil in 1920 below Fort Norman on the Mackenzie River, changed this. Mindful of the experience of the Yukon gold rush, when good luck and the dedication and competence of a tiny police force and civil service had maintained Canadian law and sovereignty in that district, the government moved quickly to establish its presence in the Mackenzie District. First, it created the Northwest Territories and Yukon Branch of the Department of the Interior to deal exclusively with the territorial North. Administrative offices were located at Fort Smith, the head of Mackenzie River navigation and, as it happened, strategically located at the southern entry point to the western NWT. Second, it arranged a treaty with the Dene along the entire length of the Mackenzie River. These events heralded further commercial development. The fur trade became well established even among the most remote Inuit, due in part to greatly improved river and coastal shipping. The airplane greatly expanded the possibilities for regular postal service and for mineral exploration in the North. Gold, silver, and radium mines were discovered and developed in the Yukon and in the Mackenzie District, and a rudimentary administrative and commercial infrastructure was also established in those areas.

The brief period of administrative activism of the 1920s was brought to an end by government austerity in the Great Depression. Government sought to minimize the pace and impact of change on northern aboriginal people, believing that if they continued to be isolated and self-sufficient hunters and trappers, the social costs of health, education, and welfare could be avoided. To this end, Canada established several large game preserves in the NWT, and restricted hunting and trapping by non-aboriginal persons. Responsibility for the Inuit, with whom no treaties were ever signed, was shifted from one agency to another in government. In the case of northern Québec, the provincial and the federal governments fought each other all the way to the Supreme Court in 1939 to shift responsibility for the Inuit to the other party. The federal government lost, and was forced to assume fiscal responsibility for Inuit health and education. As in other parts of the North, however, these services were actually provided by the Roman Catholic and Anglican missions which, during the preceding decades, had become well established in the North, gaining influence among Native peoples. In the twentieth century, the colonial administration of the North came to be personified by the trader, the police officer, and the missionary.

Inuit hunters

This period of more intensive commercial penetration from within Canada brought many changes to the native inhabitants of the North. Disease and trade-induced demographic changes continued and in some areas intensified. With the end of the fur monopoly, increasing incursions of white trappers and hunters led to direct competition for resource harvesting, and in some cases resulted in temporary depletion. The intensification of commercial whaling in the Arctic in the late nineteenth century resulted in the near extirpation of bowhead whale populations. There were also changes in the material culture and observable behaviour of aboriginal people. They had come to adopt many items of Euro-American technology, dress, and food, while certain traditional skills and practices atrophied. More people spoke English and accepted Christian rites. Aboriginal people had become dependent on material goods they could not themselves produce, and which could only be obtained through a trade relationship in which, with few exceptions, it was the traders who controlled information and set prices.

Despite these changes, the Indians and Inuit of the North continued to engage in hunting and fishing for subsistence, and trapping for trade. In some areas, unskilled seasonal wage labour became common, but this was for the most part easily integrated with hunting, fishing, and trapping. Work, cutting cordwood or at trading posts, and in guiding and survey parties, became another way of obtaining trade goods at the store during a season when trapping was not possible or profitable. The new economy allowed people to stay on the land and to live by their traditional skills and values, and indeed often rendered that life easier and more secure.

Phase Two: Administrative Colonialism and the Welfare State

The second phase of metropolitan incorporation came with the central government's full assumption of sovereignty. As before, change occurred largely in response to external pressures. World War II brought a new appreciation of the strategic significance of the North, which expanded with the Cold War of the 1950s. Airfields and bases were established in the northwest to link the United States with Alaska, and in the northeast to provide staging points for Europe. The northwestern link was strengthened by the construction of the Alaska Highway in 1942. In the postwar years, joint Canadian–American weather stations were established in the High Arctic. In the mid 1950s, two major radar lines were built across the North. The Mid-Canada line followed roughly the fifty-fifth parallel, and the Distant Early Warning (DEW) line the seventieth. While the threat of attack on Canada came from Germany, Japan, and later the Soviet Union, it was the United States, which built and controlled most of the airbases and radar stations, whose presence most threatened Canadian sovereignty in the Arctic.

The construction of these facilities brought a level of technology, investment, and productive organization previously unknown to the North. At about the same time the military arrived, there developed a crisis in the fur trade, due chiefly to a long, steady decline in fur prices accompanied by a sharp increase in the cost of imported goods. Metropolitan fur interests cut their investment in the northern trade by reducing credit both to trappers and to local independent traders, and by closing smaller or less profitable posts. No longer able to outfit themselves by trapping, many Native peoples gravitated to the new military installations in search of employment, or, in more desperate situations, in search of food and other goods.

The collapse of the fur trade in the postwar years was accompanied by an increasing awareness (promoted by writers, journalists, and government reports) on the part of southern Canadians that there were fellow citizens in the North who suffered from economic want, and who also lacked the modern conveniences and opportunities enjoyed by most Canadians. The postwar sentiment against traditional colonialism, along with a rising ethic of "equality of opportunity," created a climate ripe for change in the Far North. The publicity surrounding a few dramatic incidents of suffering and starvation, particularly among the Keewatin Inuit in the 1950s, demanded action. The more subtle, but just as pervasive, social effects of the Alaska Highway, the war-time air bases, and the DEW line could not be ignored. These opened up the most isolated parts of the continent, sometimes, unfortunately to the spread of disease, violence, and abuse by servicemen and civilians, and in the Yukon added hunting pressure on already scarce food resources. The federal government could no longer maintain a laissez-faire approach to the North and its inhabitants. That policy was clearly failing aboriginal people, and in the context of the dawning welfare state, the government was widely seen as having no alternative but to intervene.

Thus began a period of significant, planned government intervention in northern life. It was initiated by the extension of family allowances and old-age pensions to native northerners in the 1940s, followed by increased government pressure on aboriginal families to send their children to boarding schools. If isolation and self-sufficiency in the old economy were no longer possible, then government saw no alternative but to bring aboriginal people into the mainstream of Canadian society as soon and as much as possible. Aboriginal health and housing conditions had become appalling in many parts of the North by the 1950s.

The government project for the next two decades was to modernize the North by providing services and infrastructure directly to aboriginal people. This required enormous public expenditures, and because these were most economically and conveniently administered in larger centres, government had first to select and develop the necessary growth points. The construction of new towns, such as Inuvik and Iqaluit (formerly Frobisher Bay), and the transformation of established mining centres such as Yellowknife into administrative capitals stemmed directly from this second phase of northern development. Public investment in the smaller communities soon followed, in the form of federal schools and nursing stations in the 1960s, and the provision of public housing and municipal services (water, power, fuel delivery, and waste disposal) over the next decade.

In retrospect, it is evident that, notwithstanding their humanitarian intent, these government programs were developed and implemented without consultation and in ignorance of how aboriginal people lived. Consequently, they did not always effectively solve problems, and sometimes created new and unanticipated ones.

The consequences of these changes for aboriginal people were profound. The two most important were a shift in residence from the temporary camps to permanent settlements (during which aboriginal northerners became not only sedentarized but also more numerous, as birth rates and infant survival increased and mortality declined), and a major upheaval in native culture and institutions. Whereas previously it had been in the interests of the fur trade for aboriginal people to live on the land and away from the settlements, it was now in the interest of the government for them to move off the land and into the settlements.

By the end of the 1960s, the predominant residence of aboriginal families was no longer the small, seasonal encampment, but centralized settlements of two types. One was the small, predominantly aboriginal community,[8] in which it was not unusual for over half the population to be under the age of 16. The other was the larger but predominantly white community, which by southern Canadian standards would be classed as a small town. While the most dramatic demographic shift over the period was from the land to the towns, there was also a trend for aboriginal people to move from the smaller centres to the larger ones.

When explaining these movements, many observers have emphasized the attractions of these central places: greater security of income and health; the opportunity to obtain cash more easily than through trapping, either through casual labour or welfare payments; and the range of goods, services, and amenities available there. For many, however, the move to the towns was not voluntary. Compulsory schooling (which meant that to stay on the land was to be separated from one's children), and the lack of alternatives to casual employment and welfare as a means of supplementing hunting and fishing, compelled many to make the move.

Subsistence production remained important (although it was temporarily disrupted in some areas), but it became virtually unnoticed and unacknowledged by policy makers. The continuing need for trade goods was met by a combination of seasonal (summer) wage employment, family allowances, pensions, and welfare payments. The problem with these transfer payments, however, was that they were obtained not by traditional skills and organization, but from an administrative system of which aboriginal people had little understanding and even less control. Although most aboriginal people remained primarily oriented to living off the land, they became increasingly dependent on the government for shelter, heat, and power, and for the education of their children.

Federal, provincial, and territorial administrations of all political persuasions had come to a common understanding of the problem of aboriginal people in the North, and what to do about it. They concluded that the old way of life was fast dying, and that the only avenue for aboriginal people was to forsake their traditional ways and take up those of the modern world. The immediate solution to the crisis was in health and welfare programs. The long-run solution was to educate aboriginal people and give them wage employment. These policies led, particularly in the larger centres, to the growth and entrenchment of an ethnic-based class structure. Good jobs, benefits, housing and services, and high cash incomes went mainly to non-aboriginal government employees and business people. There were few aboriginal people in the new resource- or transport-based towns.

Phase Three: Megaprojects and Public Administration

The third phase, which began in the 1950s and 1960s, was characterized by the ascendancy of the "megaproject" as the expected engine of northern economic and political development, and by the growth of public administration in the North itself. Megaprojects — giant, industrial-scale extraction and transport of natural resources to metropolitan markets — require enormous sums of capital, innovative technology, and sophisticated management. This requires the initiative of large, multinational corporations, or Crown corporations, as well as the encouragement and cooperation of governments and the resident small business sector.[9] During the 1960s, megaprojects became part of a national economic development strategy, if not also a matter of national security, especially during the years of the Cold War. Thus, as in the earlier phases of northern development, external pressures played a leading role in metropolitan expansion.

Until this third phase, northern mineral developments had been the result of small-scale private initiatives that often received only marginal attention from the industry as a whole and little support from governments. These earlier developments were rarely seen as strategic to the national economic or military interest, although when they were, as in the case of the Port Radium mine at Great Bear Lake during World War II, the central government did not hesitate to take command.

From the late 1940s to the 1970s, the demand for and investment in northern resource development came chiefly from the United States. In more recent years, markets and investment sources have expanded to include the European Economic Community and Japan. Northern megaproject development began in the provincial north, in the energy, mineral, and forestry sectors. Notable examples included the iron mines and railway in Nouveau-Québec and Labrador, and hydroelectric power generating developments at Churchill Falls, Labrador; La Grande Riviére, Québec; the Churchill and Nelson Rivers in Manitoba; and on the Peace River in British Columbia (some of which were among the largest generating installations in the world).

In the northwest, megaprojects were frequently related to fossil-fuel extraction. These included the sustained exploration, delineation, and in some cases development, of oil and gas fields in northern Alberta (including tar sands developments), British Columbia, the Mackenzie Valley, the Beaufort Sea, and the High Arctic. Although only the more southerly of these were brought into production, the exploration and planning phases alone of hydrocarbon development in the territorial North brought unprecedented investment and change to that region. Major mineral developments in the territorial North consisted mostly of lead-zinc mines: Pine Point south of Great Slave Lake in 1961, Faro in the central Yukon, and in the Arctic (Nanisivik on northern Baffin Island in 1976 and Polaris on Little Cornwallis in 1981), although gold production also extended north from Yellowknife to the Lupin mine.

Megaprojects typically required innovative means of large-scale transport. The Pine Point mine resulted in a major railway extension from the Peace River area to Hay River and Pine Point. During the 1960s and 1970s, the most important transport proposals in the territorial North involved the movement of oil and gas from the Arctic. The voyage of the *Manhattan*, an American tanker, through the Northwest Passage in 1969 heralded the possibility of a commercial shipping route to bring Alaskan oil, and potentially oil from the Canadian Arctic, to the eastern United States. Later, there were proposals to transport liquified natural gas by ship from the High Arctic to southern markets. However, pipelines are generally considered the safest and most cost-efficient mode of transport, and three major large-

Oil drill site,
Cameron
Island, NWT

diameter lines were proposed in the 1970s: the Mackenzie River Valley route to bring gas from Prudhoe Bay in Alaska and the Mackenzie Delta in Canada to the American Midwest; the Alaska Highway route through the Yukon, designed for the same purpose; and a route from the High Arctic along the west coast of Hudson Bay, and south through Ontario to the Great Lakes region. None, however, was built. The only major petroleum project completed in the NWT was a smaller diameter oil pipeline south from Norman Wells along the Mackenzie Valley. That line, based on the expansion of the existing Norman Wells oil field that had already been operating since World War II, was put into operation in 1985.[10] The sharp decline in world oil prices in 1986 resulted in an immediate slow-down in northern petroleum development.

Megaprojects attracted much public attention from the 1960s onward, and led many to believe that a "breakthrough" in the exploitation of northern resources was close at hand. Often, one major project such as a gas pipeline or oil tanker route was seen as the wedge that would quickly open the door to many other developments, by providing essential transport and other infrastructure requirements. The federal government, as land owner and manager in the territorial North, saw megaprojects as the means of financing northern development and contributing to the national economy. In the outcome, however, megaproject-led development did not occur as rapidly or extensively as expected, although government and infrastructure development in anticipation of it changed the North markedly. Of the numerous major proposals for mineral and energy developments and related transport systems, few were actually brought into production or operation. Even those that at one time had the full backing of the federal government, as a national priority, were abandoned or delayed. Although two new mines opened in the Arctic Islands, most mining activity continued to focus on the south central Yukon and around Great Slave Lake.

Nonetheless, this third phase of northern development saw steady progress in both establishing an inventory of the North's mineral and energy resources, and in actual production. Seismic exploration and exploratory drilling greatly intensified during the 1970s, peaking first on land and later on the continental shelf. The value of mineral production in the two territories approximately doubled during the 1970s and again during the 1980s.

The fact remains that the North is a high-cost and comparatively disadvantaged environment in which to work. Physically, its harsh climate, permafrost, and marine ice conditions add significant costs to development, and as the 1980s demonstrated, oil, gas and mineral exploration and production are highly price sensitive. In human terms, the North remains an unfamiliar environment situated far from the core of settlement. Distance from manufacturing and consuming centres, and the difficulties of overcoming that distance, whether on land or on sea, remain powerful limitations on northern resource development.

Although industrial development in the territorial North did not reach the proportions that occurred in neighbouring Alaska, or in the northern parts of some of the provinces, both the expansion in exploration activity and the anticipation of future developments promoted rapid growth in the government and resident small business sectors. In the early 1960s, Yellowknife was a mining town, in which almost the entire federal administration was housed in a modest, two-storey wooden building. After being designated the territorial capital in 1967, matters formerly administered in Ottawa, such as financial management, education of Native peoples, and social services, were gradually transferred to the territorial government in Yellowknife. While the federal Department of Indian Affairs and Northern Development remains the owner and manager of public lands in the territorial North, the actual administration of lands and resources has been largely moved to Yellowknife. Now, numerous federal and territorial government agencies operate out of modern high-rise office buildings. The same process has occurred in the Yukon, where Whitehorse was designated the territorial capital in 1953. Government became by far the leading source of employment, especially in the Northwest Territories.

The regulation of development activity also became more complex. For example, in the 1970s, proposals to develop and transport oil and gas were subject to extensive government review processes, not only by the National Energy Board, but also by the Federal Environmental Assessment Review process and even by public inquiries established to determine regional socio-economic impacts. The best known of the latter was the Mackenzie Valley Pipeline Inquiry which, in 1977, recommended that no pipeline be built along that route for at least ten years to provide time for the settlement of native claims and the development of programs and institutions that could minimize the risks and maximize the benefits of development.

This third phase of northern development saw the establishment in the North of a full range of welfare state machinery, including pensions, unemployment insurance, and government programs for job training, employment creation, labour allocation, and regional economic development. No longer were government functions concentrated in one department of northern administration, a characteristic feature of the era of administrative colonialism. There was also rapid development of a banking and financial infrastructure, and especially of communications. By the 1980s, every northern community received telephone service and television by satellite. It was these developments, rather than simply increased industrial resource extraction activity, that integrated the North much more completely into the national economy.

One result was that both Yellowknife and Whitehorse changed in character from mining towns to administrative and business centres, and became by far the dominant and most populous central places in their respective territories (Table 10.2). Growth in the non-aboriginal population of the North occurred almost entirely in the two capitals (and in regional administrative centres such as Inuvik and Iqaluit), or in resource centres such as Faro and Pine Point.

Table 10.2 Growth of the Territorial Capitals, 1961, 1991

		Whitehorse	Yellowknife
1961	Total population	5 031	3 245
	Proportion of territorial population	34.4%	14.1%
1991	Total population	17 805	15 120
	Proportion of territorial population	64.4%	26.3%

Summary

The North was slowly but progressively tied into the global economy and the Canadian nation in three distinct phases. The era of discovery, exploration, and trade brought the North and its peoples into the sphere of European knowledge and influence, and later under Canadian sovereignty. It was then administered for several decades as a colony by the Dominion government in Ottawa, in effect as a territory in waiting. Since the 1950s, the planned development of the North's economy, infrastructure, and government has integrated the two territories much more fully into national life. But if the North was no longer simply a colony of the

Dominion, it certainly remained a distant hinterland, dependent on the south for investment both public and private, and with little effective voice in its own affairs, let alone those of the nation. The northern economy remained a staples economy, highly dependent on the production of a few primary resources for export. The old staples were replaced by new ones (fur was the leading single product of the NWT for the last time in 1946, since then far eclipsed by minerals), but near complete dependency on external markets, and a weakly integrated local economy, were as much a fact of northern life late in the twentieth century as one hundred years before. If non-aboriginal northerners felt powerless and disenfranchised, aboriginal northerners remained effectively excluded from economic and political power altogether, being merely the subjects of public policy making. The development model that prevailed in the third phase — of active government promotion of industrial development and social change — ignored the views and visions of aboriginal northerners and eventually provoked their resistance.

THE CONTEMPORARY NORTH

The single most important factor in the course of northern development in the last twenty-five years has been the assertion of power by aboriginal northerners, and the accommodation of Canada and its two northern territories to that assertion. Underlying this reversal was the simple but underestimated (by both government and industry) survival of aboriginal distinctiveness; of economic, social, and cultural persistence of a way of life and of a commitment to a different vision of the future. The incorporation of the North into Canada did not lead inevitably to the assimilation of its people. This persistence was expressed at first as resistance (again unanticipated by government and industry) to the course of economic and political development that was occurring in the 1960s and 1970s, and later, as that development slowed, articulation and assertion of a distinctive aboriginal future.

The ability of aboriginal northerners to so alter the course of northern political development since the 1970s was in significant part due to the limited success of the third-phase strategy of megaproject-led growth, which led to neither a major influx of southern Canadians, nor the dominance of major outside industrial influence over territorial and municipal governments. Either of those outcomes would have greatly restricted aboriginal political influence. The persistence of a distinctive aboriginal way of life can be seen in the distribution of the North's population, the patterns of land and resource use, and the character of the northern economy.

Population

The North has not become the expanding frontier of settlement in Canada, as was anticipated by some through much of the twentieth century, and the geographical centre of Canada's population has not moved northward. The territorial North's approximately 0.3 per cent share of the national population is unchanged over several decades, although the rate of population increase in the NWT has been higher than in the Yukon. The aboriginal population has experienced a high rate of natural increase in recent decades, but migration has been largely within the territories, from smaller settlements to larger ones, rather than to or from other parts of Canada. The non-aboriginal population has a lower rate of natural increase, but is much more mobile. A high proportion of the non-aboriginal population in each territory are short-term rather than permanent residents, and the rate of, and balance between, in-migration and out-migration is substantially dependent on economic conditions.

Resource developments no longer require a large, resident labour force. For example, new mining operations now tend to rely on flying in the bulk of the labour force on a rotational basis, rather than establishing permanent communities. It now seems unlikely that the North's population will increase significantly by virtue of the establishment of single-industry communities. Between 1981 and 1991, the number of such communities in the Northwest Territories fell from seven to two, and their proportion of the population from 6.7 per cent to 1.6 per cent. On the other hand, there has been a very substantial growth in the population of the capital cities as central places in each territory (Table 10.2), in line with the growth of government administration in the North.

For the foreseeable future, population growth in the territories is more likely to occur from the continued natural increase in the aboriginal population, and from growth in administration (especially in Nunavut) and business, requiring immigration of skilled personnel, than from growth in primary production. Overall, the population of the territorial North is about evenly divided between non-aboriginals and aboriginals, although the former predominate in the Yukon and Mackenzie Valley, the latter in the northern and eastern parts of the Northwest Territories. The NWT is the only jurisdiction in North America with a predominantly aboriginal population, in contrast to the situation in the northern parts of the provinces and Alaska, where regional concentrations of aboriginal people in rural and remote areas are greatly outnumbered by urban centres to the south.

The territorial North is also distinguished by the relative distribution of its aboriginal and non-aboriginal populations. Formerly, aboriginal people lived in small, widely dispersed groups throughout the North. Each group moved camp several times annually, using tens of thousands of square kilometres of land and water for sustenance. In recent decades they have become more sedentarized, moving into nearby communities established by the fur trade and expanded by government to provide housing and services in central locations. Non-aboriginal people have generally established year-round settlements, usually with a specific economic motivation, whether these were fur trade posts, whaling stations, military bases, or mining camps. When these were no longer needed or profitable, people moved on.

Today there are about seventy-three permanently inhabited communities in the territorial North, which may be classified functionally in four categories: the capital cities, regional administrative centres (highway service centres in the Yukon), resource communities, and historic aboriginal communities (Table 10.3). Aboriginal and non-aboriginal people are distributed very differently among these categories.

The population of the capital cities is in each case over 80 per cent non-aboriginal, although the proportion of the aboriginal population residing in these centres is rising, particularly in the Yukon where nearly 50 per cent of the aboriginal population resides in Whitehorse. The regional administrative or highway service centres are somewhat more balanced (58 per cent aboriginal in the NWT, 30 per cent in the Yukon). Aboriginal people account for a low proportion of residents of resource communities, especially in the Yukon, which reflects their low participation in the industrial labour force, especially mining. Of the 73 territorial communities, 58 are predominantly aboriginal and have populations of less than 1000. Typically these are located on the main rivers and coasts and originated as fur trade posts. The proportion of non-aboriginal residents in these communities has actually declined in recent decades and in the Northwest Territories is now less than 10 per cent. While most aboriginal northerners reside in these small communities, over 60 per cent of non-aboriginal residents live in the two capitals, and another 20 per cent live in the major regional administrative and transport centres such as Iqaluit, Fort Smith, Hay River, and Inuvik. Single-industry resource communities now only account for less than 5 per cent of the non-aboriginal population.

Two Ways of Life in the North

The distribution of the northern population is perhaps the most obvious manifestation of the persistence of a distinctive aboriginal communal and economic life that relies on continuing extensive use of the land. There continue to be two distinctive ways of life, economic orientation, or "mode of production"[11] in the North, even though each has changed considerably in the last several decades. Non-aboriginals have been oriented to trade, resource extraction, and later, wage employment in a town setting. Hunting, fishing, and trapping, or "living off the land" has been the exception, and for those few non-aboriginals who did so, it was primarily to earn cash income, as in trapping and commercial fishing. The smaller communities where most aboriginal northerners live are characterized by a mixed, subsistence-based economy.[12] In this economy, the household acts not only as a unit of consumption, but also as the basic unit of production (see Figure 10.1). Some members of the household draw income from wage employment (which may be seasonal or occasional rather than full-time), some may hunt and fish, providing the bulk of the family's food and perhaps selling some of this produce. Household income is thus

Table 10.3 Ethnic Composition of Populated Centres by Function, Northwest Territories and Yukon, 1991

	Northwest Territories				
	Capital (Yellowknife)	Other Administrative Towns	Resource Communities	Historic Aboriginal	Total
Number of places	1	6	2	48	57
Total population	15 120	15 380	920	25 345	57 435*
Aboriginal population	2 965	8 860	260	23 030	35 390*
Non-aboriginal population	12 155	6 520	660	2 315	22 045*
Average population	15 120	1 087	460	528	—
Aboriginal as % of total population	19.6	57.6	28.3	90.9	61.6
Percentage of total NWT aboriginal population	8.3	25.0	0.7	65.1	99.2
Percentage of total NWT non-aboriginal population	55.1	29.6	3.0	10.5	98.2

*Includes 670 persons (275 aboriginal and 395 non-aboriginal) residing outside of organized communities.

	Yukon				
	Capital (Whitehorse)	Highway Service Centres	Resource Communities	Historic Aboriginal	Total
Number of places	1	4	1	10	16
Total population	17 805	1 730	1 220	3 750	27 655*
Aboriginal population	3 110	520	110	2 020	6 385*
Non-aboriginal population	14 695	1 210	1 110	1 730	21 270*
Average population	17 805	433	1 220	375	—
Aboriginal as % of total population	17.5	30.1	9.0	53.9	23.1
Percentage of total Yukon aboriginal population	48.7	8.1	1.7	31.6	90.2
Percentage of total Yukon non-aboriginal population	69.1	5.7	5.2	8.1	88.1

*Includes 3150 persons (625 aboriginal and 2525 non-aboriginal) residing outside of organized communities.

Source: Census of Canada, 1991

derived from a combination of wages, commodity sales, and income-in-kind. Where wage-earning opportunities are limited, unemployment insurance and welfare may also contribute, and in former years, family allowances and old-age pensions were a significant source of cash.

The role of harvesting, as hunting, fishing, and trapping are often referred to, is both significant and distinctive in aboriginal communities, although for a long time it was largely unrecognized because conventional economic indicators focussed primarily on employment and income. The continuing high rate of aboriginal participation in harvesting was masked by the fact that many aboriginal harvesters also participate in wage employment. Modern hunting and fishing techniques allow many people to assume casual or even full-time employment and still produce much of their own food. Income-generating functions within the family, or even among households, are also becoming more specialized, so that a few people can hunt for many. The importance of domestic food production was often underestimated by both incomplete data-gathering by government fish and wildlife agencies, and because this production, which did not enter the marketplace, was either not evaluated or undervalued by conventional economic systems of measurement. With effective replacement values for country food now running from $5.00 to $20.00 per kilogram in the North, country food production contributes hundreds and perhaps thousands of dollars to real per capita income. This estimate also does not include the imputed value of domestically consumed furs, hides, fuelwood, building materials, and the like. There are also important nutritional values in country food. Beyond these measurable economic facts are the important social and cultural values that Native peoples attach to both the production and consumption of country food, and to the way of life those activities represent.

FIGURE 10.1 The Household in the Mixed, Subsistence–based Economy

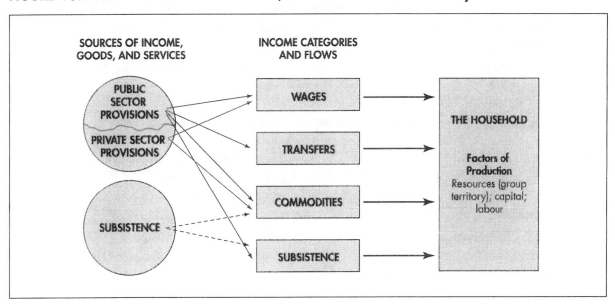

Traditional Arctic fishing

Relations Between the Two Modes of Production

The relation between the two ways of life in the North changed over time, and as they became more interconnected, certain frictions arose. During the first phase of northern development, in which furs were the staple export commodity, aboriginal people were an integral part of the production process. The trading companies encouraged them to retain their ties to the land and their way of life. During the second phase of development, the locus of contact between whites and aboriginals was restricted to the administrative and trade centres. The intrusion of non-aboriginal society and authority, massive as it was, was limited in geographical extent to a few tiny dots on the enormous map of the Northland (Map 10.3). Life on the land remained basically unchanged, and Native peoples continued to enjoy relatively free access to the land and its resources.

When large-scale resource development and modern government centres first appeared in the North, many observers and policy makers regarded the two economies as independent because, although they existed in close geographic proximity, there were few linkages between them in terms of cash or commodity flow, transport, labour, or technology. The chief point of contact appeared to lie with the new economy's employment of aboriginal peoples, and even this was restricted largely to the resident business sector (chiefly tourism and recreation) and to the government sector (chiefly unskilled labour). Such employment was mostly available in the larger communities where aboriginal people were incorporated into the lowest and least secure levels of the occupational structure.

The apparent independence of the two modes of production gave rise to the notion of a dual economy. Since economic growth (or measures to promote it) in the industrial or modern sector failed to stimulate growth in the aboriginal or traditional sector, it was assumed that the two sectors were functionally separate. In the third-phase model of northern development, the solution to the problems of the "moribund" native economy was to move people out of it and into the modern economy (Map 10.4). Governments of all political persuasions came to believe that mineral and energy development was not only the way the North would make the greatest possible contribution to the national or regional economy, but also that such development offered the only viable economic option to an apparently "impoverished and unproductive" aboriginal population.

There is a considerable overlap between traditionally used areas and those undergoing non-renewable resource exploration and development. In the third phase of northern development, widespread conflict arose over land use, resource management, the allocation of public and private funds, and the concerted encouragement of Native peoples to join the industrial labour force. The extensive and intrusive nature of some resource development activities introduced a new element of conflict, and highlighted to aboriginal people for the first time the practical consequences of non-aboriginal ownership and control of the land. For example, oil and gas exploration was widely dispersed and had broad environmental implications. Seismic exploration crews in the Western Arctic and Mackenzie Delta, their use of explosives and heavy equipment, poor trail construction, and garbage disposal practices, and their occasional interference with traplines and bush camps, raised fears among aboriginal people that the land — their chief source of livelihood — was being threatened. In the Yukon and upper Mackenzie Valley, as roads were built and non-aboriginal residents increased, there was more competition for limited fish and wildlife resources, and aboriginal harvesting was more strictly regulated. This progressive encroachment and restriction on aboriginal land and livelihood was an important factor in aboriginal political development in the early 1970s. The concern of aboriginal people for the maintenance of their land and resource base, and for the way of life it sustains, has been expressed at major public hearings and inquiries across northern Canada, and has been documented in numerous recent studies of northern development conflicts.[13]

The direct experience of aboriginal people with development impacts was paralleled by a growing environmental awareness and concern on the part of the Canadian public. As a result of these and other events, the key developments of the last two decades have included the ongoing settlement of aboriginal claims through negotiations, the development of environmental management regimes, and a trend toward self-government.

Map 10.3 Settlements and Land Use

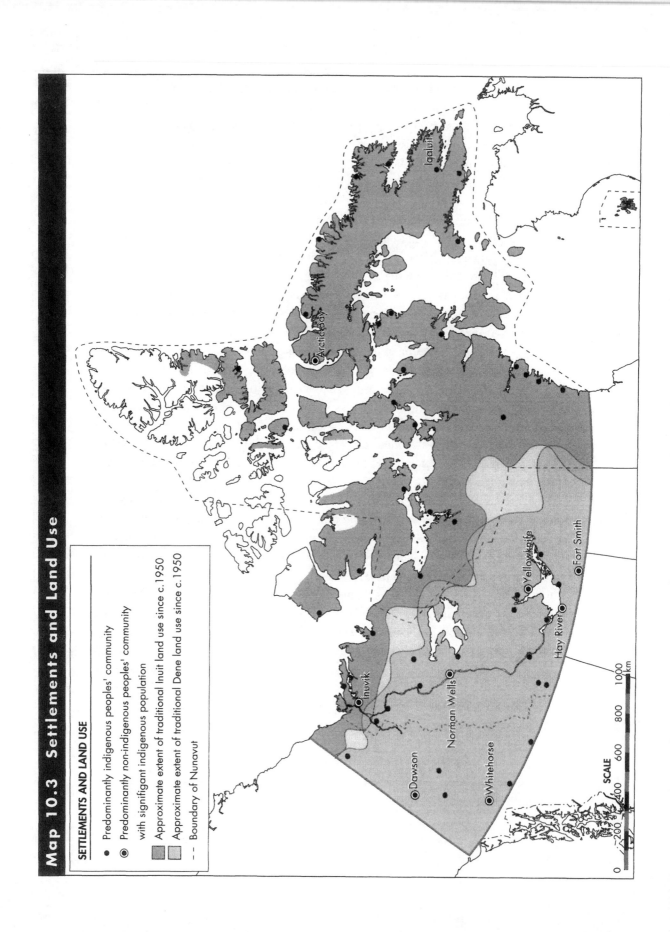

SETTLEMENTS AND LAND USE

- Predominantly indigenous peoples' community
- ⊙ Predominantly non-indigenous peoples' community with signifigant indigenous population
- Approximate extent of traditional Inuit land use since c.1950
- Approximate extent of traditional Dene land use since c.1950
- -- Boundary of Nunavut

Iqaluit

Arctic Bay

Fort Smith

Yellowknife

Hay River

Inuvik

Norman Wells

Whitehorse

Dawson

SCALE

0 200 400 600 800 1000 km

Map 10.4 Economic Activity, 1996

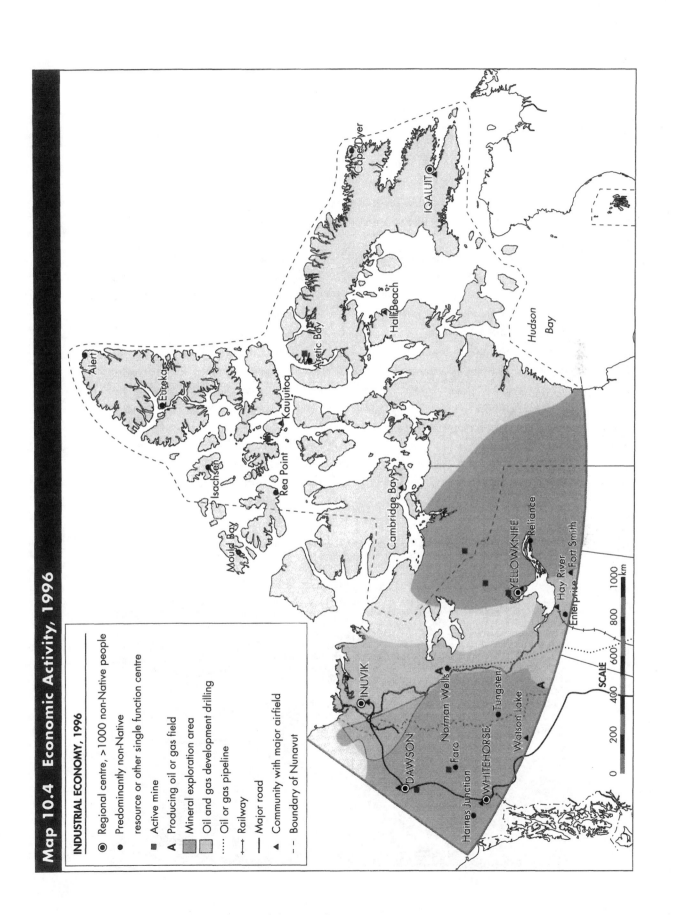

INDUSTRIAL ECONOMY, 1996

- ◉ Regional centre, >1000 non-Native people
- ● Predominantly non-Native
  resource or other single function centre
- ■ Active mine
- ▲ Producing oil or gas field
- ▣ Mineral exploration area
- ▣ Oil and gas development drilling
- ⋯ Oil or gas pipeline
- ⊢⊢ Railway
- — Major road
- ▲ Community with major airfield
- -- Boundary of Nunavut

Cape Dyer

IQALUIT

Hall Beach

Hudson Bay

Alert

Eureka

Arctic Bay

Kaujuitoq

Rea Point

Isachsen

Mould Bay

Cambridge Bay

YELLOWKNIFE

Reliance

Hay River

Fort Smith

Enterprise

INUVIK

Norman Wells

Tungsten

Watson Lake

DAWSON

Faro

Haines Junction

WHITEHORSE

SCALE

0   200   400   600   800   1000
km

Aboriginal Claims

The initial response of aboriginal northerners was simply to oppose exploration and development. The several northern aboriginal organizations founded at about this time, however, soon formulated a more comprehensive strategy. They sought first to establish a legal claim to their territories based on aboriginal title and then to negotiate a settlement of land claims with the federal government. The claim of aboriginal title to northern lands was based on the fact that the Native peoples of the territorial North had never formally surrendered the lands they traditionally occupied.

The principle of aboriginal title and rights was in fact grounded in British colonial doctrine, and articulated in the Royal Proclamation of 1763. The assertion of sovereignty by Britain was actually a declaration of intent to acquire aboriginal lands. Britain acquired underlying title in the sense that aboriginal nations with whom the Crown formed an alliance or offered protection could cede land title only to the Crown, not to other imperial powers, or to individuals. However, a lawful process of acquisition was established whereby "Indian territories" (in which the tribes were to be "unmolested") could be converted to "general lands" in which the land and property regime of the colony would apply. Only after such conversion could the Crown dispose of lands to settlers. Formal treaty-making was the normal legal procedure by which the Crown's underlying title to aboriginal lands was converted to a full and unencumbered title. Until this occurred, aboriginals were legally entitled to the full and free use of their lands, and their internal arrangements of property, tenure, and management were undisturbed. This doctrine was

Yellowknife

the basis of treaty-making until the 1920s, when Canada lapsed the practice. When the issue was revived in the 1970s, the weight of judicial opinion indicated that the principle survived and that Canada was obligated to respect it.[14]

In an historic statement of policy in August, 1973, the federal government acknowledged these outstanding obligations, pledging to negotiate the settlement of claims wherever an aboriginal group could document its traditional use and occupancy of lands not previously surrendered by treaty or other legal means. Since establishing a documentation and negotiation process in 1975, several land claims have been ratified and implemented, chiefly in the territorial North. With respect to lands and resources, the modern land claims agreements provide for the following elements, in exchange for the extinguishment of aboriginal title in each settlement region: title to a limited amount of land (including subsurface rights to a small proportion thereof), preferential or exclusive access to fish and wildlife, participation in resource and environmental decision-making in the form of "comanagement," and cash compensation for surrendered lands. These arrangements include some significant practical benefits either not provided for or not implemented under the historic Indian treaties.

The amount of land retained in the territorial North has varied from about ten to thirty per cent of the original land base (see Map 10.5). Unlike Indian reserves in the provinces, which are federal lands held "for the benefit" of Indians, aboriginal lands in the modern claims agreements are private lands in fee simple title held directly by aboriginal land-holding corporations. Aboriginals are now by far the largest private landowners in the territorial North, as there are virtually no other private lands outside of the communities themselves. Most aboriginal lands are in surface title only, but a developer with subsurface rights must negotiate an impact benefits agreement with the aboriginal owners.

Unlike the old treaties, the hunting and fishing rights provided are exclusive or preferential, and to some extent compensable in respect of damage or loss. Cash compensation for land loss is generally in the tens or hundreds of millions of dollars per settlement area, and is paid to an aboriginal corporate entity on a finite schedule, rather than as small annuities to individuals as in the case of the historic treaties.

The principle of comanagement is perhaps the most innovative and yet least understood element of the modern treaties. It applies not only to wildlife and fisheries — the so-called "traditional" resources — but also to environmental protection and regulation, and land-use planning. The basic structure of comanagement consists of boards or committees responsible for specific management areas such as wildlife, fisheries, impact screening and review, land-use planning, and water management. Members are usually appointed in equal numbers by governments and aboriginal organizations. Geographically, the jurisdiction of these boards covers the entire settlement area, whether in aboriginal, Crown, or third party tenure. The boards are technically advisory to the appropriate minister, and do not replace

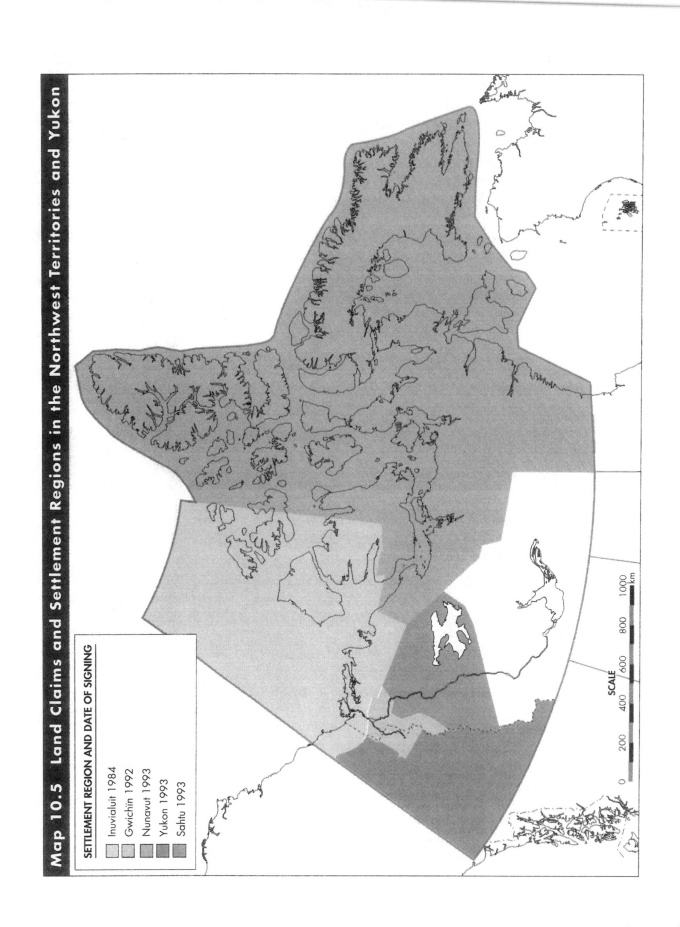

Map 10.5 Land Claims and Settlement Regions in the Northwest Territories and Yukon

SETTLEMENT REGION AND DATE OF SIGNING

Inuvialuit 1984
Gwich'in 1992
Nunavut 1993
Yukon 1993
Sahtu 1993

SCALE

0   200   400   600   800   1000
km

existing government agencies. They are, however, intended to guide the overall direction of policy, and their powers range from making binding decisions, to providing approvals, advice, and research direction.

Resource Management and Environmental Protection

The settlement of aboriginal land claims is part of a longer process of land allocation and management, and environmental assessment and regulation, that has developed in the territorial North since the early 1970s. The federal government continues to be the owner and manager of Crown lands in both territories, and is also responsible for management and control of the marine environment. Statutes regulating the use of land and of renewable and non-renewable resources, and providing for environmental protection, have been enacted or substantially revised in the last twenty-five years, and several new national parks have been created. Major resource developments are now subject to environmental review under the framework of the Canadian Environmental Assessment Act.

Some federal authority over these matters has been delegated to the respective territories, but more significantly, power sharing arrangements have been developed involving the territorial governments as well as aboriginal peoples through the comanagement regimes established under the land claims settlements. As a result, local involvement in, and influence over, decisions relating to land, resources, and environment is substantial.

Self-government

The transfer of government powers and functions from Ottawa to Yellowknife and Whitehorse responded largely to the demands of the local settler populations for responsible government and progress toward provincial status. The key issues at stake between the federal and territorial governments up to the end of the 1970s were the proper timing and appropriate threshold population for the establishment of responsible government institutions and devolution of powers, and the adequacy of the local tax base, or whether and when the territories would have the administrative and fiscal capacity to exercise certain constitutional powers, in line with the other provinces.

The scope of land claims negotiations as originally agreed to by Canada did not include self-government provisions. The Inuit, however, succeeded in tying their land claim to a separate agreement with Canada to create a third territory by dividing the Northwest Territories. The new eastern Arctic territory of Nunavut, because Inuit will constitute a very large majority of its population, will effectively provide for a form of Inuit self-government, even though the fundamental political rights of all residents will be equal regardless of ethnic status. The special rights of Inuit are those specified in the land claim itself, which is separate from the creation of the Nunavut Territory.

Full provincial status for any of the three territories still appears to be a distant step. None has more than the population of a small city in southern Canada, and none is financially self-supporting. Both the federal and territorial governments spend far more on services, administration, and capital expenditures in the North than they receive in the form of taxes and royalties, and this seems unlikely to change in the near future. Both territorial governments have traditionally relied on block transfers from the federal government, but these have declined in recent years, forcing reductions in government expenditure in the North. These are the fiscal conditions under which Nunavut will be established, and the additional challenge will be to create an effective administration drawn primary from a resident population whose educational levels to date have been low.

The North is emerging from its colonial political status within Canada. However, in contrast to the situation at the beginning of the twentieth century when the Prairie provinces were created out of the old Northwest Territories, the northern territories are unlikely to receive significant settler populations, or to achieve provincial status in the foreseeable future, even if there is significant growth in the mining, oil, and gas sectors. The distinctive place of aboriginal peoples, their economy, and their institutions may continue to set the territorial North apart from the rest of Canada as much as its physical, climatic, and ecological characteristics.

NOTES

1. The three major aboriginal populations of the territorial North are Inuit, Dene, and Métis. Inuit is the term that the people formerly known as Eskimos in Canada use to describe themselves. Dene is the term that Athapaskan Indians use to describe themselves (all Indians native to the territorial North are Athapaskans except those of the extreme southern Yukon). The terms Dene and Inuit are the ones now in general use in the North, and are used throughout this chapter, except where the word Indian is, in the context, more inclusive than Dene. There are also people in the territorial North who identify themselves as Métis. This identity is not simply a matter of mixed parentage, as there are many Native people in the North who consider themselves fully Dene or Inuit despite some non-Native ancestry. The Métis have a distinctive culture and history, dating from the early fur trade era, and have tended to play an intermediary economic and cultural role between whites and Indians.

2. Hence the title of the report of the Mackenzie Valley Pipeline Inquiry: T. R. Berger, *Northern Frontier, Northern Homeland* (Ottawa: Supply and Services, 1977).

3. See also R. M. Bone, *The Geography of the Canadian North* (Toronto: Oxford University Press, 1992). For more detailed bioregionalization, see E. Wiken, *Terrestrial Ecozones of Canada* (Ottawa: Lands Directorate, Environment Canada, 1986), and Ecoregions Working Group, *Ecoclimatic Regions of Canada* (Ottawa: Canadian Wildlife Service, Environment Canada, 1989).

4. For a concise discussion of permafrost in northern Canada, and its implications for construction with special reference to pipelines, see P. J. Williams, *Pipelines and Permafrost* (London: Longmans, 1979).

5. This description of the aboriginal peoples of the North applies to the period of discovery from about 1600 to 1900. The prehistoric record of peoples in the North, as revealed by archaeological research, is marked by successive cultures and migrations, but by and large there has been a clear difference between those who inhabited the Arctic and Subarctic environments.

6. L.-E. Hamelin has devised a complex scale of "Nordicity" based on ten criteria, of which six are environmental and four relate to human settlement and development (chiefly accessibility, density of population, and economic activity). On this basis, he has divided the North into three regions: the middle, far, and extreme norths. (He also identified a region called the near North, which includes such agricultural fringe areas as the Peace River and the Clay Belt, and a number of the mining and forestry towns on or near the northernmost transcontinental railway.) These human criteria, however, relate to non-Native perceptions and objectives. Degrees of isolation and inaccessibility are entirely relative. Inuit would no more have thought of themselves as living in the far North than Nova Scotians would think of themselves as living in the far northeast, although that would be an apt description from a Chinese or Australian perspective. See L.-E. Hamelin, *Canadian Nordicity: It's Your North, Too*, trans. William Barr (Montreal: Harvest House, 1978).

7. Cited in Morris Zaslow, *The Opening of the Canadian North, 1870–1914* (Toronto: McClelland and Stewart, 1971), p. 252.

8. Such communities are usually located on Indian Reserves in the provincial north, but reserves were almost never created in the territorial North, and not at all for Inuit, to whom the Indian Act did not apply.

9. Some energy projects, especially large hydroelectric developments in the provinces, were promoted and financed by public utilities, and for a brief period in the 1970s and 1980s, state-owned petroleum corporations played a significant role in Arctic oil and gas exploration, although they did not actually bring any megaprojects to fruition. See also R. M. Bone, 1992, op.cit. 135–137.

10. For an extended discussion of this project and its impacts, see R. M. Bone, 1992, op. cit., 145–155.

11. A mode of production encompasses not only the resources and technology by which a people make their living, but also the social organization and ideological system which combine the factors of production — land (including resources), labour (including skills, knowledge and technique), and capital (fixed investment in tools, buildings, transport, i.e. the means of production) — into a functioning system. It is the means by which a people provide not only for their material needs — food, clothing, and shelter — but also for their social, cultural, and spiritual needs. To understand the dynamics of a mode of production, we must examine not only the factors of production, but also the ways in which human groups manage labour tasks, organize the ownership and use of production factors, and handle the distribution, exchange, and consumption of product and surplus. The concept of "mode of production," and its application to the North, was elaborated in detail in the previous edition of this chapter.

12. For further reading, see N. C. Quigley and N. J. McBride, "The Structure of an Arctic Microeconomy: The Traditional Sector in Community Economic Development, *Arctic* 40 (1987), 204–210; R. J. Wolfe and R. J. Walker, "Subsistence Economies in Alaska: Productivity, Geography, and Development Impact," *Arctic Anthropology*, 24 (1987), 56–81; Wenzel, *Animal Rights, Human Rights: Ecology, Economy and Ideology in the Canadian Arctic* (Toronto: University of Toronto Press) 1991; and P.J. Usher and M. S. Weinstein, *Towards Assessing the Effects of Lake*

Winnipeg Regulation and Churchill River Diversion on Resource Harvesting in Native Communities in Northern Manitoba, Canadian Technical Report of Fisheries and Aquatic Sciences, Report 1794, 1991, Winnipeg.

13. For a review of these, see P. J. Usher, "Northern Development, Impact Assessment, and Social Change," in *Anthropology, Public Policy and Native Peoples in Canada*. N. Dyck and J. B. Waldram, eds. (Montréal and Kingston: McGill-Queen's University Press, 1993), pp. 98–130.

14. For further reading on aboriginal rights, treaties, and land claims, see W.R. Morrison, "Aboriginal Land Claims in the Canadian North," in *Aboriginal Land Claims in Canada*, K. Coates, ed. (Toronto: Copp Clark Pitman, 1992), pp. 168–194; and P. J. Tough and R. M. Galois, "Reclaiming the Land: Aboriginal Title, Treaty Rights, and Land Claims in Canada, *Applied Geography* 12(2) (1992) 109–132.

chapter 11

Cole
Harris

Regionalism
and the Canadian
Archipelago

This chapter has three parts. It deals first with the underlying structure of the territory settled by Europeans which eventually was defined as Canada. This is done to establish the frame with which the conceptualization of this country has had to contend. The second part deals with sentiment and regionalism. It is, really, an inventory and perhaps, to some extent, an explanation of feelings about the parts and the whole of this country, including those between core and peripheral regions. Finally, the chapter deals with the relationship between these regional sentiments and the future course of Canadian development. Together, these parts stand in summary as a perspective on Canadian regionalism, and of some of the roles played by core and periphery in shaping Canada's evolving regional character.[1]

THE CANADIAN ARCHIPELAGO

ecumene = inhabited land

The political map of North America sustains the illusion that Canada is a continental giant spanning 70 degrees of longitude and some 40 degrees of latitude; whereas on any long, clear-night flight, this Canada dissolves into an oceanic darkness spotted by occasional islands of light. These lights mark the lived-in Canada, the Canadian ecumene, an island archipelago spread over 7200 east-west kilometres. Between the islands in winter are snow and ice, and in summer are rock, muskeg, endless tangles of black spruce, and black flies — little possibility for the innumerable cabotage that over the centuries served the islands, peninsulas, and coastal plains of Braudel's Mediterranean.[2] There have been routes through the rock: first rivers navigated by canoes, then railways, and finally highways; but railways began to make connections just over a hundred years ago and highways considerably more recently, while canoe routes, never followed by many people, preceded substantial white settlement. For the most part, the spaces between these islands have been hard to traverse and harder to use. Like Newfoundland outports, the islands back into rock.

The problem, if that is what it is, began to come into European focus in the sixteenth century. Verrazano, sailing off Carolina, thought he glimpsed Arcadia to the west; Cartier, along the shore of the Gulf of St. Lawrence, found himself face to face, as he said, with the land God gave as his portion to Cain. He had met the crystalline edge of the Canadian Shield. Over the years little enough would temper this stark, geographical reality. The Shield edge looms over Québec, is on the near horizon in Montréal, and is only just beyond it in Toronto and Winnipeg. On the plains its limiting place is taken by the short summer of a severe continental climate and, farther west, by the Cordillera. To the south, as time went by, a political boundary evolved, most of its length the negotiated balance, at different dates, between American perception of settlement opportunities and British interest in the fur trade. North and south, these were the confines of almost all the Canadian archipelago: together they defined its territory and constricted its possibilities.

Early European settlement of these northern islands had a particular inadvertency. On return voyages to Europe, fish were more profitable than fishermen, and men were left behind to struggle through bleak winters — the origin of Newfoundland's shifting, male population of the seventeenth and eighteenth centuries. Descended from a few immigrant families, a considerable Acadian population created cultivable niches out of tidal marshlands around the Bay of Fundy, lived in hamlets of close kin, and until its expulsion by the British early in the Seven Years' War, was little affected by the fluctuations of French and British jurisdiction to which its exposed location made it vulnerable. Along the lower St. Lawrence there was more bureaucracy and more of Europe. Québec and Montréal approximately reflected the social gradients of French provincial towns, but most people lived in the countryside and depended on a primarily subsistence agriculture. They had little to do with the towns, with the fur trade for which the colony had been created but which could not employ its growing population, or with the larger world of the North Atlantic.[3] Overall, there was no master plan, no vision, as in New England, of Old World regeneration overseas, and next to no interaction among these somewhat adventitious French and British beginnings in the northwestern Atlantic.

In the early nineteenth century the islands received a massive British migration, principally Irish and Scottish, propelled by clearances, famine, and rural pauperization, and by the technological and demographic changes of early

industrialization. Most of these immigrants passed through the seigneuries bordering the lower St. Lawrence to settle in southern Ontario and establish an English-
speaking, largely Protestant population that viewed its Roman Catholic,
French-speaking neighbours with a large measure of contempt. Others stayed on
the Atlantic islands to create the Irish and English shores of Newfoundland, the
predominant Scottishness of northern Nova Scotia, and the pockets of Irish,
English, and Scots that together with returned Acadians comprised the populations
of New Brunswick and Prince Edward Island. All of these newcomers, save the
Acadians, came to British colonies from a British hearth. Their nostalgia and, when
they could write, their letters, returned to Britain. Other islands of the Canadian
archipelago were by and large outside their experience.

Their descendants, together with those of the older French-speaking population along the lower St. Lawrence, soon faced a common predicament. The islands
were small, their agricultural possibilities circumscribed, and as numbers multiplied in still rural, pre-industrial societies there was soon a shortage of land. The
pioneer fringe ran into rock. There was no western safety valve; there were only the
granite Shield and the other already settled British North American islands. The
surplus young faced the choice of striking north into the land God had given to
Cain or south into the United States.

Early in the nineteenth century the drain began from Québec to New England's
first factory towns and then, later in the century, to the lumber camps of the
American Midwest. Driven by demographic pressures, this was an annual migration first of hundreds, then of thousands. Worried nationalists could not stop it.
With the blessing of priests, other young French Canadians, who would make the
North serve for the West Québec did not have, went to the lumber camps, sawmills,
and podzols of the Canadian Shield — a brutal encounter of defenceless people
with the devastating realities of an almost impossible agricultural land and of early
industrial capitalism. In Ontario, population pressures built up a few years later.
Government and manufacturers promoted northward expansion. Colonization roads
were built into the Shield. By the 1850s there was official interest in the Red River
settlement, 1600 kilometres away, but ordinary people had neither means nor
desire to go there, and few would tackle the Shield. Rather, they slipped across the
international border, creating by 1900 a plume of Ontarians and their descendants,
as numerous as the Ontarians themselves, stretching westward through the United
States to the High Plains. From the Atlantic islands, migration went to the "Boston
States" as, vicariously, New England became their missing West. None of these
migrations mixed the different populations of British North America. South of the
border, immigrants were being absorbed into a larger America; north of it, island
societies that now exported people bypassed the mixing effects of the migrations
they had launched.

Then, about the time the American historian Frederick Jackson Turner
lamented the closing of the American frontier,[4] a Canadian West finally opened for

settlement. When the Canadian Pacific Railway (CPR) reached Winnipeg in 1883, the fur trade was already gone from most of the prairies. Although a long way off, expensive to reach, and in a different physiographic realm of uncertain agricultural potential, a large new island, a possible counterpoise to the attraction of the United States, was available for settlement. In both Québec and Ontario, this West found dedicated promoters, although enthusiasm dwindled among French-speakers in Québec after the Protestant outburst over the Riel Rebellion and the collapse in Manitoba of French educational and linguistic guarantees. By 1900, leaders in Québec were almost unanimous that Québec should keep its young, or, failing that, should colonize a contiguous northern Ontario (or even, some felt, a contiguous New England); they turned deaf ears to western pleas for immigrants who might have stabilized the region's Catholic, French-speaking future.[5]

As it turned out, neither Québec nor Ontario would be reproduced on the Canadian prairie. There would be more of the latter than of the former — some enclaves, such as southwestern Manitoba, were overwhelmingly Ontarian — but the Canadian prairie was settled over a short generation before World War I, when migrants from the eastern Canadian islands mixed with immigrants coming directly from the British Isles, with the northern fringe of the late nineteenth- and early twentieth-century peasant migration to North America from central Europe, and with a wave of American settlement that pushed northward along the eastern flank of the Rockies. Languages spoken in the new ethnic ghettoes in American cities appeared on the Canadian prairie, an element in a new mix of eastern Canadians, Britons, Americans, Germans, Scandinavians, Russians, Ukrainians, and Poles; of Lutherans, Eastern Orthodox, Mennonites, Doukhobors, and Mormons as well as

Saskatchewan landscape

Roman Catholics and Protestants of British origin. In British Columbia the mix was different again: much more of the British Isles, particularly of England, much less of continental Europe, a good deal of Ontario, something of the Atlantic islands, and, there on the Pacific Ocean, elements from the Orient.

Thus were settled the islands between an implacable north and the United States. There was no continuous, expansive Canadian experience with the land. What was common was the lack of continuity imparted by the close limits of confined lands. Settlement proceeded in patches, island by island. One island would fill up, then people would emigrate, south more often than north because the United States was more inviting than the Shield. Until the last hundred years, there was no settlers' West. The next Canadian island was inaccessible or occupied, and when a West finally opened, the eastern islands would be partially represented and much diluted there. The process of Canadian settlement has imparted striking discontinuities. Different islands were settled at different times with different technologies and economies by people from different backgrounds. Considered overall, the archipelago was settled island by island from Europe; it did not expand westward from an Atlantic beginning. Ranald Macdonald, Ralph Connor's Man from Glengarry who left his native Ontario to chop Douglas fir and extended a new Dominion in British Columbia, represents one strand of the Canadian settlement process; but Maria Chapdelaine, torn between the lure of the United States and the rock of Peribonka, represents its overriding dilemma.

This pattern of settlement sharply differentiated the Canadian experience from the American. In the United States, the land was perceived as a garden as readily as wilderness, and it attracted far more settlers, and focussed European dreams.[6] There, eastern seaboard beginnings could migrate westward across the piedmont plain, across the narrow and substantially inhabitable Appalachians, across the rich soils of the Mississippi basin to the desert margins, situated 3200 kilometres inland. These were the first major environmental obstacles to an expanding agrarian civilization. There the West — unoccupied land to the west suitable for cultivation — was a stimulant for three hundred years, to the point that the essence of the American experience could be plausibly interpreted as a succession of waves on a succession of westward-moving frontiers. There a Lockian liberalism, popularized by Thomas Paine and reinforced generation after generation by individual opportunities (for many if not for all) in a bounteous, on-going land, could become a pervasive ideology because the experience of so many seemed to attest to the rewards and the virtues of individual enterprise. And there, as different streams from the initial settlements along the colonial seaboard, augmented by newcomers from Europe, moved west across the Appalachians, different ways met and substantially merged. As it gathered momentum in the late eighteenth and nineteenth centuries, the American occupation of an essentially welcoming land had the capacity to mold different peoples into a relatively homogeneous culture as it spread them over an astonishing area. In Canada, all of this was checked by the land's ineluctable niggardliness.

In the United States there was also more North American time, more temporal as well as spatial continuity.[7] Behind the Republic lay almost two centuries of colonial America, ample time for English ways to evolve in different, New World directions that would be marked by the great nation-building events of revolution and independence. In Canada, only the French-speaking reach back like this in North America, and their early North American evolution was capped by the deportation of almost 10 000 Acadians, and by the conquest of Canada — by the imposition of limits rather than by the opening of autonomous beginnings. English Canada is a product of the nineteenth century, of the Victorian age. In the West some beginnings are still remembered. Such recency harbours lingering memories of homes elsewhere, and blurs new circumstances. Hence, of all Canadians it is the French Canadians who have recognized themselves most clearly as North American, and who have had fewest doubts about who and where they were.

However little they recognized what was going on, the different peoples on the British North American islands soon ceased to be European. Emigration undertaken with no radical intent by people who, for the most part, sought no more than a living, had unexpected implications for Europeans in non-European settings. Put briefly, these settings tended to accept some European ways and to reject others, and thereby to create societies that were selections reduced from the European social whole. Characteristically this selection pared back complex European hierarchies of honour, status, rank, and deference; and obliterated the fine spatial texture that in pre-industrial Europe was reflected in local cultures.[8]

For many years, land on these islands was relatively inexpensive, and because ordinarily people could substitute land for labour, labour was relatively dear. The development of new resources required strong backs, practical ingenuity and, in some cases, capital, but not social polish or refined learning. These circumstances provided opportunity for peasants, artisans, and a few of the middle class, but discouraged gentility. At the same time, new and often strikingly uniform work environments diminished European occupational labels and their implicit social sorting. European social hierarchies were being drastically simplified. Something of European gentility surfaced in the military, in civil bureaucracies, among the commercial elite and even, here and there, in the countryside — as with Susanna Moodie's incongruous presence in the Ontario bush and with orchardists who struggled to grow fruit and play polo on British Columbia's mountainsides. Something of the European occupational range survived in the towns. Overall, European gentility had been deprived of a landed base and of a context for its manners, and ordinary people had lost sight of most of the complex European hierarchy of status and deference.

On the other hand, a northwestern European sentiment of the family, a concept of private property, and the commercial values in the middle class, had found congenial settings. Where markets were poor and there was little to attract capital, as for years on the marshlands around the Bay of Fundy and on most seigneuries along the St. Lawrence, there emerged strikingly egalitarian rural societies of semisubsistent families. Over the years, such societies developed dense networks of kin and local traditions that amalgamated elements of the different regional backgrounds of founding populations into distinctive folk cultures. Wherever commercial opportunities were greater, as they were in Newfoundland from the beginning, or as they became in habitant Québec early in the nineteenth century, the social gradient steepened, but this gradient was defined by the market rather than by custom. The same transformation, "the great transformation" Karl Polanyi called it,[9] was affecting European society, but here the pace of change was accelerated and the social dismemberment was more complete. The few deported Acadians who reached France and were settled on the Île de Re near the port of La Rochelle from which some of their ancestors had probably embarked, found themselves amid alien people. Like the other people on the Canadian islands, they were no longer Europeans, and most of them soon left for Louisiana.

While diverging socially from Europe, these North American islands all became British possessions. They were parts of the same empire, each connected by governors and colonial administrations to Westminster and the Crown, that is, to a European core. Each evolved political institutions derived from British models, and each depended on the presence of the British military. Whether a source of pride or of consternation, the British connection was a constant. In some expatriate minds, empire acquired a certain fanaticism, as it did on the 12th of July or, more gently, in the garden that Emily Carr's father molded into a patch of England overseas. But, as we shall see, the imperial fervour that led harmlessly back to Britain rasped across the grain of British North America. What was empire to some was colonialism to others. Islands that would rather have danced to different tunes were on strings held by the same puppeteer.

And finally, there were people on all the islands, particularly people of some eminence, who would not believe that the surrounding rock and climate were real. Settlement would have a larger outlet and some of the islands would merge. The mistake was easily made when there had been little scientific surveying, when climatology was more theoretical than empirical, when the air was full of the American West, and when one lived in the towns. Indeed, there was evidence that marginal environments could be settled. Irishmen from the bogs and rocks of Connemara, Highlanders from two- or three-acre crofts and a collapsing kelp industry, and French Canadians were all driven into the Shield where, given their prior experience and the unknown limits of untested land, long-term prospects could seem reasonably bright. The mirage of abundant agricultural land would persist into the twentieth century. Most spectacularly and most erroneously, it would accompany the National Transcontinental Railway (the Grand Trunk Pacific) across Canada from Moncton, New Brunswick, to a new town, Prince Rupert, hewn out of an Indian reserve at the mouth of the Skeena River, a railway created by visions of a northwest passage to the Orient and of a continuous band of agricultural settlement linking the islands of British North America.

'Such was not to be. The geography of Canadian settlement remained disjointed and discontinuous. This is the underlying structure of Canada on which, as Braudel would have said, economic circumstances and political events would work their more ephemeral passage. Another Frenchman, the anarchist Elisée Reclus, surveyed the map of Canadian settlement and introduced the Canadian chapters of his *Géographie Universelle* (written c. 1890) in the following way: "The vast stretch of lands occupying all the northern section of North America and politically defined as the 'Dominion of Canada' constitutes no distinct geographical unit." Canada, he wrote, had a "fantastic frontier" with the United States. Only in the St. Lawrence–Great Lakes area was the population "dense enough to constitute really independent groups and autonomous centres of political and social life." Little political importance could be attached to a "precarious political frontier liable to be effaced by the least change of equilibrium."

THE GEOGRAPHY OF CANADIAN REGIONALISM

Elisée Reclus found no relationship between the political map of Canada and the pattern of Canadian settlement. The one was transcontinental, the other was local. The one was a geopolitical vision, the other was the frame of ordinary lives. Indeed, as has been argued, Canada is a composition of islands. From island vantage points, outlooks have been bounded, local feelings intense, and ignorance of other circumstances considerable. Emotional lines of attachment often led back to distant homelands. But over and against this island Canada are the power, spatial range, and integrating capacity of modern technologies; a transcontinental political territory that has survived far longer than Reclus thought it would; and on all the islands, if in varying intensity, a good deal of sentiment for Canada. Another level has been superimposed on the islands, and the mix has created a tangle of sentimental attachments. To deal with the components of this tangle, we shall start with the smallest units of regional feeling.

In our electronic age, it is worth remembering that local feeling in this country did not first develop with the province or, earlier, the colony, both political abstractions well removed from daily life, but with the settlement, the place where people lived and whose horizons they knew. In Newfoundland and much of Nova Scotia these settlements were outports; in New Brunswick and on Prince Edward Island they were small towns or farm communities; in Québec they were the rangs and parishes that were the principal units of sociability beyond the family; in Ontario they were farmhouses along concession-line roads and a local service centre; on the prairie they were rural neighbourhoods, often strengthened by ethnicity, of farm families on quarter or half sections; in British Columbia they were fishing, logging, and mining camps. From Buckler's Annapolis Valley to Hodgins' Vancouver Island, this grain abounds in Canadian letters for it was the immediate horizon of everyday experience.[10] On this scale, nature and people were known. In the older settlements of the Maritimes, the rhythms of the land, the traditional ways that earned a living, and the people who lived nearby were the context of most experience. Even today, genealogical conversation is a Maritimes staple, a reflection of communities whose people have known each other through the generations. In the West such conversation is rarer for the local texture has been different, having less of custom and the generations and more of movement, technology, markets, and memories of other places. In either case the settlement has been a tangible world, the home of one's people and of one's peoples' people or, at the other extreme, a place to boost because it was one's point of attachment in a new land and because its fortunes were substantially one's own. Beyond the local settlement lay more amorphous scales of attachment, not so readily experienced or accepted, usually somewhat abstract and often threatening.

Cities provided the earliest and most direct connections beyond the local settlements. At first these ties were with a European core, but after a time some cities on the Canadian islands organized internal hinterlands of considerable size. Toronto's rise to metropolitan dominance as it displaced Kingston and other lakefront towns reflected the growing integration of the Ontario peninsula. Imported goods and services passed through Toronto's warehouses, insurance brokers, and merchants to be distributed throughout the province; politicians and bankers congregated there; secondary manufacturing concentrated in Toronto and in Hamilton nearby; and many an ambitious young man, with or without his cat, would go to Toronto as Dick Whittington had once gone to London. Broadly similar developments took place elsewhere. Early in the nineteenth century, fish merchants from

Highway 407 in Ontario uses transponder technology

Poole and Dartmouth, their operations disrupted by the Napoleonic Wars, joined by general trading partnerships from Greenock and Liverpool, moved to St. John's, the city that henceforth would dominate the Newfoundland fishery and the outport economy. In the West, the first transcontinental railway would dictate urban primacy. After the CPR arrived in 1883, Winnipeg dominated the eastern prairie, and on the Pacific the railway created a western port that quickly displaced Victoria and soon would service most of the isolated settlements of the British Columbian coast and interior.

Regional metropolitan dominance brought about an increased circulation of goods, people, and ideas, thereby creating a framework for the emergence of regional awareness. The settlement acquired a context, an intermediate frame of reference between its own localness and a distant outside. The city might be visited now and then; its news, and perhaps its newspapers reached far afield; and it provided a channel for contact between the various places it served. On the other hand, urban dominance could easily appear as exploitation, as grandeur bought at local cost. As lesser towns in the urban system slipped into relative stagnation, as local enterprise was eclipsed and local economies became increasingly dependent, as lines of credit extended out from the city and perquisites seemed to be bestowed on it, this feeling was almost inescapable. The perception of injustice acquired a spatial dimension from which none of the regional metropolitan centres was immune. Halifax would not be popular on Cape Breton Island, nor Saint John along the Gulf Shore.

Provincial boundaries have corresponded poorly to the hinterlands of regional cities. In the West they were arbitrary lines drawn before substantial settlement. In the East they bore some relationship to colonial territories and to prior settlements; but with the exception of a small part of the southern boundary of Québec none of these political lines had been drawn to reflect cultural regions. They were lines of cartographic or administrative convenience suggested, usually, by the configuration of the land and the border with the United States. Still, local settlements developed within a province, and this meant, as time went on, that they were exposed to the same provincial politicians and laws; to the same provincial capital; and to an identifying name. Over time such exposure would foster a sense of provincial identity.

Provincial feeling would be strongest among French-speaking Québecers. They lived where their ancestors had lived two hundred years before Confederation; were well aware that their civil law, language, and religion were unprotected outside the province; and were confronted by an alien population whose presence only reinforced their definition of themselves. If sometimes they have been tempted by larger visions, when the chips of survival were down, Québec has stood out as the largest political unit where a people were a majority and where institutions for collective defence were at hand. Inside Québec, therefore, French-speaking sentiment readily crystallized around the province. In

The Shield

Newfoundland, where separate Dominion status is well remembered, where after the Irish and southern English migrations of the early nineteenth century there has been little subsequent immigration, and where the elemental context of rock, settlement, and sea is everywhere apparent, the province corresponds to a strong sense of separate identity. Much the same is true of Prince Edward Island. Elsewhere, provincial feeling has been weaker. It has been diluted in New Brunswick and Nova Scotia by ethnicity, isolation, and the tension of local metropolis–hinterland relationships; in Ontario by a tendency to equate the centre with the whole — to think of English Canada rather than Ontario; and in the West by recency. British Columbians, for example, are recent immigrants from different backgrounds who have converged on a complex physiographic realm dominated by different, rapidly changing local economies. Inevitably, their sense of themselves as British Columbians has been inchoate — a surrounding natural magnificence, a certain frontier optimism ("western spirit" some writers called it), and a certain unconventionality in a new and relatively benign setting, but nothing like the unconscious, generational identification of Prince Edward Islanders and Newfoundlanders with their islands, or of most Québécois with Québec.

Beyond the provinces are the larger regions: Central Canada, the Maritimes, the Prairies, the West, the East, the North. Clear enough from afar — Canadians apply these names to parts of the country where they do not live — these regions are less apparent from inside. They have not been reinforced by political organization, functional economic integration, or common settlement history. Yet the names locate blocks of the country in relation to each other and are part of the country's vocabulary of spatial ambition and resentment. In the Maritimes and on the Prairies, common landscapes, common experiences with land and economy, and common perceptions of an impinging outside give these regions additional meaning for those who live there. From time to time, as during the Maritime Rights movement of the early 1920s, there has been talk of regional political consolidation. Overall, however, such movements have been diffused by more local frames of reference, by the lack of functional economic integration at the regional scale, and by the growing visibility of the provinces. Of the other broad regions, the West, Central Canada, and the East have only fuzzy locational meanings. Although home of a three-centuries-old, tri-racial society created by the fur trade, the North sprawls across provincial boundaries; and until very recently its small, scattered settlements have been largely unknown to each other. As identity develops in the North, it probably will have more to do with race than with "northernness," as the impending creation of Nunavut suggests.

Finally, the terms French and English Canada are clear enough when used to indicate language, but are ambiguous foci of sentiment. French Canada comprises Canadiens and Acadiens — people with the same European roots but separated by

Igloo

more than three hundred different North American years — and French speakers in the West, most of whom are many generations from Québec. Accents and memories differ, although outside Québec there is a common sense of being part of a French-speaking minority in Canada. English-speaking Canada is more diffuse. From its long-established centre in Ontario and Montréal, it can look relatively firm, but from anywhere else English Canada looks like a collection of different English-speaking peoples with a common unease about the centre. Because neither French nor English Canada has a political base (though increasingly Québec serves as the base for the former), the terms are unrepresented abstractions in Canadian political life.

Of singular importance is the fact that, among all these different scales of Canadian sentiment, the provinces have become increasingly dominant. In effect, they are now the repositories of the country's fragmented structure. At Confederation the local settlement was still the predominant scale of Canadian life, but settlements that once provided definition and defence for traditional ways have been overridden by modern transportation and communications. Their isolation and stability have largely gone; they survive in some urban shadow of an urbanized and industrialized society. In such a society, horizons are broadened and the local defence of custom is superseded. The state assumes a growing symbolic and practical importance. In this situation, the Canadian province, with its constitutionally defined power, its growing political history, and a location that bears some relationship to the fragmented structure of the country, replaces both the local settlements that no longer support Canadian life, and the broader but amorphous regions that have no clear political definition. The provinces are crystal clear. Their territorial boundaries are precise. For all the arguments, their powers are explicit. Their scale is supportable within modern technology. As political territories that reflect something of the country's island structure, they enormously simplify Canadian reality, and it is this simplified and thereby politically more powerful regionalism that now confronts the concept and the sentiment of Canada.

In considering Canada as a whole we must begin, as George Grant has reminded us, with the relationship between technology and empire.[11] There can be no doubt about this connection. The political boundary with the United States grew out of the transcontinental momentum of the fur trade as contained by the transcontinental momentum of the American settlement frontier, and was sustained by railways and factories: the Maritimes and the Intercolonial; the West and the CPR; factories in central Canada and markets across a continent; eventually, airplanes, a highway, and the CBC. To say that Canadian confederation requires this technological arsenal is to understate the case. Confederation became conceivable within an industrial technology and, in good part, is one of its byproducts.

Industrial technology has the capacity to integrate the bulky products of a large area within a single market. Associated with such spatial integration are metropolitan centres where there are clear economies of agglomeration and distribution, and extensive resource and market hinterlands. The railway and the factory would impose this structure on the Canadian archipelago. In the late nineteenth century, a continental market (no tariffs along the United States border) would impose a north-south economic orientation and would strengthen primary and weaken secondary industry throughout the Canadian Islands. Market areas on the scale of the individual British North American colonies (tariffs between the colonies as well as along the United States border) would thwart grander entrepreneurial ambitions, undermine the scale economies of modern technology, and, in relation to any larger market area, depress average living standards. Implicit in Confederation and explicit in the National Policy that followed, was the decision to create a Canadian market, a decision that would shape the spatial structure of the Canadian economy for the next century. Protected from the United States by tariffs, the metropolitan centre — the core — would stabilize in the St. Lawrence–Great Lakes Lowlands where most of the market was located and where there was optimum overall access to the hinterlands — the peripheries — to east and west. The rest of the country would consume the manufactures of the core, and would supply it with some raw materials. This structure could be intensified by public or private policy (for example, by changing freight rates) but, given the pattern of Canadian settlement at Confederation and the spatial character of an industrial economy, it followed in all essentials from the one decision to create a Canadian market.

Such a spatial economy would have an obvious bearing on regional sentiment. Those at the core would tend to feel expansive about the country on which their economy relied and over which their institutions exerted much influence. Those on the peripheries would be suspicious of the core, their suspicion stemming from jealousy, from a sense that local circumstances were controlled by the uninformed from afar ("Ottawa," said a mayor of Vancouver in the 1930s, "is 2500 miles from Vancouver, but Vancouver is 25,000 miles from Ottawa"), and from the conviction that, by being forced "to sell cheap and buy dear," they are subsidizing the Industrial Heartland and absorbing the cost of Confederation. What was a "National Policy" in central Canada would easily be interpreted in the Maritimes as "Upper Canadian imperialism," and in the West as the dark manipulation of James and Bay streets.

From a Prairie vantage point, "big interests" and "special privilege" lived in the Industrial Heartland. They were assumed to control the CPR and hence freight rates, grain elevators, and boxcars — the lifeline to the outside. They were assumed to control the insurance companies, the banks, the grain exchanges — the financial infrastructure of the wheat economy. Such talk often filled the prairie air. In a dependent primary resource economy, class conflict assumed a clear spatial direction. Eventually collective action would build grain elevators and, when goaded by depression and drought, would create radical political movements, but in an economy focussed on a distant core, power would not be transferred easily to the periphery. Neither capital nor votes was there. The lesson of the reciprocity election of 1911, when Ontario voted for protection, would echo through the West for the next seventy years.

For a time, the relationship of the Maritimes with central Canada seemed less lopsided. In 1867 the Maritime economy was still vigorous, relying on fish, forests, agriculture, and a wooden merchant fleet largely locally built, that made the new Dominion the fourth largest shipping nation in the world. From this base and with abundant coal and iron and ice-free Atlantic ports, the Intercolonial Railway and the National Policy seemed to offer the Maritimes a continental market. During the 1880s, secondary industry — cotton mills, sugar refineries, rope works, iron and steel mills — developed rapidly in Nova Scotia and New Brunswick. Even so, trade from central Canada to the Maritimes was twice that in the other direction, and by the 1890s the Maritime provinces' vulnerability to their new continental connection was becoming apparent. The underpinning of the traditional regional economy — staple production — was threatened by American tariffs, by resource depletion, and by technological change, while the new industries scattered through the local maze of the Maritimes and remote from continental markets were not able to take up the slack. The railway had not imposed a primate city on the Maritimes as it had on British Columbia. Maritime enterprises were scattered and undercapitalized, and by 1895 many had been bought out by Montréal competitors. Control of the banks shifted west as well, and workshops and craftsmen, the life of many Maritime settlements, gradually succumbed to the competition of mass-produced goods from central Canada. With the creation of the British Empire Steel Corporation in 1920, Maritimers had lost control of every significant element of their secondary industry. Symbolizing the change all too clearly, the head office of the Intercolonial, the railway that once had brought a vision of industrial growth, was subsumed within the new Canadian National Railway and moved to Toronto. Even the Acadians, usually more interested in culture than in railways, felt that "la pacte de la confederation" had been broken. More gradually than Westerners, Maritimers, too, had come to perceive that the "big cities" and "big interests" were in the growing Industrial Heartland, a perception that circumstances would not alter.[12]

Not everyone in the core would feel expansive about a larger Canada. French Canadians, who worked in the factories but did not own them, would have no entrepreneurial enthusiasm for a transcontinental country and a good deal of cultural suspicion of it. The commercial empire of the St. Lawrence was not in their hands. Many clerics denounced a Protestant world of rationalism, individualism, and materialism, and some of them tended to reach out not to a Canada of railways and progress but, in an ecclesiastical dream that returned to the continental territory of the French fur trade, to the collapse of Protestantism and triumph of Roman Catholicism that Louisiana and Québec, as Spanish and French America, embraced on the banks of the Mississippi.[13]

But for the English-speaking population in Montréal and Ontario, a British Dominion from sea to sea would do two very satisfactory things simultaneously: it would reinforce their traditions and it would expand their markets. For them, the West, as it came into focus in the second half of the nineteenth century, was to be settled. Sent from Canada in 1857 to report on the prairie environment, the geologist H. Y. Hind found both a northern extension of the Great American Desert that "is permanently sterile and unfit for the abode of civilized man," and also "a broad strip of fertile country" north of the desert. It was, he concluded, "a physical reality of the highest importance to the interests of British North America that this continuous belt can be settled and cultivated from a few miles west of the Lake of the Woods to the passes of the Rocky Mountains."[14] Twenty years later, John Macoun, botany teacher from Belleville, Ontario, asserted that Hind had found a desert only because "enormous herds of buffalo" had eaten the grass. According to Macoun, much of Hind's sterile area was arable and most of the rest was good pasture. Far more than a fertile belt, there was a vast tract of land, almost 250 000 square miles (647 500 km^2) in Macoun's calculation, fit for agriculture. This was what Ontario and the CPR wanted to hear. There was, after all, a British garden to the west. Canada could expand from sea to sea as a British Dominion, part of a British empire and of all that implied for cultural continuity and economic growth.

For most English speakers in the core, an image of a British Canada from sea to sea seemed authentic and just. It was rooted in a Loyalist tradition; in the trans-Atlantic migrations that had settled Ontario and the Maritimes; in a widely felt sense of belonging to a British parliamentary tradition and empire; and in unease about the United States. It defined Canadians as separate in North America, made sense of the border, and held out a transcontinental vision of the Canadian future. A few argued for an imperial federation of Canada and Britain in which, they thought, power would shift from the Mother Country to her vigorous offspring. In such a Canada the crass pursuit of gain would be checked by men of breeding, leaders of a stable, conservative society that stood apart from the acquisitive, republican rabble to the south.[15] Far from this Tory vision, but equally British in sympathy, were the many thousands who knelt at their initiation into the Orange Lodge to swear "that I will be faithful and bear true allegiance to Her Majesty Queen Victoria and to her lawful heirs and successors . . . so long as she or they shall maintain the Protestant Religion," and who went on to declare that they had never been Papists, that they would never educate their children in the Roman Catholic faith, and that they would "steadily maintain the connection between the Colonies of British America and the Mother Country, and be ever ready to resist all attempts to weaken British influence or dismember the British Empire." "So help me God," they concluded, "and keep me steadfast in this my Orangeman's obligation."[16] The Lodge consistently advocated that Newfoundland enter Confederation, and worked vigorously to establish lodges in the West. Like other British imperialists, Orangemen had found the Canadian identity; their challenge was to impose it on Canada.

Fishing village in the Maritimes

Such sentiments linger to the present, but as time went on English Canadians — especially those in central Canada — began to replace them with a vision of Canada's northern character and destiny. We were not so much British as northern, a people invigorated by arctic ozone in a bracing environment that generated freedom.[17] Decadence and slavery were southern. Such views reflected Darwinian strands in late nineteenth century thought, the contemporaneous American discovery of the wilderness, and the growing urbanization of Canada; and they were enormously convenient. They differentiated Canadians from Americans while placing Canadians on the side of virtue. They incorporated all Canadians, even Catholic French-speaking ones. Who were French Canadians, after all, but Normans, and who were Normans but Vikings? Northern credentials could not be better. And, as Canada moved toward independence from Britain, they provided an autonomous vision that for two full generations — until sometime well beyond World War II — would give many English Canadians their most explicit sense of Canada.

The discovery of "northernness" could take many forms. Artists who eventually formed the Group of Seven were inspired by the fringe of the Canadian Shield, and would leave their Toronto jobs as commercial artists for weekends in Algonquin Park and for vacations on the north shore of Lake Superior — the land settlers refused. Bold canvases depicted the wilderness; and their painters, and eventually many of their viewers, felt that they captured something at the heart of Canada. On the canvases of A. Y. Jackson, the farm buildings of the Laurentides merged into the geometry of a larger land, while Lawren Harris reworked the rocky headlands of Lake Superior and the glaciers of Baffin Island into a transcendentalist's vision of hope. Scholars, particularly Harold Adams Innis, emphasized the distinctiveness of an economy built on the northern staples of cod and beaver. Canada had grown out of the fur trade, out of its rivers and canoe routes, out of the Shield and the larger north where the best furs were to be had. Eventually the patterns of the fur trade would be reproduced in the grain trade. City and hinterland, civilization and wilderness, northern land and nation state — here was an encompassing theory of Canada, a Laurentian interpretation.

The problem with British and Laurentian visions was their fit with reality. Many of the polyglot settlers on the prairie and almost all French Canadians, who could readily equate Britishness with conquest, would not include themselves in an imperial and essentially Protestant conception of Canada. A British Canada from sea to sea implied the recasting of French Canadians in another cultural mold, a possibility intermittently envisaged but not so easily accomplished. The Laurentian vision had been culturally neutralized by shifting the Canadian focus from the islands of settlement to the rocks between. It was a dehumanized vision, expressed in beaver, maple leaves, and vibrant wilderness canvases, not in people. Academically, it moved the fur trade and transcontinental enterprise to centre stage, and island populations to the wings. This disembodied Canada could present an illusion of unity, for the country was certainly northern, wilderness was everywhere at hand, and the fur trade had left traces across the land. But to turn a part into the whole, which in their different ways both the British and Laurentian visions attempted, was either to impose a cultural tyranny or to ignore culture altogether. The price of Canadian definition was a severely distorted understanding of Canadian reality.

The fact was that a satisfying national definition would not be found. If some islands could be defined easily enough, the archipelago could not. Definitions that satisfied here, grated there. Symbols like the Union Jack or the fleur-de-lis were island rallying points and national battle cries. More neutral symbols turned to nature, and no benign Uncle Sam pervaded the national mythology. But there could be, and there was, an accumulating national experience.

Confederation had launched a transcontinental state that, over the years, has become part of the consciousness of the people who live in it. For newcomers there have been immigration laws and immigration officers to deal with, and Canadian citizenship to obtain, a tangible definition of new circumstances. For French speakers, on whose North-American past Confederation was grafted, there have been, at the very least, members of parliament to send to Ottawa and Canadian laws to obey. For imperialists there was prospect of empire. People connected to Canada in innumerable ways, but connections overlapped here and there, and cumulative experience grew. Some politicians have become known nationally, and the same political arguments have been joined, however differently, across the country. Federal legislation has been lived with and federal institutions — before radio, the post office was the most ubiquitous example — have reached into settlements across the land. The World Wars generated conscription crises, but also Canadian feeling as the magnitude of the national contribution became apparent. The country, Canadians knew, had become visible internationally. However badly and with whatever biases, something of Canada has been taught in Canadian schools. A host of private institutions from churches to corporations have been organized nationally. There were hockey games and Grey Cups, occasional Olympic successes, pensions and family allowance cheques, Crown corporations and royal commissions. A political space has been institutionalized across the archipelago, creating a web of experiences that are part of Canadian life and, much more on some islands than others but nowhere entirely absent, a sense of being Canadian.

Such experience has developed with the influence of modern communications. There have been many media and many messages with diverging implications for the geography of regionalism. The newspaper has probably reinforced provincialism for there has been no truly national press, while small-town newspapers have been victims of regional metropolitan dominance. Magazines have been more nearly national, but do not cross the language divide. Electronic communication tends to organize space linguistically. On the English-speaking islands, particularly, north–south signals contend with and often obliterate east–west ones. The National Hockey League televises its all-star games in American cities and our children soak up American "sitcoms." The idea of Canada is squeezed between our long-standing localism and a continentalism reinforced by the electronic media. Yet these media also report the ongoing national experience, and regulation deflects part of the barrage from the south. The islands are better known to each other than ever before; considering their location, an astonishingly small number of Canadians would contemplate a continental political realignment.

Thus, an amalgam of sentiment attaches to Canada. For some there is a lingering sense of Britain, for others a northern vision, and for many more a sense of Canada that has grown out of the continuing experience of being here. Not everyone, of course, has appreciated the experience of Canada. As the Métis lad, Jules Skinner, told Morag in Margaret Laurence's *The Diviners*: "The Prophet and his guys and the Indians and their guys, they'd just beat the shit out of the Mounties at someplace, and everybody was feeling pretty fine." But, for all the current acrimony, it is likely that among English Canadians, national feeling outweighs provincial — except perhaps in Newfoundland. Western separatism still runs into a western wall of Canadian feeling. Among French speakers, Canada is approached more cautiously although, paradoxically, the viability of French outside Québec now depends on the federal defence of bilingualism. As the last twenty years have shown only too clearly, Québec sentiment is predominant among French speakers in Québec but, as polls, referenda, and many less categoric indicators continue to show, Canadian sentiment is not absent. Of course, any concept of Canada must reckon with the profound structural localism of which this country is composed, with the regional tensions generated by a national economy, and with the provinces' growing visibility.

ON THE FUTURE OF CANADA

In an earlier form of this essay, written in 1981, I concluded my analysis with these still essentially optimistic thoughts. Given the location of the Canadian archipelago, the manner of its settlement, and the tensions inherent in a core–periphery economy, the overwhelmingly remarkable Canadian quality has been stability. On the face of it, Elisée Reclus should have been right. The country seems conceptually impossible. It is not remarkable that there have been and continue to be sharp regional tensions. They are built into the country's structure. It is remarkable that they have not been even stronger, that Canada has held together while generating a considerable reservoir of transcontinental feeling.

Stability has had much to do with isolation. Differences have not been perceived or, at least, have not had to be dealt with day by day because they were regionally compartmentalized. French speakers were conveniently tucked away on long-lot farms, in working-class Montréal, or in the Clay Belt. They did not obtrude nationally. Newfoundlanders kept to their fishing or went to Boston. For many years, stability also had a good deal to do with standard of living. For all the trauma of relocation, most immigrants were better off here than in their homelands, and this bred satisfaction not only with personal circumstances but also with a new country. In the long run, economic oppression has been tempered by a resource base relatively abundant for a small population; by a productive industrial system and its associated institutions of working-class defence; and by the fact that Canada has managed to achieve a fair measure of regional and social redistribution of wealth — less than desirable but enough to diffuse a great deal of tension. Programs for radical change have encountered these muting realities. There is probably a relationship, too, between the rather quiet, undefined nature of Canadian nationalism and national stability. Clear cultural symbols are rallying points, but they are also targets. Canada is sustained by nationalism based on experience, and weakened by nationalism based on cultural belief. In sum, the country has been stable, until recently, because most Canadians are not deeply angered by Canada as it is. Admitting the Conquest, the orange-green head-knocking in the Ottawa Valley, the Métis rebellions, the rabid anti-orientalism in British Columbia, the depossession of Native peoples, and much raw exploitation in sweatshop and industrial camp, the country does not have a tradition of violence. To a considerable extent cultural conflict has been diffused by distance and ignorance and by the failure of national definition. Economic conflict has been diffused by relatively high standards of living, and by a tradition of government that has mitigated some of the more blatant inequities of unregulated industrial capitalism. Overall, most Canadians have not lived with a festering memory of overwhelming injustice, as do many Irish, but rather with the all-too-smug and often too myopic satisfaction that accompanies the feeling that life here is fairly good.

But this is changing. There is now far more regional tension in Canada than there was a generation ago. Canadians live with the possibility that Canada may disintegrate and, for all the fissionable tendencies built into the structure of this country, this is a relatively new apprehension. Earlier generations of Canadians wondered whether Canada had an identity apart from Britain or could withstand the continental pressure of the United States. We wonder whether the Canadian archipelago will hold together. Canada is suddenly not as stable as before. This change has only superficially to do with the combative personalities of our politicians. It reflects the challenge raised by the Parti Québécois, but that challenge itself is part of a structural change in the relationship between the various parts of this country. Heightened internal tension is the almost inevitable corollary of the growing role and visibility of the provinces and of provincial governments, especially when accompanied by the first major challenge in a century to the spatial character of the Canadian economy.

The consolidation of island sentiment in the provinces has taken place while governments were assuming a larger role in Canadian life, and while the evolution of the Canadian economy was changing the significance of some of the terms of the British North America Act. In an age of water power and steam, control of natural resources did not mean what it does today. Provincial governments have been assuming a growing role in the economy, and the province a growing role in

Old church
in Québec

Canadian feeling. A host of activities that once were organized at many regional scales are now organized provincially. At the same time federal power has also increased, as Ottawa, too, has expanded its services and economic presence. In short, provincial and federal governments have grown simultaneously at the expense of a multitude of other levels of Canadian interaction. This is a recipe for conflict. It reduces a whole gamut of different relationships on different scales to a polarized struggle between the provinces and the federal government, and between the politically defined hinterlands and the national heartland.

The effects of such reorganization have been felt across the country, but are seen most clearly in Québec. There, the cultural defence of a French-speaking, Roman Catholic people once was mounted by the local community, by a variety of nationalistic societies, and, above all, by the Roman Catholic Church. For some, the clearest defence of culture was a rural life and a high birth rate, and from this perspective the provincial government could do little more than encourage colonization. The province was only one of the scales of cultural defence that ranged from the family and parish to French-speaking, Roman Catholic North America. The Church was not the church of Québecers but of Roman Catholics, and it could never be altogether comfortable within provincial boundaries. In recent years, the defence of culture has shifted toward government. The effects are obvious. One level of cultural defence is mounted by a provincial government that is prepared, as the Roman Catholic Church could never be, to write off French speakers outside Québec. The other level is mounted federally, and would provide more protection for French speakers outside Québec and less for those within. Two modes of cultural defence — both with long pedigrees — become the focus of rivalry between the provinces and the federal government. Elsewhere the issues have been different, having more to do with economics than culture, but across the country the tendency toward polarization has been essentially the same.

While this has been going on, the location of oil and gas fields, the spiralling world price of these commodities, and the rapidly growing economic importance of the Pacific basin have challenged the spatial economy that Canadians assumed for almost a hundred years. Core and peripheries seem to have come unstuck. Suddenly there is oil on the Grand Banks. In the West, the tar sands may have oil for two centuries, and there are massive coal deals with the Japanese and the highest residential land prices in the country. In the centre, there is the strong possibility of de-industrialization in the face of competition from the very Japanese products that increasingly will be manufactured from western Canadian resources. Whether the Canadian spatial economy will be turned inside out is another question; in a country like Canada it takes only the possibility of spatial change to raise a clamour as the ghosts of the last hundred years of spatial conflict reappear.

Seen from the centre, the old peripheries of wheat farmers and fishers would now seem to be infested with oil sheiks who wield the ominous power to turn off the tap. Seen from the peripheries, the centre is finally tasting some of its own medicine. Their turn would seem to have come — as long as the natural momentum of different circumstances is left to run its course. But if political power is in the centre, and if the British North America Act leaves ample opportunity for federal influence on resource policy, then the economic advantages of the peripheries can be compromised by the protective instincts of the centre. Hence the Western provinces' and Newfoundland's paranoia over resource control. Hence the struggle between Alberta and Ottawa over the price of oil. As long as the federal government is elected in central Canada, as given the distribution of population it is likely to be, conflict over the economy is immediately translated into conflict beween the provincial hinterlands and the centre.

So it seemed to me when I wrote the earlier form of this essay in 1981. I thought then that Canadians would probably have to bear with a fractious, tense Canada but, hopefully, a Canada that they could appreciate for what it was. I offered three simple observations that, if we could keep them in mind, seemed to bode well for our inevitably somewhat cantankerous future. First, however much we may have recognized our debt to institutions and traditions from across the Atlantic, we were no longer British or French. A social change as old as Canadian settlement took care of that. But if North Americans, we were not Americans. The circumstances of Canadian and American life had always been different, and different societies resulted. We would not easily define this cultural divide between adjacent New World peoples, but we experienced it, and I thought that perhaps we could relax with that. Finally, we would do well to remember that in a society of islands the majority could be culturally destructive. For this reason minorities would need protection that majorities would not. This, I held, applied particularly to French-speaking Canadians, without whom this country would never have been, and who, through more than two hundred years, had held out against an anglicizing continent. The same could be said of Native peoples. This country worked when it displayed understanding and humanity, and was bottled up by dogma. I ended with a quote from Margaret Laurence (in *The Diviners*) "I never connected it with that," Jules told Morag, "because my dad's version was a whole lot different."

Here, I thought, was a recipe for tolerance. Morag was not less interested in Jules because their families told different stories. But most of the claims I made in 1981 about the divisive, centrifugal tendencies in this country have only been emphasized over the intervening years. Now, however, the context has changed. The balance between the idea of Canada and of other geographical scales of attachment has shifted. I cannot be as confident as I was then that, for all the tensions inherent within it, Canada remains a stable creation. Quite the contrary. As we near the end of the century, Canada seems to hang by a thread. We could easily become two countries, the smaller based on the territory that Jacques Cartier explored in the 1530s, and the larger, ironically, carrying the name he gave to that area.

In a general way developments here have reflected worldwide trends. On the one hand, scales of attachment are becoming more transnational and global, and on the other more tribal and local. The intermediate space occupied by a country like Canada is squeezed. When the federal government is short of and curtails its programs, the effects of that squeezing are compounded. The balance of power between federal and provincial governments tilts towards the provinces. When increasingly salient provinces have different ideas about the cultures they represent and defend, the result is a fundamental challenge to the idea of Canada — which, as I argued in 1981, cannot survive cultural nationalisms. Canada is pressured to reconfigure itself around a nationalism focussed on Québec, based on cultural belief, and drawing on centuries of French experience in North America; and around another much more contrived nationalism in the rest of Canada, "English" Canada — a nationalism that hardly exists (English Canadian nationalism turns around Canada rather than around English Canada) and that would have to be made. Were this to happen, as well it could, a very awkward and unpredictable future would lie ahead, particularly perhaps for English Canada, which would seem to be an inherently unstable creation. Some precious Canadian qualities would be lost.

Québec is *the* Canadian issue now. Nothing else, in the foreseeable future, has the capacity to tear the country apart. From the perspective of many Québecers, recent developments have strengthened the case for independence. Increasingly global flows of information, capital, and trade resituate Québec in global rather than Canadian space. NAFTA opens up north–south markets, and the old east-west economy, originally built on steam power and protective tariffs, is weakened. Montréal is no longer the Canadian metropolis. At the same time, Canada's linguistic geography is polarizing: a higher percentage of Québecers speak French at home, and a lower percentage of Canadians outside Québec than in 1981. There has been a steady net interprovincial migration from Québec, some of it francophone but more of it not, while a surprising number of Jacques Parizeau's Montréal "ethnics" have returned to their countries of origin. Outside Québec, French-speaking minorities are being assimilated rapidly; the French language appears secure only in New Brunswick and, perhaps, in parts of Ontario. In such circumstances, Québec rather than Canada seems the more obvious protector of the French language in North America. Not less important has been the continuing refusal of the rest of Canada to entrench the distinctiveness of Québec in the constitution. Any provincial society is distinctive, but most Québecers claim a particular distinctiveness by right of language, history, and constitution. I thought at the time, and think now, that it was a national tragedy that we did not quite ratify the Meech Lake Accord.

The reluctance of English Canada (that is, largely English-speaking and polyethnic Canada outside Québec) to recognize the distinctiveness of Québec rests, finally, on the liberal assumption that all people should be equal before the law. The balance between individual and collective rights is a fundamental issue in any democracy. In Canada under Pierre Trudeau the balance shifted toward individual rights, a position entrenched in the Constitution Act of 1982, which included the *Canadian Charter of Rights and Freedoms*. In Canada, the emphasis on individual rights particularly exposes two groups of people: French speakers (who battle the tide of an English-speaking continent backed by an ascendant international English), and Native peoples (who have been the victims of Canada and, for the most part, have shown no wish to assimilate into a larger Canadian society). Scattered across the country, Native peoples do not have the power to divide Canada, but theirs is another claim, rooted in the long trajectory of this country, that some cultures require a measure of collective defense to survive in this world. If Canadian society as a whole cannot meet these claims, then Québec will leave and Native peoples will become ever more acrimonious. The rest of Canada, "English" Canada, hardly needs equivalent protection. It is no longer English, except in language, and the English language needs no safeguards. Immigrants have not come to preserve cultures so much as to get on in the world.

In the near term, then, the future of Canada somewhat as we know it comes down to the ability of the rest of Canada to work out an accommodation with Québec. Québecers will not assume that, when the chips are down, a government in Ottawa will look after their linguistic and cultural interests. Given the demographic and political realities, how could they? The distinctive society clause in the Meech Lake Accord seemed a minimal step in the right direction. It is probably insufficient now. The recent suggestion from Professor John Richards, economist at Simon Fraser University, that linguistic rights in Canada should be territorial rather than, as now, attached to individuals, warrants serious consideration. Québec, he thinks, should receive broad linguistic powers, and the pretense of bilingual opportunity for everyone throughout Canada should be dropped. Such ideas are important and need to be on the table. We don't have much time. Most of Canada, inured to Québec, sleepwalks toward the next Québec referendum, when it will be a bystander.

If we cannot find a solution and Québec separates, the repercussions will be enormous. One fraction of the country will have created a nation, while the other will have lost one. English Canadian nationalism, surprisingly deeply felt, embraces the whole country, not the English-speaking parts of it. For most in the West, Newfoundlanders are not more Canadian than Québecers because they speak English. This larger Canada will have been taken away, and the grief and anger associated with its going will be compounded by jubilation in parts of Québec — hardly the circumstances in which to work out a new and amicable relationship. Living standards will probably fall across the country, most in Québec. The ease of movement of people across new international borders will diminish, by how much no one can say. There will be uncertainty everywhere, from the division of the debt to the division of collections in the National Gallery. The most precious casualty will be the idea that this country can provide respectful support for its own fundamental cultural differences. Canada is not, and never has been, anything like a homogeneous society. It cannot sustain a strident cultural nationalism because it invariably contradicts much of the rest of Canadian reality. Ours is a multivalent, unspecified nationalism, the product of our inductive past and fragmented geographical composition. The Canadian virtues, which are also requirements if this country is to be, are those of tolerance and respect. If we cannot hold this country together, these qualities will have diminished a little on the world stage, both Canada and Québec will be meaner, narrower places, and the space, worldwide, between globalism and tribalism will have shrunk considerably.

NOTES

1. A version of this essay was originally given as a Walter L. Gordon lecture at McGill University on March 17, 1981, and is published here with the kind permission of the Walter L. Gordon Lecture Series and the Canadian Studies Foundation.

2. Fernand Braudel, *The Mediterranean and the Mediterranean World in the Age of Philip II*, trans. Siân Reynolds (New York: Harper and Row, 1972).

3. Some of these points are discussed more fully in R. Cole Harris, "The Extension of France into Rural Canada," in *European Settlement and Development in North America: Essays on Geographical Change in Honour and Memory of Andrew Hill Clark*, ed. J. R. Gibson (Toronto: University of Toronto Press, 1978), pp. 27–45.

4. Frederick Jackson Turner, "The Significance of the Frontier in American History," first published in 1893 and reprinted in F. J. Turner, *The Frontier in American History* (New York: 1920).

5. A.-N. Lalonde, "L'intelligentsia de Québec et la migration des canadiens français vers l'ouest canadien, 1870–1930," *Revue d'histoire de l'amérique française*, 33 (1979), pp. 163–85.

6. See, for example, Henry Nash Smith, *Virgin Land: The American West as Symbol and Myth* (Cambridge, Mass.: Harvard University Press, 1950); and Leo Marx, *The Machine in the Garden: Technology and the Pastoral Ideal in America* (New York: Oxford University Press, 1964).

7. This point is made by Northrop Frye in a splendid essay that concludes *Literary History of Canada: Canadian Literature in English*, ed. C.F. Klinck (Toronto: University of Toronto Press, 1965), p. 826.

8. I have developed this argument in "The Simplification of Europe Overseas," *Annals of the Association of American Geographers*, 67 (1977), pp. 469–83.

9. Karl Polanyi, *The Great Transformation* (New York: Farrar and Rinehart, 1944).

10. Ernest Buckler, *The Mountain and the Valley* (New York: Holt, 1952); and Jack Hodgins, *The Invention of the World* (Toronto: Macmillan, 1977).

11. George Grant, "In Defense of North America," in idem, *Technology and Empire: Perspectives on North America* (Toronto: House of Anansi, 1969).

12. Ernest R. Forbes, *The Maritime Rights Movement, 1919–1927: A Study in Canadian Regionalism* (Montréal: McGill-Queen's University Press, 1979), especially chapter 2.

13. Christian Morissonneau, *La Terre Promise: le mythe du nord Québecois* (Montréal: Hurtubise, 1978), especially chapters 2 and 3.

14. The most relevant of Hind's and Macoun's writing, including the parts quoted here, are in John Warkentin, *The Western Interior of Canada: A Record of Geographical Discovery*, The Carleton Library, No. 15 (Toronto: McClelland and Stewart, 1964).

15. Carl Berger, *A Sense of Power: Studies in the Ideas of Canadian Imperialism, 1867–1914* (Toronto: University of Toronto Press, 1970).

16. Quoted in C. J. Houston and W. J. Smyth, *The Sash Canada Wore: A Historical Geography of the Orange Order in Canada* (Toronto: University of Toronto Press, 1980), p. 120.

17. Carl Berger, "The True North Strong and Free," in *Nationalism in Canada*, Peter H. Russell ed. (Toronto: McGraw-Hill, 1966), pp. 3-26.

Photo Credits

Index

Note: Bolded page numbers refer to drawings or illustrations; italicized page numbers refer to maps or tables.

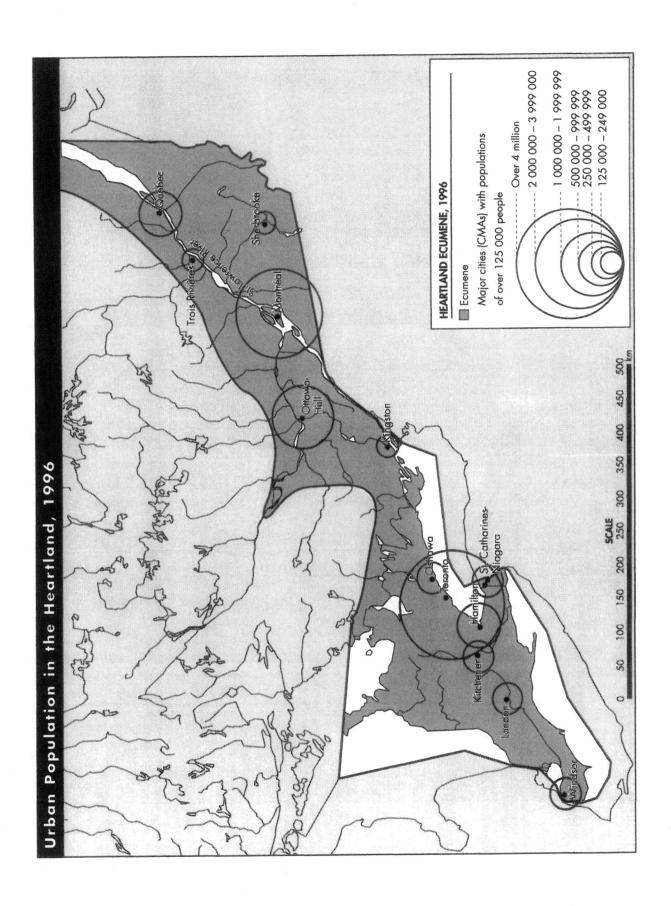

Urban Population in the Heartland, 1996

HEARTLAND ECUMENE, 1996

Ecumene

Major cities (CMAs) with populations
of over 125 000 people

------- Over 4 million
------- 2 000 000 – 3 999 000
------- 1 000 000 – 1 999 999
------- 500 000 – 999 999
------- 250 000 – 499 999
------- 125 000 – 249 000

St. Lawrence River

Québec

Sherbrooke

Trois Rivières

Montréal

Ottawa-Hull

Kingston

Oshawa

Toronto

Hamilton

St. Catharines-Niagara

Kitchener

London

Windsor

SCALE

0   50   100   150   200   250   300   350   400   450   500
km